Workbook for Mathematical Economics Part 2

Michael Sampson

Workbook for Mathematical Economics Part 2.

ISBN: 978-0-9947930-3-4

If you find typos or errors, *please tell me!* I value your fresh eyes. Any and all comments can be forwarded to *comments@loglinear.com.*

Acknowledgment

Thanks to my former students for struggling through earlier versions. For pointing out typos and errors, particular thanks go to: Dana Al-Aghbar, Kimia Amouzegar, Viyasan Asokanathan, Todor Atanassov, Audrey Autmezguine, Ramzi Bahsoun, Philippe Baril Lecavalier, Hamza Benzekri, Andras A. Birkus, Katherine Boisvert, David Bourque, Elizabeth Brown, Jonathan Camarda, Noah Campbell, Ysa Cao, Carlo Catino, Alan Lai Yuk Chuen, Mark Costello, Herby Damas, Alejandro Fuenmayor, Sabrina Gilbert, Jonas Graham, Mathieu Grenier, Alexis Grondin, Xinzhu Han, Wendy Ip, Ryan Joyce, Vladimir Khripunov, Mohammad Kureishy, Rucktooa Mitheela Lay Ling, Luigi Francesco Lento, Tamarha Louis-Charles, Alain Lumbroso, Bogdan Malaelea, Christopher Martin, Curtis McArdle, Jennifer Meyer, Alassane Anand Ndour, Eduard Nenciu, Ted Nehme, Ha Dao Ngoc, Alexia Parhas, Alexandre Perrault, Edi Salem, Samnang Om, Stefano Patone, Jeffrey Penney, Maya Prince, Simon Rivard, Patrick Rizzetto, Sarah Saleh, Thomas Suchorski, Juri Takubo, Andrea Vallejos, Andrija Vasic, Olivia I. Vasiliscu, Apira Vigneshwaran, Yang Wang, Melanie Wodzicki, Zijiang Xi, Mingyuan Yang, Sebastian Yeung, Ianitt Yoo, Bülent Yurtsever, and Irina.

Michael Sampson is an Associate Professor in the Department of Economics, Concordia University, Montreal.

Cover image taken from John Tenniel's illustrations for Lewis Carroll's *Alice in Wonderland.*

Preface

Practice is the key to learning mathematics, or just about anything for that matter. Here is a large collection of problems for you to practice on. Answers are included, but you should only look at them *as a last resort.* First *struggle* with the problem. A good rule might be to *only look at the answer the next day.* It is the act of struggling that will prepare your mind, not simply reading the answer. The best outcome is to struggle, and get the answer yourself. The next best outcome is to struggle, fail after a long period, and only then look at the answer. Then when you see the answer you're likely to have that 'Aha!' experience, and you'll remember that particular technique for the rest of your life. Otherwise if you simply read the answer each time you get a little frustrated, you'll forget very soon afterwards (typically by the next class).

It is also important to work on the *right kind of problems.* This is depends on where *you are.* There is no point working on A problems if you are currently working at a D level. Work on D and C problems, try to get yourself to a C level, and only then go for the B problems. Assuming you are still ambitious, once you've mastered the B problems, go for the A problems.

Some problems are more important than others. Some are absolutely essential: they deal with skills (for example the chain rule) that you simply *must* master. Other problems are less important but still useful; some are not important at all. Some are there simply because I find them interesting.

With these distinctions in mind I have attempted to categorize the problems according to their level, that is whether they are A, B, C, D problems, and by their importance. The importance is indicated as follows:

*	Not very important.
**	Moderately important.
***	Absolutely essential.

For example *[C***]* would indicate a C level problem that is absolutely essential to master if you want to get a C, while *[A***]* would indicate an A level problem that is absolutely essential to master if you want to get an A . *[B**]* would indicate a moderately important B level problem, and *[A*]* would indicate an unimportant A level problem, although perhaps important for other courses, or simply an interesting or fun problem.

Chapter 1

Mathematical Tools

1.1 Homogeneous Functions

Problem 1.1 *[D***] Show that the function*

$$f(x_1, x_2) = \left(2x_1^{\frac{1}{2}} + 3x_2^{\frac{1}{2}}\right)^{\frac{4}{3}}$$

is homogeneous and determine the degree of homogeneity k.

Answer: We have

$$\begin{aligned} f(\tau x_1, \tau x_2) &= \left(2(\tau x_1)^{\frac{1}{2}} + 3(\tau x_2)^{\frac{1}{2}}\right)^{\frac{4}{3}} = \left(\tau^{\frac{1}{2}}\left(2x_1^{\frac{1}{2}} + 3x_2^{\frac{1}{2}}\right)\right)^{\frac{4}{3}} \\ &= \tau^{\frac{2}{3}}\left(2x_1^{\frac{1}{2}} + 3x_2^{\frac{1}{2}}\right)^{\frac{4}{3}} = \tau^{\frac{2}{3}} f(x_1, x_2) \end{aligned}$$

so the function is homogenous of degree $k = \frac{2}{3}$.

Problem 1.2 *[D***] Show that*

$$Q = F(L, K) = L^{\alpha} K^{\beta}$$

is homogeneous and determine k.

Answer: We have

$$F(\tau L, \tau K) = (\tau L)^{\alpha} (\tau K)^{\beta} = \tau^{\alpha} \tau^{\beta} (L)^{\alpha} (K)^{\beta} = \tau^{\alpha+\beta} (L)^{\alpha} (K)^{\beta} = \tau^{\alpha+\beta} F(L, K)$$

so that $F(L, K) = L^{\alpha} K^{\beta}$ is homogeneous of degree $\alpha + \beta$.

Problem 1.3 *[C***] Show that*

$$Q = F(L, K) = L^{\frac{1}{2}} K^{\frac{1}{2}} + 10$$

is not homogeneous.

Answer: We have

$$F(\tau L, \tau K) = (\tau L)^{\frac{1}{2}} (\tau K)^{\frac{1}{2}} + 10 = \tau^{\frac{1}{2}} \tau^{\frac{1}{2}} L^{\frac{1}{2}} K^{\frac{1}{2}} + 10 = \tau L^{\frac{1}{2}} K^{\frac{1}{2}} + 10 \neq \tau \left(L^{\frac{1}{2}} K^{\frac{1}{2}} + 10 \right)$$

since $10 \neq 10\tau$ for $\tau \neq 1$. We conclude that $F(L, K) = L^{\frac{1}{2}} K^{\frac{1}{2}} + 10$ is **not** homogeneous.

Problem 1.4 *[A***] Show that if $f(x_1, x_2)$ is homogenous of degree 1 then the Hessian of $f(x_1, x_2)$ is singular for all non-zero x_1, x_2.*

Answer: If $f(x_1, x_2)$ is homogenous of degree 1 then both $\frac{\partial f(x_1, x_2)}{\partial x_1}$ and $\frac{\partial f(x_1, x_2)}{\partial x_2}$ are homogeneous of degree 0. By Euler's theorem

$$\begin{aligned} 0 \times \frac{\partial f(x_1, x_2)}{\partial x_1} &= \frac{\partial^2 f(x_1, x_2)}{\partial x_1^2} x_1 + \frac{\partial^2 f(x_1, x_2)}{\partial x_1 \partial x_2} x_2 = 0 \\ 0 \times \frac{\partial f(x_1, x_2)}{\partial x_2} &= \frac{\partial^2 f(x_1, x_2)}{\partial x_1 \partial x_2} x_1 + \frac{\partial^2 f(x_1, x_2)}{\partial x_2^2} x_2 = 0 \end{aligned}$$

or in matrix notation

$$\underbrace{\begin{bmatrix} \frac{\partial^2 f(x_1, x_2)}{\partial x_1^2} & \frac{\partial^2 f(x_1, x_2)}{\partial x_1 \partial x_2} \\ \frac{\partial^2 f(x_1, x_2)}{\partial x_1 \partial x_2} & \frac{\partial^2 f(x_1, x_2)}{\partial x_2^2} \end{bmatrix}}_{H} \begin{bmatrix} x_1 \\ x_2 \end{bmatrix} = \begin{bmatrix} 0 \\ 0 \end{bmatrix}$$

which is of the form $Hx = 0$ where H is the Hessian. Since $x \neq 0$ it follows that H is a singular matrix.

Problem 1.5 *[D***] Show that*

$$Q = F(L, K) = \left(\frac{2}{3} L^{-3} + \frac{1}{3} K^{-3} \right)^{-\frac{2}{3}}$$

is homogeneous of degree 2.

Answer: Since

$$\begin{aligned} F(\tau L, \tau K) &= \left(\frac{2}{3} (\tau L)^{-3} + \frac{1}{3} (\tau K)^{-3} \right)^{-\frac{2}{3}} = \left(\tau^{-3} \left(\frac{2}{3} L^{-3} + \frac{1}{3} K^{-3} \right) \right)^{-\frac{2}{3}} \\ &= \tau^2 \left(\frac{2}{3} L^{-3} + \frac{1}{3} K^{-3} \right)^{-\frac{2}{3}} = \tau^2 F(L, K). \end{aligned}$$

Problem 1.6 *[B***] Show that the Constant Elasticity of Substitution or CES production function*

$$Q = F(L, K) = (\alpha L^{\rho} + (1 - \alpha) K^{\rho})^{\frac{\gamma}{\rho}}$$

is homogeneous of degree γ.

Answer: Since

$$\begin{aligned} F(\tau L, \tau K) &= \left(\alpha (\tau L)^{\rho} + (1-\alpha)(\tau K)^{\rho}\right)^{\frac{\gamma}{\rho}} = \left(\tau^{\rho}\left(\alpha L^{\rho} + (1-\alpha) K^{\rho}\right)\right)^{\frac{\gamma}{\rho}} \\ &= \left(\tau^{\rho}\right)^{\frac{\gamma}{\rho}} \left(\alpha L^{\rho} + (1-\alpha) K^{\rho}\right)^{\frac{\gamma}{\rho}} = \tau^{\gamma}\left(\alpha L^{\rho} + (1-\alpha) K^{\rho}\right)^{\frac{\gamma}{\rho}} = \tau^{\gamma} F(L, K). \end{aligned}$$

Problem 1.7 *[A***] Use Euler's theorem to show that if $Q = F(L, K)$ is homogeneous of degree k and if the firm pays labour and capital according to its marginal product so that $\frac{\partial F(L,K)}{\partial L} = \frac{W}{P}$ and $\frac{\partial F(L,K)}{\partial K} = \frac{R}{P}$, then the firm receives positive, 0, and negative profits according to whether $k < 1$, $k = 1$, or $k > 1$.*

Answer: By Euler's theorem

$$kQ = \frac{\partial F(L,K)}{\partial L} L + \frac{\partial F(L,K)}{\partial K} K.$$

Since $\frac{\partial F(L,K)}{\partial L} = \frac{W}{P}$, $\frac{\partial F(L,K)}{\partial K} = \frac{R}{P}$ we have

$$kQ = \frac{W}{P} L + \frac{R}{P} K \Longrightarrow kPQ = WL + RK$$

and so profits π are given by

$$\pi = PQ - (WL + RK) = PQ - kPQ = (1-k) PQ.$$

Thus if $0 < k < 1$ (there are decreasing returns to scale) then $\pi > 0$ while if $k = 1$ (constant returns to scale) then $\pi = 0$. If $k > 1$ then profits must be negative, which is indicative of the fact that increasing returns to scale are not consistent with perfect competition.

Problem 1.8 *[B***] Show that*

$$f(x_1, x_2, x_3) = x_1^{\alpha}\left(x_2^{\alpha} + x_3^{\alpha}\right)^{\frac{1-\alpha}{\alpha}}$$

is homogeneous and determine the degree of the homogeneity.

Answer: We have

$$\begin{aligned} f(\tau x_1, \tau x_2, \tau x_3) &= (\tau x_1)^{\alpha}\left((\tau x_2)^{\alpha} + (\tau x_3)^{\alpha}\right)^{\frac{1-\alpha}{\alpha}} \\ &= \tau^{\alpha} x_1^{\alpha}\left(\tau^{\alpha} x_2^{\alpha} + \tau^{\alpha} x_3^{\alpha}\right)^{\frac{1-\alpha}{\alpha}} = \tau^{\alpha} x_1^{\alpha}\left(\tau^{\alpha}\left(x_2^{\alpha} + x_3^{\alpha}\right)\right)^{\frac{1-\alpha}{\alpha}} \\ &= \tau^{\alpha}\left(\tau^{\alpha}\right)^{\frac{1-\alpha}{\alpha}} x_1^{\alpha}\left(x_2^{\alpha} + x_3^{\alpha}\right)^{\frac{1-\alpha}{\alpha}} = \tau^{\alpha}\tau^{1-\alpha} x_1^{\alpha}\left(x_2^{\alpha} + x_3^{\alpha}\right)^{\frac{1-\alpha}{\alpha}} \\ &= \tau^{1} x_1^{\alpha}\left(x_2^{\alpha} + x_3^{\alpha}\right)^{\frac{1-\alpha}{\alpha}} = \tau^{1} f(x_1, x_2, x_3) \end{aligned}$$

so $f(x_1, x_2, x_3)$ is homogeneous of degree $k = 1$.

Problem 1.9 *[C***] Show that*

$$f(x_1, x_2) = \left((x_1)^{\frac{1}{2}}(x_2)^{\frac{1}{4}} + (x_1)^{\frac{1}{4}}(x_2)^{\frac{1}{2}}\right)^{\frac{1}{3}}$$

is homogeneous and determine k.

Answer: We have

$$\begin{aligned}
f(\tau x_1, \tau x_2) &= \left((\tau x_1)^{\frac{1}{2}}(\tau x_2)^{\frac{1}{4}} + (\tau x_1)^{\frac{1}{4}}(\tau x_2)^{\frac{1}{2}}\right)^{\frac{1}{3}} \\
&= \left(\tau^{\frac{1}{2}}(x_1)^{\frac{1}{2}}\tau^{\frac{1}{4}}(x_2)^{\frac{1}{4}} + \tau^{\frac{1}{4}}(x_1)^{\frac{1}{4}}\tau^{\frac{1}{2}}(x_2)^{\frac{1}{2}}\right)^{\frac{1}{3}} \\
&= \left(\tau^{\frac{1}{2}+\frac{1}{4}}(x_1)^{\frac{1}{2}}(x_2)^{\frac{1}{4}} + \tau^{\frac{1}{4}+\frac{1}{2}}(x_1)^{\frac{1}{4}}(x_2)^{\frac{1}{2}}\right)^{\frac{1}{3}} \\
&= \left(\tau^{\frac{3}{4}}\left((x_1)^{\frac{1}{2}}(x_2)^{\frac{1}{4}} + (x_1)^{\frac{1}{4}}(x_2)^{\frac{1}{2}}\right)\right)^{\frac{1}{3}} \\
&= \tau^{\frac{1}{4}}\left((x_1)^{\frac{1}{2}}(x_2)^{\frac{1}{4}} + (x_1)^{\frac{1}{4}}(x_2)^{\frac{1}{2}}\right)^{\frac{1}{3}} = \tau^{\frac{1}{4}} f(x_1, x_2)
\end{aligned}$$

so that $f(x_1, x_2)$ is homogeneous of degree $\frac{1}{4}$.

Problem 1.10 *[D***] Consider the production function*

$$Q = F(L, K) = \left(L^{-\frac{1}{3}} + K^{-\frac{1}{3}}\right)^{-\frac{3}{2}}.$$

*Show that the production function is homogeneous and determine the degree of the homogeneity. [C***] Write Euler's theorem for this function. Show that if the firm is competitive it will make a positive profit. Find the marginal product of labour and show that the marginal product of labour is homogeneous and determine the degree of its homogeneity.*

Answer: We have

$$\begin{aligned}
F(\tau L, \tau K) &= \left((\tau L)^{-\frac{1}{3}} + (\tau K)^{-\frac{1}{3}}\right)^{-\frac{3}{2}} = \left((\tau)^{-\frac{1}{3}}\left(L^{-\frac{1}{3}} + K^{-\frac{1}{3}}\right)\right)^{-\frac{3}{2}} \\
&= \left((\tau)^{-\frac{1}{3}}\right)^{-\frac{3}{2}}\left(L^{-\frac{1}{3}} + K^{-\frac{1}{3}}\right)^{-\frac{3}{2}} = \tau^{\frac{1}{2}}\left(L^{-\frac{1}{3}} + K^{-\frac{1}{3}}\right)^{-\frac{3}{2}} = \tau^{\frac{1}{2}} F(L, K)
\end{aligned}$$

so that $F(L, K)$ is homogeneous of degree $\frac{1}{2}$. Euler's theorem then states that

$$\frac{1}{2}F(L, K) = \frac{\partial F(L, K)}{\partial L}L + \frac{\partial F(L, K)}{\partial K}K.$$

If the firm is competitive then $\frac{\partial F(L,K)}{\partial L} = \frac{W}{P}$, $\frac{\partial F(L,K)}{\partial K} = \frac{R}{P}$ so that

$$\begin{aligned}
\frac{1}{2}F(L, K) &= \frac{W}{P}L + \frac{R}{P}K \\
&\Longrightarrow \frac{1}{2}PQ = WL + RK \\
&\Longrightarrow \pi = PQ - WL - RK = PQ - \frac{1}{2}PQ = \frac{1}{2}PQ > 0.
\end{aligned}$$

Using the chain rule the marginal product of labour is

$$\frac{\partial F(L, K)}{\partial L} = \frac{\partial}{\partial L}\left(L^{-\frac{1}{3}} + K^{-\frac{1}{3}}\right)^{-\frac{3}{2}} = \frac{1}{2}\left(L^{-\frac{1}{3}} + K^{-\frac{1}{3}}\right)^{-\frac{5}{2}} L^{-\frac{4}{3}}$$

which is homogeneous of degree $\frac{1}{2} - 1 = -\frac{1}{2}$ since $F(L,K)$ is homogenous of degree $\frac{1}{2}$.

Problem 1.11 *[C***] If $f(x_1,x_2)$ is homogeneous of degree k then what is $f(\tau x_1, \tau x_2)$? [B***] What then does Euler's theorem state? [A***] Prove that $\frac{\partial f(x_1,x_2)}{\partial x_1}$ is homogeneous of degree $k-1$. [D***] Show that*

$$f(x_1,x_2) = \left((x_1)^{\frac{1}{2}}(x_2)^{\frac{1}{4}} + (x_1)^{\frac{1}{4}}(x_2)^{\frac{1}{2}}\right)^{\frac{1}{3}}$$

is homogeneous and determine k.

Answer: By definition $f(\tau x_1, \tau x_2) = \tau^k f(x_1,x_2)$. Euler's theorem states then that

$$kf(x_1,x_2) = \frac{\partial f(x_1,x_2)}{\partial x_1}x_1 + \frac{\partial f(x_1,x_2)}{\partial x_2}x_2.$$

Differentiating both sides of $f(\tau x_1,\tau x_2) = \tau^k f(x_1,x_2)$ with respect to x_1 and using the chain rule we have

$$\begin{aligned}
\tau\frac{\partial f(\tau x_1,\tau x_2)}{\partial x_1} &= \tau^k\frac{\partial f(x_1,x_2)}{\partial x_1} \\
\implies \frac{\partial f(\tau x_1,\tau x_2)}{\partial x_1} &= \tau^{k-1}\frac{\partial f(x_1,x_2)}{\partial x_1}
\end{aligned}$$

and so $\frac{\partial f(x_1,x_2)}{\partial x_1}$ is homogeneous of degree $k-1$.

For

$$f(x_1,x_2) = \left((x_1)^{\frac{1}{2}}(x_2)^{\frac{1}{4}} + (x_1)^{\frac{1}{4}}(x_2)^{\frac{1}{2}}\right)^{\frac{1}{3}}$$

we have

$$\begin{aligned}
f(\tau x_1,\tau x_2) &= \left((\tau x_1)^{\frac{1}{2}}(\tau x_2)^{\frac{1}{4}} + (\tau x_1)^{\frac{1}{4}}(\tau x_2)^{\frac{1}{2}}\right)^{\frac{1}{3}} = \left(\tau^{\frac{1}{2}}(x_1)^{\frac{1}{2}}\tau^{\frac{1}{4}}(x_2)^{\frac{1}{4}} + \tau^{\frac{1}{4}}(x_1)^{\frac{1}{4}}\tau^{\frac{1}{2}}(x_2)^{\frac{1}{2}}\right)^{\frac{1}{3}} \\
&= \left(\tau^{\frac{1}{2}+\frac{1}{4}}(x_1)^{\frac{1}{2}}(x_2)^{\frac{1}{4}} + \tau^{\frac{1}{4}+\frac{1}{2}}(x_1)^{\frac{1}{4}}(x_2)^{\frac{1}{2}}\right)^{\frac{1}{3}} = \left(\tau^{\frac{3}{4}}\left((x_1)^{\frac{1}{2}}(x_2)^{\frac{1}{4}} + (x_1)^{\frac{1}{4}}(x_2)^{\frac{1}{2}}\right)\right)^{\frac{1}{3}} \\
&= \tau^{\frac{1}{4}}\left((x_1)^{\frac{1}{2}}(x_2)^{\frac{1}{4}} + (x_1)^{\frac{1}{4}}(x_2)^{\frac{1}{2}}\right)^{\frac{1}{3}} = \tau^{\frac{1}{4}}f(x_1,x_2)
\end{aligned}$$

so that $f(x_1,x_2)$ is homogeneous of degree $\frac{1}{4}$.

1.2 Multivariate Taylor Series

Problem 1.12 *[B*] Construct the Taylor series for*

$$f(x_1,x_2) = 4x_1^{\frac{1}{4}}x_2^{\frac{1}{2}}$$

around the point $x_1 = 1, x_2 = 1$. Evaluate the error at $x_1 = 1, x_2 = 1$ and $x_1 = 1.2, x_2 = 1.3$. Can you predict the sign of the error without making any calculations?

Answer: We have

$$\begin{aligned}
f(x_1, x_2) &= 4x_1^{\frac{1}{4}}x_2^{\frac{1}{2}} \\
&\Longrightarrow f(1,1) = 4 \times 1^{\frac{1}{4}}1^{\frac{1}{2}} = 4 \\
\frac{\partial f(x_1, x_2)}{\partial x_1} &= x_1^{-\frac{3}{4}}x_2^{\frac{1}{2}} \\
&\Longrightarrow \frac{\partial f(1,1)}{\partial x_1} = 1^{-\frac{3}{4}}1^{\frac{1}{2}} = 1 \\
\frac{\partial f(x_1, x_2)}{\partial x_2} &= 2x_1^{\frac{1}{4}}x_2^{-\frac{1}{2}} \\
&\Longrightarrow \frac{\partial f(1,1)}{\partial x_2} = 2 \times 1^{\frac{1}{4}}1^{-\frac{1}{2}} = 2
\end{aligned}$$

so the gradient at $x_1 = 1, x_2 = 1$ is

$$\nabla f(1,1) = \begin{bmatrix} 1 \\ 2 \end{bmatrix}.$$

The first-order Taylor series then is

$$\begin{aligned}
\bar{f}(x) &= f(x^0) + \nabla f(x^0)^T (x - x^0) \\
&= 4 + \begin{bmatrix} 1 \\ 2 \end{bmatrix}^T \begin{bmatrix} x_1 - 1 \\ x_2 - 1 \end{bmatrix} \\
&= 4 + \begin{bmatrix} 1 & 2 \end{bmatrix} \begin{bmatrix} x_1 - 1 \\ x_2 - 1 \end{bmatrix} \\
&= 4 + 1 \times (x_1 - 1) + 2 \times (x_2 - 1) = 1 + x_1 + 2x_2.
\end{aligned}$$

The Taylor series is shown in the figure below where $\bar{f}(x_1, x_2)$ is a flat plane in black just touching $f(x_1, x_2) = 4x_1^{\frac{1}{4}}x_2^{\frac{1}{2}}$ at $x_1 = 1, x_2 = 1$.

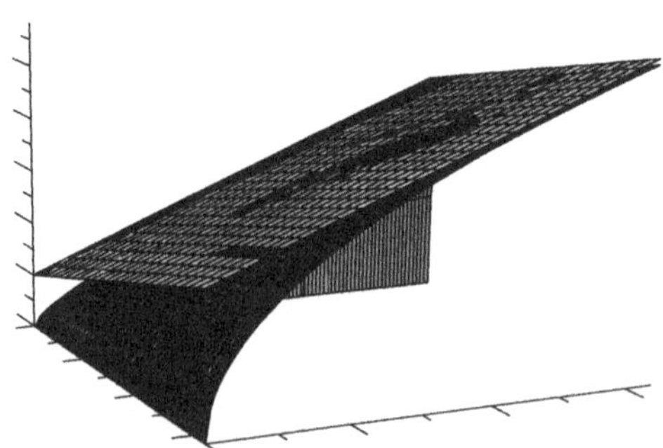

We can write the Taylor series approximation as

$$4x_1^{\frac{1}{4}}x_2^{\frac{1}{2}} \approx 1 + x_1 + 2x_2.$$

For $x_1 = 1, x_2 = 1$ the approximation is exact as

$$\begin{aligned} f(1,1) &= 4x_1^{\frac{1}{4}}x_2^{\frac{1}{2}} = 4 \times 1^{\frac{1}{4}}1^{\frac{1}{2}} = 4 \\ \bar{f}(1,1) &= 1 + x_1 + 2x_2 = 1 + 1 + 2 \times 1 = 4. \end{aligned}$$

If we move away from $x_1 = 1, x_2 = 1$ to $x_1 = 1.2, x_2 = 1.3$ the function and its Taylor series differ as

$$\begin{aligned} f(1.2, 1.3) &= 4x_1^{\frac{1}{4}}x_2^{\frac{1}{2}} = 4 \times (1.2)^{\frac{1}{4}}(1.3)^{\frac{1}{2}} \approx 4.77 \\ \bar{f}(1.2, 1.3) &= 1 + x_1 + 2x_2 = 1 + 1.2 + 2 \times 1.3 = 4.8 \end{aligned}$$

and so the error is

$$4.77 - 4.8 = -0.03.$$

Since $4x_1^{\frac{1}{4}}x_2^{\frac{1}{2}}$ is strictly concave it follows that

$$4x_1^{\frac{1}{4}}x_2^{\frac{1}{2}} < 1 + x_1 + 2x_2$$

for $x_1 \neq 1$ or $x_2 \neq 1$ and so the error will in general be negative. This can be seen in the figure above where the function falls below the plane given by the Taylor series.

Problem 1.13 *[B*] Construct the Taylor series for*

$$f(x_1, x_2) = e^{-\frac{1}{2}\left(x_1^2 + x_2^2\right)}$$

around the point $x_1 = \frac{1}{2}, x_2 = \frac{1}{2}$.

Answer: We need to evaluate the function and its first-order partial derivatives as

$$\begin{aligned} f(x_1, x_2) &= e^{-\frac{1}{2}\left(x_1^2 + x_2^2\right)} \\ &\Longrightarrow f\left(\frac{1}{2}, \frac{1}{2}\right) = e^{-\frac{1}{4}} \\ \frac{\partial f(x_1, x_2)}{\partial x_1} &= -x_1 e^{-\frac{1}{2}\left(x_1^2 + x_2^2\right)} \\ &\Longrightarrow \frac{\partial f\left(\frac{1}{2}, \frac{1}{2}\right)}{\partial x_1} = -\frac{1}{2}e^{-\frac{1}{4}} \\ \frac{\partial f(x_1, x_2)}{\partial x_2} &= -x_2 e^{-\frac{1}{2}\left(x_1^2 + x_2^2\right)} \\ &\Longrightarrow \frac{\partial f\left(\frac{1}{2}, \frac{1}{2}\right)}{\partial x_2} = -\frac{1}{2}e^{-\frac{1}{4}} \end{aligned}$$

so the gradient at $x_1 = \frac{1}{2}, x_2 = \frac{1}{2}$ is

$$\nabla f\left(x^0\right) = \nabla f\left(\frac{1}{2}, \frac{1}{2}\right) = \begin{bmatrix} -\frac{1}{2}e^{-\frac{1}{4}} \\ -\frac{1}{2}e^{-\frac{1}{4}} \end{bmatrix}.$$

The first-order Taylor series around $x_1 = \frac{1}{2}, x_2 = \frac{1}{2}$ then is

$$\begin{aligned} \bar{f}(x) &= f\left(x^0\right) + \nabla f\left(x^0\right)^T\left(x - x^0\right) = e^{-\frac{1}{4}} + \begin{bmatrix} -\frac{1}{2}e^{-\frac{1}{4}} \\ -\frac{1}{2}e^{-\frac{1}{4}} \end{bmatrix}^T \begin{bmatrix} x_1 - \frac{1}{2} \\ x_2 - \frac{1}{2} \end{bmatrix} \\ &= e^{-\frac{1}{4}} + \begin{bmatrix} -\frac{1}{2}e^{-\frac{1}{4}} & -\frac{1}{2}e^{-\frac{1}{4}} \end{bmatrix} \begin{bmatrix} x_1 - \frac{1}{2} \\ x_2 - \frac{1}{2} \end{bmatrix} \\ &= e^{-\frac{1}{4}}\left(1 - \frac{1}{2}\left(x_1 - \frac{1}{2}\right) - \frac{1}{2}\left(x_2 - \frac{1}{2}\right)\right) = e^{-\frac{1}{4}}\left(\frac{3}{2} - \frac{1}{2}x_1 - \frac{1}{2}x_2\right). \end{aligned}$$

The Taylor series is the black plane touching the bell-shaped function at $x_1 = \frac{1}{2}, x_2 = \frac{1}{2}$ in the figure below.

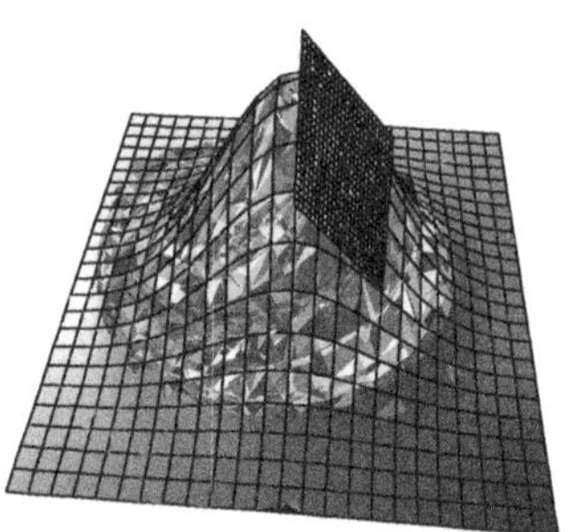

Problem 1.14 *[A*] Consider a first-order Taylor series around x^0 for $f(x)$ given by*

$$f(x) = \bar{f}(x) + E(x) = f\left(x^0\right) + \nabla f\left(x^0\right)^T\left(x - x^0\right) + E(x)$$

where $E(x)$ is the error. Here $f(x)$ is globally concave iff $E(x) < 0$ for $x \neq x^0$, and $f(x)$ is globally convex iff $E(x) > 0$ for $x \neq x_0$. Show that if $f(x)$ is globally concave and x^* satisfies the first-order conditions $\nabla f(x^*) = 0$, then $f(x)$ has a unique global maximum at x^*. Similarly show that if $f(x)$ is globally convex and x^* satisfies the first-order conditions $\nabla f(x^*) = 0$, then $f(x)$ has a unique global minimum at x^*.

Answer: Let x^* satisfy $\nabla f(x^*) = 0$ and suppose $x \neq x^*$ is any other x. To show x^* is a global maximizer we need to show that $f(x) < f(x^*)$. From a Taylor series around x^* and $\nabla f(x^*) = 0$ we have

$$f(x) = f(x^*) + \nabla f(x^*)^T(x - x^*) + E(x) = f(x^*) + E(x) < f(x^*)$$

since concavity of $f(x)$ implies that $E(x) < 0$. Thus $f(x) < f(x^*)$ and so $f(x)$ has a unique global maximum at x^*.

Similarly if $f(x)$ is globally convex we have $E(x) > 0$ and

$$f(x) = f(x^*) + \nabla f(x^*)^T (x - x^*) + E(x) = f(x^*) + E(x) > f(x^*)$$

so that $f(x) > f(x^*)$ and $f(x)$ has a unique global minimum at x^*.

Problem 1.15 *Consider the problem of maximizing or minimizing $f(x)$ subject to the constraint $g(x) = 0$. The Lagrangian then is*

$$\mathcal{L}(\lambda, x) = f(x) + \lambda g(x)$$

and the first-order conditions are

$$\begin{aligned} g(x^*) &= 0 \\ \nabla f(x^*) + \lambda^* \nabla g(x^*) &= 0. \end{aligned}$$

Show that if 1) $f(x)$ is globally concave (convex) and 2) the constraint is linear as $g(x) = a - b^T x$, then x^* corresponds to a constrained global maximum (minimum).

Answer: Let $x \neq x^*$ be any other x which satisfies $g(x) = 0$. Suppose that $f(x)$ is globally concave so that $E(x) < 0$ in the Taylor series around x^*

$$f(x) = f(x^*) + \nabla f(x^*)^T (x - x^*) + E(x).$$

To show x^* corresponds to a constrained global maximum we need to show that: $f(x) < f(x^*)$.

Since $g(x) = a - b^T x$ we have $\nabla g(x) = -b$ so that

$$\begin{aligned} \nabla f(x^*) + \lambda^* \nabla g(x^*) &= 0 \\ &\Longrightarrow \nabla f(x^*) - \lambda^* b = 0 \\ &\Longrightarrow \nabla f(x^*) = \lambda^* b. \end{aligned}$$

Both x and x^* satisfy the constraints so that $g(x) = 0$ and $g(x^*) = 0$. Thus we have

$$\begin{aligned} a - b^T x &= 0 \text{ and } a - b^T x^* = 0 \\ &\Longrightarrow b^T (x - x^*) = 0. \end{aligned}$$

From the Taylor series around x^* we have

$$f(x) = f(x^*) + \overbrace{\nabla f(x^*)^T}^{\lambda^* b^T} (x - x^*) + E(x) = f(x^*) + \lambda^* \overbrace{b^T (x - x^*)}^{0} + E(x) = f(x^*) + E(x) < f(x^*)$$

since $E(x) < 0$ by the concavity of $f(x)$. Thus $f(x) < f(x^*)$ and so x^* corresponds to a constrained global maximum. If $f(x)$ is convex then the proof is the same but with $E(x) > 0$ so that $f(x) > f(x^*)$.

Problem 1.16 *[A*] Consider the problem of maximizing or minimizing $f(x)$ subject to the constraint $g(x) = 0$. A solution to the first-order conditions from the Lagrangian*

$$\mathcal{L}(\lambda, x) = f(x) + \lambda g(x)$$

is given by

$$\begin{aligned} g(x^*) &= 0 \\ \nabla f(x^*) + \lambda^* \nabla g(x^*) &= 0. \end{aligned}$$

Show that if 1) $f(x) = a + b^T x$ is linear, 2) $g(x)$ is globally concave, and 3) $\lambda^* > 0$ then x^* corresponds to a constrained global maximum. That is show that if $x \neq x^*$ and $g(x) = 0$ then $f(x) < f(x^*)$. Similarly show if $g(x)$ is globally convex then x^* corresponds to a constrained global minimum.

Answer: Let $x \neq x^*$ be any other x which satisfies $g(x) = 0$. To show x^* corresponds to a constrained global maximum we need to show that $f(x) < f(x^*)$.

From the first-order conditions and $\nabla f(x) = b$ we have

$$\begin{aligned} \nabla f(x^*) + \lambda^* \nabla g(x^*) &= 0 \\ &\Longrightarrow b + \lambda^* \nabla g(x^*) = 0 \\ &\Longrightarrow b = -\lambda^* \nabla g(x^*). \end{aligned}$$

From a Taylor series around x^* for $g(x)$ we have using $g(x) = g(x^*) = 0$ that

$$\begin{aligned} g(x) &= g(x^*) + \nabla g(x^*)^T (x - x^*) + E_g(x) \\ &\Longrightarrow \nabla g(x^*)^T (x - x^*) + E_g(x) = 0 \\ &\Longrightarrow \nabla g(x^*)^T (x - x^*) = -E_g(x) > 0 \end{aligned}$$

since $g(x)$ is globally concave.

Using $f(x) = a + b^T x$ and $b = -\lambda^* \nabla g(x^*)$ we have

$$f(x) - f(x^*) = \left(a + b^T x\right) - \left(a + b^T x^*\right) = b^T (x - x^*) = -\lambda^* \nabla g(x^*)^T (x - x^*) < 0$$

since $\lambda^* > 0$ and

$$\nabla g(x^*)^T (x - x^*) > 0.$$

Thus

$$f(x) - f(x^*) < 0 \Longrightarrow f(x) < f(x^*)$$

and x^* corresponds to a constrained global maximum.

If $g(x)$ is globally convex the proof is the same except $E(x) > 0$ leading to $f(x) > f(x^*)$ and x^* corresponds to a constrained global minimum.

Problem 1.17 *[A*] A solution to the first-order conditions from the Lagrangian*

$$\mathcal{L}(\lambda, x) = f(x) + \lambda g(x)$$

is given by

$$\begin{aligned} g(x^*) &= 0 \\ \nabla f(x^*) + \lambda^* \nabla g(x^*) &= 0. \end{aligned}$$

Show that if 1) $f(x)$ is globally concave, 2) $g(x)$ is globally concave, and 3) $\lambda^* > 0$ then x^* corresponds to a constrained global maximum. Show that x^* in fact maximizes $f(x)$ for all x such that $g(x) \geq 0$ (and not just all x such that $g(x) = 0$).

Answer: Let $x \neq x^*$ be any other x which satisfies $g(x) = 0$. To show x^* corresponds to a constrained global maximum we need to show that $f(x) < f(x^*)$.

From the first-order conditions we have

$$\begin{aligned} \nabla f(x^*) + \lambda^* \nabla g(x^*) &= 0 \\ &\implies \nabla f(x^*) = -\lambda^* \nabla g(x^*). \end{aligned}$$

From a Taylor series around x^* for $g(x)$ we have using $g(x) = g(x^*) = 0$ that

$$\begin{aligned} g(x) &= g(x^*) + \nabla g(x^*)^T (x - x^*) + E_g(x) \\ &\implies \nabla g(x^*)^T (x - x^*) + E_g(x) = 0. \end{aligned}$$

Since $g(x)$ is globally concave we have

$$\begin{aligned} \nabla g(x^*)^T (x - x^*) + E_g(x) &= 0 \text{ and } E_g(x) < 0 \\ &\implies \nabla g(x^*)^T (x - x^*) > 0. \end{aligned}$$

Thus using $\nabla g(x^*)^T (x - x^*) > 0$ and $\lambda^* > 0$ in

$$f(x) = f(x^*) - \lambda^* g(x^*)^T (x - x^*) + E_f(x) < f(x^*)$$

since $E_f(x) < 0$ by the concavity of $f(x)$. Thus x^* corresponds to a constrained global maximum.

If $g(x) \geq 0$ then from a Taylor series around x^* for $g(x)$ we have using $g(x^*) = 0$ that

$$\begin{aligned} g(x) &= g(x^*) + \nabla g(x^*)^T (x - x^*) + E_g(x) \\ &\implies \nabla g(x^*)^T (x - x^*) + E_g(x) \geq 0 \\ &\implies \nabla g(x^*)^T (x - x^*) \geq -E_g(x) > 0 \end{aligned}$$

so that

$$f(x) = f(x^*) - \lambda^* g(x^*)^T (x - x^*) + E_f(x) < f(x^*)$$

since $E_f(x) < 0$ by the concavity of $f(x)$. Thus x^* is a constrained global maximizer for all x which satisfy $g(x) \geq 0$.

1.3 Positive and Negative Semi-Definite Matrices

Problem 1.18 *[A**] If X is an $n \times p$ matrix, show that $A = X^T X$ is $p \times p$, symmetric and positive semi-definite. Suppose that X is an $n \times p$ matrix and rank $[X] = p$; that is for any vector y: $Xy = 0 \Longrightarrow y = 0$. Show that $A = X^T X$ is positive definite. Notice that $X^T X = X^T IX$ where I is positive definite. [A*] Consider generalizing this result: Assuming that B is positive definite, show that $A = X^T BX$ is positive semi-definite, and that if rank $[X] = p$ then $A = X^T BX$ is positive definite.*

Answer: With these kind of general problems we can't use eigenvalues of determinants to prove the result. Rather we have to go to the original definitions of positive semi-definite and positive definite matrices. If y is any $p \times 1$ vector (we are not using x because the letter x is already used in $A = X^T X$) then A is positive semi-definite if and only if $y^T Ay \geq 0$. Since $A = X^T X$ we have

$$y^T Ay = y^T \left(X^T X\right) y = \left(y^T X^T\right)(Xy) = z^T z$$

where $z = Xy$ and $z^T = (Xy)^T = y^T X^T$. Since X is $n \times p$ and y is $p \times 1$ it follows that z is an $n \times 1$ vector so that

$$\begin{aligned} y^T Ay &= z^T z = \begin{bmatrix} z_1 & z_2 & \cdots & z_n \end{bmatrix} \begin{bmatrix} z_1 \\ z_2 \\ \vdots \\ z_n \end{bmatrix} \\ &= z_1^2 + z_2^2 + \cdots + z_n^2 \geq 0 \end{aligned}$$

since squares are never negative and hence $z_1^2 \geq 0, z_2^2 \geq 0, \cdots, z_n^2 \geq 0$.

To prove that A is positive definite we need to show that $y^T Ay > 0$ for $y \neq 0$ or that

$$y^T Ay = 0 \Longrightarrow y = 0.$$

We have

$$y^T Ay = y^T \left(X^T X\right) y = \left(y^T X^T\right)(Xy) = z^T z = z_1^2 + z_2^2 + \cdots + z_n^2$$

where $z = Xy$. It follows that

$$\begin{aligned} z_1^2 + z_2^2 + \cdots + z_n^2 &= 0 \\ &\Longrightarrow z_1 = 0, z_2 = 0, \cdots, z_n = 0 \Longrightarrow z = 0. \end{aligned}$$

But $z = 0 \Longrightarrow Xy = 0 \Longrightarrow y = 0$. Therefore y is positive definite.

If y is any $p \times 1$ vector then

$$y^T Ay = y^T X^T BXy = z^T Bz \geq 0$$

where $z = Xy$ and $z^T = (Xy)^T = y^T X^T$ and where the inequality follows from the fact that B is positive semi-definite. Since $y^T Ay \geq 0$ it follows that A is positive semi-definite.

Now suppose that $rank[X] = p$. This means that

$$z = Xy = 0 \Longrightarrow y = 0$$

Thus

$$y^T Ay = 0 \Longrightarrow z^T Bz = 0 \Longrightarrow z = 0 \Longrightarrow Xy = 0 \Longrightarrow y = 0.$$

It follows then that $y^T Ay > 0$ for $y \neq 0$ and so A is positive definite.

Problem 1.19 *[A**] If A is positive semi-definite and B is positive semi-definite, show that AB is not necessarily positive definite.*

Answer: Here is a counter-example. Both

$$\begin{aligned} A &= \begin{bmatrix} 5 & -4 \\ -4 & 5 \end{bmatrix} \\ &\Longrightarrow M_1 = 5 > 0, M_2 = 9 > 0 \\ B &= \begin{bmatrix} 1 & 2 \\ 2 & 6 \end{bmatrix} \\ &\Longrightarrow M_1 = 1 > 0, M_2 = 2 > 0 \end{aligned}$$

are positive definite. But

$$AB = \begin{bmatrix} 5 & -4 \\ -4 & 5 \end{bmatrix} \begin{bmatrix} 1 & 2 \\ 2 & 6 \end{bmatrix} = \begin{bmatrix} -3 & -14 \\ 6 & 22 \end{bmatrix}$$

is not positive definite since it has a negative diagonal element, or if we go to the definition since if

$$y = \begin{bmatrix} 2 \\ 0 \end{bmatrix} \neq \begin{bmatrix} 0 \\ 0 \end{bmatrix}$$

then

$$y^T ABy = \begin{bmatrix} 2 & 0 \end{bmatrix} \begin{bmatrix} -3 & -14 \\ 6 & 22 \end{bmatrix} \begin{bmatrix} 2 \\ 0 \end{bmatrix} = -12 < 0$$

and so AB cannot be positive definite. Note AB is not even in general a symmetric matrix since $(AB)^T = B^T A^T = BA \neq AB$. With the example above $BA \neq AB$ since

$$BA = \begin{bmatrix} 1 & 2 \\ 2 & 6 \end{bmatrix} \begin{bmatrix} 5 & -4 \\ -4 & 5 \end{bmatrix} = \begin{bmatrix} -3 & 6 \\ -14 & 22 \end{bmatrix}.$$

Problem 1.20 *[A**] If A and B are both positive semi-definite, show $A + B$ is positive semi-definite.*

Answer: If y is any $p \times 1$ vector with $y \neq 0$ then

$$y^T (A + B) y = y^T Ay + y^T By \geq 0$$

since if A and B are positive semi-definite: $y^T Ay \geq 0$ and $y^T By \geq 0$.

Problem 1.21 *If tr* $[A] < 0$ *can* A *be positive semi-definite?*

Answer: Since $tr\,[A]$ is the sum of the eigenvalues or

$$tr\,[A] = \lambda_1 + \lambda_2 + \cdots + \lambda_n < 0$$

it follows that at least one eigenvalue must be negative and hence A cannot be positive semi-definite.

Problem 1.22 *[B**] If* A *is given by*

$$A = \begin{bmatrix} 8 & -4 \\ -4 & 2 \end{bmatrix}$$

show that A *is positive semi-definite using eigenvalues.*

Answer: The eigenvalues are the solutions to

$$\begin{aligned} \det\left(\begin{bmatrix} 8 & -4 \\ -4 & 2 \end{bmatrix} - \lambda \begin{bmatrix} 1 & 0 \\ 0 & 1 \end{bmatrix} \right) &= 0 \\ &\Longrightarrow \lambda^2 - 10\lambda = 0 \end{aligned}$$

so that $\lambda_1 = 0, \lambda_2 = 10$. Since $\lambda_1 \geq 0, \lambda_2 \geq 0$ is satisfied, it follows that A is positive semi-definite.

Problem 1.23 *[A**] Suppose that* A *is symmetric and positive semi-definite. Consider the problem of calculating the square root of* A *or finding a matrix* B *such that* $A = BB$*. (This turns out to be an important problem in econometrics.) Show that such a matrix* B *exists, and that* B *is symmetric and positive semi-definite.*

Answer: Since A is symmetric we can write A as $A = C\Lambda C^T$ where C is orthogonal or $C^T = C^{-1}$ and Λ is diagonal with the eigenvalues of A along the diagonal or

$$\Lambda = \begin{bmatrix} \lambda_1 & 0 & \cdots & 0 \\ 0 & \lambda_2 & \cdots & 0 \\ \vdots & \vdots & \ddots & \vdots \\ 0 & 0 & 0 & \lambda_n \end{bmatrix}$$

and where $\lambda_i \geq 0$ since A is positive semi-definite. Since $\lambda_i \geq 0$ we can take the positive square root of each of the eigenvalues. Therefore define B by $B = C\Lambda^{\frac{1}{2}}C^T$ where

$$\Lambda^{\frac{1}{2}} = \begin{bmatrix} \lambda_1^{\frac{1}{2}} & 0 & \cdots & 0 \\ 0 & \lambda_n^{\frac{1}{2}} & \cdots & 0 \\ \vdots & \vdots & \ddots & \vdots \\ 0 & 0 & 0 & \lambda_n^{\frac{1}{2}} \end{bmatrix}.$$

You can then verify that B is symmetric, positive semi-definite and that

$$BB = C\Lambda^{\frac{1}{2}}C^T C\Lambda^{\frac{1}{2}}C^T = C\Lambda^{\frac{1}{2}}\Lambda^{\frac{1}{2}}C^T = C\Lambda C^T = A.$$

Problem 1.24 *[A**] For scalars a and b if $a - b \geq 0$ then $\frac{1}{b} - \frac{1}{a} \geq 0$. For example if $a = 5, b = 3$ then $a - b = 2 \geq 0 \Longrightarrow \frac{1}{3} - \frac{1}{5} = \frac{2}{15} \geq 0$. Consider generalizing this to matrices (this turns out to be important for econometrics). Prove that if A and B are positive definite then $A - B$ is positive semi-definite if and only if $B^{-1} - A^{-1}$ is positive semi-definite. Hint: prove that $A - B$ is positive semi-definite if and only if* $\det[A - \lambda B] = 0 \Longrightarrow \lambda \geq 1$.

Answer: Since B is positive definite we can write it as $B = DD$ where $D = D^T$ is symmetric. We then have

$$\begin{aligned}
\det[A - \lambda B] &= \det[A - B - (\lambda - 1) B] = \det[A - B - (\lambda - 1) DD] \\
&= \det\left[D\left(D^{-1}(A - B) D^{-1} - (\lambda - 1) I\right) D\right] \\
&= \det[D] \det\left[D^{-1}(A - B) D^{-1} - (\lambda - 1) I\right] \det[D] \\
&= \det[DD] \det\left[D^{-1}(A - B) D^{-1} - (\lambda - 1) I\right] \\
&= \det[B] \det\left[D^{-1}(A - B) D^{-1} - (\lambda - 1) I\right] \\
&= \det[B] \det[E - (\lambda - 1) I]
\end{aligned}$$

where $\det[B] > 0$ and

$$E = D^{-1}(A - B) D^{-1}.$$

Therefore $\det[A - \lambda B] = 0$ if and only if $\lambda - 1$ is an eigenvalue of the matrix E

Now suppose that $A - B$ is positive semi-definite. It follows then that $E = D^{-1}(A - B)\left(D^{-1}\right)^T$ is positive semi-definite since E has the form $C\Omega C^T$ where $\Omega = A - B$ is positive semi-definite and $C = D^{-1}$. Therefore the eigenvalues of E are non-negative so that $\det[A - \lambda B] = 0$ implies that $\lambda - 1 \geq 0$ or $\lambda \geq 1$.

Now suppose that $\det[A - \lambda B] = 0$ implies that $\lambda \geq 1$. Since E is positive semi-definite

$$x^T(A - B)x = x^T DD^{-1}(A - B)\left(D^{-1}\right)^T D^T x = y^T E y \geq 0$$

where $y = D^T x$. It follows then that $A - B$ is positive semi-definite.

We then have

$$\begin{aligned}
\det[A - \lambda B] &= \det\left[A\left(B^{-1} - \lambda A^{-1}\right) B\right] \\
&= \det[A] \det\left[B^{-1} - \lambda A^{-1}\right] \det[B]
\end{aligned}$$

so that

$$\det[A - \lambda B] = 0 \Longleftrightarrow \det\left[B^{-1} - \lambda A^{-1}\right] = 0.$$

Therefore

$$\begin{aligned}
A - B \text{ is positive semi-definite} &\Longleftrightarrow \det[A - \lambda B] = 0 \Longrightarrow \lambda \geq 1 \\
&\Longleftrightarrow \det\left[B^{-1} - \lambda A^{-1}\right] \Longrightarrow \lambda \geq 1 \\
&\Longleftrightarrow B^{-1} - A^{-1} \text{ is positive semi-definite.}
\end{aligned}$$

Problem 1.25 *[A**] If A and B are symmetric positive semi-definite matrices, show that ABA is positive semi-definite.*

Answer: A is positive semi-definite if and only if $x^T Ax \geq 0$ for all x, where x is an $n \times 1$ vector. Furthermore

$$x^T ABAx = y^T By \geq 0$$

where $y = Ax$ and $y^T = x^T A^T = x^T A$ and $y^T By \geq 0$ since B is positive semi-definite.

Problem 1.26 *[A**] If P is symmetric and idempotent ($PP = P$) show that the eigenvalues of P are all either 1 or 0. Show that P is positive semi-definite. What can you say about P if it is also positive definite?*

Answer: If P is idempotent then an eigenvector $x \neq 0$ and eigenvalue λ satisfy $Px = \lambda x$. Multiplying both sides by P we find that

$$\begin{aligned} PPx &= \lambda Px \Longrightarrow \lambda x = \lambda^2 x \\ &\Longrightarrow \lambda (1 - \lambda) x = 0 \end{aligned}$$

since $Px = \lambda x$ and $PPx = Px = \lambda x$. Since $x \neq 0$, $\lambda (1 - \lambda) = 0$ so that $\lambda = 0$ or $\lambda = 1$. Since $\lambda = 0$ or $\lambda = 1$, $\lambda \geq 0$ and so P is positive semi-definite.

If P is positive definite then all eigenvalues must be greater than 0 or all eigenvalues must be 1. Then from the representation $P = C\Lambda C^T$ where $C^T = C^{-1}$ and Λ is a diagonal matrix with eigenvalues along the diagonal it follows that $\Lambda = I$ and so $P = C\Lambda C^T = CIC^{-1} = CC^{-1} = I$.

1.4 Advanced Concavity and Convexity

Problem 1.27 *[A**] Show that $y = f(x) = |x|$ is convex, but not strictly convex.*

Answer: If x_1 and x_2 are both negative or positive then since $\lambda \geq 0$ we have

$$|\lambda x_1 + (1 - \lambda) x_2| = \lambda |x_1| + (1 - \lambda) |x_2| .$$

For example with $\lambda = \frac{1}{5}, x_1 = 10, x_2 = 20$ we have

$$\left| \frac{1}{5} \times 10 + \left(1 - \frac{1}{5}\right) \times 20 \right| = \frac{1}{5} |10| + \left(1 - \frac{1}{5}\right) |20| = 18$$

while if $\lambda = \frac{1}{5}, x_1 = -10, x_2 = -20$ we have

$$\left| \frac{1}{5} \times (-10) + \left(1 - \frac{1}{5}\right) \times (-20) \right| = \frac{1}{5} |-10| + \left(1 - \frac{1}{5}\right) |-20| = 18.$$

If x_1 and x_2 have different signs then

$$|\lambda x_1 + (1 - \lambda) x_2| < \lambda |x_1| + (1 - \lambda) |x_2| .$$

For example with $\lambda = \frac{1}{5}, x_1 = 10, x_2 = -20$ we have

$$\left|\frac{1}{5} \times 10 + \left(1 - \frac{1}{5}\right) \times (-20)\right| = 14 < \frac{1}{5}|10| + \left(1 - \frac{1}{5}\right)|-20| = 18$$

while if $\lambda = \frac{1}{5}, x_1 = -10, x_2 = 20$ we have

$$\left|\frac{1}{5} \times (-10) + \left(1 - \frac{1}{5}\right) \times 20\right| = 14 < \frac{1}{5}|-10| + \left(1 - \frac{1}{5}\right)|20| = 18.$$

In general

$$\begin{aligned}
|\lambda x_1 + (1-\lambda) x_2| &= \lambda |x_1| + (1-\lambda)|x_2| \text{ if } x_1 \geq 0 \text{ and } x_2 \geq 0 \\
|\lambda x_1 + (1-\lambda) x_2| &< \lambda |x_1| + (1-\lambda)|x_2| \text{ if } x_1 \geq 0 \text{ and } x_2 < 0 \\
|\lambda x_1 + (1-\lambda) x_2| &< \lambda |x_1| + (1-\lambda)|x_2| \text{ if } x_1 < 0 \text{ and } x_2 \geq 0 \\
|\lambda x_1 + (1-\lambda) x_2| &= \lambda |x_1| + (1-\lambda)|x_2| \text{ if } x_1 < 0 \text{ and } x_2 < 0.
\end{aligned}$$

Thus

$$\begin{aligned}
f(\lambda x_1 + (1-\lambda) x_2) &= |\lambda x_1 + (1-\lambda) x_2| \\
&\leq \lambda f(x_1) + (1-\lambda) f(x_2) = \lambda |x_1| + (1-\lambda)|x_2|
\end{aligned}$$

and so by definition $f(x) = |x|$ is convex.

But $f(x) = |x|$ is **not** strictly convex since strict convexity requires that

$$f(\lambda x_1 + (1-\lambda) x_2) = |\lambda x_1 + (1-\lambda) x_2|$$

be less than

$$\lambda f(x_1) + (1-\lambda) f(x_2) = \lambda |x_1| + (1-\lambda)|x_2|$$

for all $x_1 \neq x_2$ and $0 < \lambda < 1$. But this is violated for any two points on a linear segment. For example if $\lambda = \frac{1}{5}$ and $x_1 = 10$ and $x_2 = 20$ then

$$\left|\frac{1}{5} \times 10 + \left(1 - \frac{1}{5}\right) \times 20\right| = \frac{1}{5}|10| + \left(1 - \frac{1}{5}\right)|20| = 18.$$

Problem 1.28 *[A**] Consider the function*

$$\mu(x) = \begin{cases} x \text{ if } x \geq 0 \\ 0 \text{ if } x < 0 \end{cases}.$$

Show that $\mu(x)$ is convex but not strictly convex.

Answer: For $0 \leq \lambda \leq 1$ if $\lambda x_1 + (1-\lambda) x_2 \geq 0$ then

$$\mu(\lambda x_1 + (1-\lambda) x_2) = \lambda x_1 + (1-\lambda) x_2 \Longrightarrow \mu(\lambda x_1 + (1-\lambda) x_2) \geq \lambda x_1 + (1-\lambda) x_2.$$

If $\lambda x_1 + (1-\lambda) x_2 < 0$ then

$$\mu(\lambda x_1 + (1-\lambda) x_2) = 0 > \lambda x_1 + (1-\lambda) x_2 \Longrightarrow \mu(\lambda x_1 + (1-\lambda) x_2) \geq \lambda x_1 + (1-\lambda) x_2.$$

Thus it is always the case that

$$\mu(\lambda x_1 + (1-\lambda) x_2) \geq \lambda x_1 + (1-\lambda) x_2$$

and so $\mu(x)$ is convex. It is not strictly convex since if $\lambda x_1 + (1-\lambda) x_2 \geq 0$

$$\mu(\lambda x_1 + (1-\lambda) x_2) = \lambda x_1 + (1-\lambda) x_2$$

and not

$$\mu(\lambda x_1 + (1-\lambda) x_2) > \lambda x_1 + (1-\lambda) x_2$$

as required by strict convexity.

Problem 1.29 *[A**] Show that if $Q = F(L, K)$ is homogeneous of degree 1 then it cannot be strictly concave. Show that if $Q = F(L, K)$ is homogeneous of degree greater than 1 then it cannot be concave.*

Answer: Suppose that $F(L, K)$ is homogeneous of degree 1. For strict concavity we require for $L_2 \neq L_1$ and $K_2 \neq K_1$ and $0 < \lambda < 1$ that

$$F(\lambda L_1 + (1-\lambda) L_2, \lambda K_1 + (1-\lambda) K_2) > \lambda F(L_1, K_1) + (1-\lambda) F(L_2, K_2).$$

We will show this is not the case for $L_2 = \tau L_1$ and $K_2 = \tau K_1$ for $\tau > 1$. Clearly $L_2 \neq L_1$ and $K_2 \neq K_1$ but

$$\begin{aligned} F(\lambda L_1 + (1-\lambda) L_2, \lambda K_1 + (1-\lambda) K_2) &= F((\lambda + (1-\lambda)\tau) L_1, (\lambda + (1-\lambda)\tau) K_1) \\ &= (\lambda + (1-\lambda)\tau) F(L_1, K_1) \\ &= \lambda F(L_1, K_1) + (1-\lambda)\tau F(L_1, K_1) \\ &= \lambda F(L_1, K_1) + (1-\lambda) F(\tau L_1, \tau K_1) \\ &= \lambda F(L_1, K_1) + (1-\lambda) F(L_2, K_2). \end{aligned}$$

Now suppose that $F(L, K)$ is homogeneous of degree $k > 1$. For concavity we require for $0 < \lambda < 1$ that

$$F(\lambda L_1 + (1-\lambda) L_2, \lambda K_1 + (1-\lambda) K_2) \geq \lambda F(L_1, K_1) + (1-\lambda) F(L_2, K_2).$$

We will show this is not the case for $L_2 = \tau L_1, K_2 = \tau K_1$ with $\tau > 1$. Now

$$\begin{aligned} F(\lambda L_1 + (1-\lambda) L_2, \lambda K_1 + (1-\lambda) K_2) &= F((\lambda + (1-\lambda)\tau) L_1, (\lambda + (1-\lambda)\tau) K_1) \\ &= (\lambda + (1-\lambda)\tau)^k F(L_1, K_1) \\ &< \left(\lambda 1^k + (1-\lambda)\tau^k\right) F(L_1, K_1) \end{aligned}$$

since x^k convex for $k > 1$ implies that $(\lambda + (1-\lambda)\tau)^k < \left(\lambda 1^k + (1-\lambda)\tau^k\right)$. Now

$$\begin{aligned} \left(\lambda 1^k + (1-\lambda)\tau^k\right) F(L_1, K_1) &= \lambda F(L_1, K_1) + (1-\lambda)\tau^k F(L_1, K_1) \\ &= \lambda F(L_1, K_1) + (1-\lambda) F(\tau L_1 \tau, K_1) \\ &= \lambda F(L_1, K_1) + (1-\lambda) F(L_2, K_2) \end{aligned}$$

and so we conclude that

$$F(\lambda L_1 + (1-\lambda) L_2, \lambda K_1 + (1-\lambda) K_2) < \lambda F(L_1, K_1) + (1-\lambda) F(L_2, K_2).$$

Problem 1.30 *If $f(x)$ is convex then (by definition) for a weighted sum of x_1, x_2 satisfies*

$$f(\lambda x_1 + (1-\lambda)x_2) \leq \lambda f(x_1) + (1-\lambda) f(x_2)$$

*for $0 < \lambda < 1$. [A**] Generalize this to n weighted sums. That is show that if $\lambda_i \geq 0$ for $i = 1, 2, \ldots, n$ satisfy*

$$\lambda_1 + \lambda_2 + \cdots + \lambda_n = 1$$

then

$$f(\lambda_1 x_1 + \lambda_2 x_2 + \cdots + \lambda_n x_n) \leq \lambda_1 f(x_1) + \lambda_2 f(x_2) + \cdots + \lambda_n f(x_n).$$

This is known as Jensen's inequality.

Answer: Consider a proof by induction. The result is true for $n = 2$ since if $\lambda_1 = \lambda, \lambda_2 = 1-\lambda$ then

$$f(\lambda_1 x_1 + \lambda_2 x_2) = f(\lambda x_1 + (1-\lambda)x_2) \leq \lambda f(x_1) + (1-\lambda) f(x_2) = \lambda_1 f(x_1) + \lambda_2 f(x_2).$$

Now suppose the result is true for up to $n-1$. The proof consists of showing that the result is true for n.

If $\lambda_n = 1$ the result is true trivially so assume $\lambda_n < 1$. Then

$$\begin{aligned} \lambda_1 x_1 + \lambda_2 x_2 + \cdots + \lambda_n x_n &= (1-\lambda_n)\left(\frac{\lambda_1}{1-\lambda_n}x_1 + \frac{\lambda_2}{1-\lambda_n}x_2 + \cdots \frac{\lambda_{n-1}}{1-\lambda_n}x_{n-1}\right) + \lambda_n x_n \\ &= (1-\lambda_n)\tilde{x}_1 + \lambda_n x_n \end{aligned}$$

where

$$\tilde{x}_1 = \frac{\lambda_1}{1-\lambda_n}x_1 + \frac{\lambda_2}{1-\lambda_n}x_2 + \cdots + \frac{\lambda_{n-1}}{1-\lambda_n}x_{n-1}.$$

Thus

$$f(\lambda_1 x_1 + \lambda_2 x_2 + \cdots + \lambda_n x_n) = f((1-\lambda_n)\tilde{x}_1 + \lambda_n x_n) \leq (1-\lambda_n) f(\tilde{x}_1) + \lambda_n f(x_n).$$

Now since the result is true for $n-1$ we have

$$\begin{aligned} f(\tilde{x}_1) &= f\left(\frac{\lambda_1}{1-\lambda_n}x_1 + \frac{\lambda_2}{1-\lambda_n}x_2 + \cdots + \frac{\lambda_{n-1}}{1-\lambda_n}x_{n-1}\right) \\ &\leq \frac{\lambda_1}{1-\lambda_n}f(x_1) + \frac{\lambda_2}{1-\lambda_n}f(x_2) + \cdots + \frac{\lambda_{n-1}}{1-\lambda_n}f(x_{n-1}) \end{aligned}$$

since

$$\frac{\lambda_1}{1-\lambda_n} + \frac{\lambda_2}{1-\lambda_n} + \cdots + \frac{\lambda_{n-1}}{1-\lambda_n} = \frac{1-\lambda_1-\lambda_2-\cdots-\lambda_{n-1}}{1-\lambda_n} = \frac{1-\lambda_n}{1-\lambda_n} = 1$$

which follows from

$$\lambda_1 + \lambda_2 + \cdots + \lambda_n = 1 \Longrightarrow \lambda_1 + \lambda_2 + \cdots + \lambda_{n-1} = 1 - \lambda_n.$$

Thus

$$\begin{aligned}
f\left(\lambda_1 x_1+\lambda_2 x_2+\cdots+\lambda_n x_n\right) &= f\left(\left(1-\lambda_n\right)\tilde{x}_1+\lambda_n x_n\right)\\
&\leq \left(1-\lambda_n\right) f\left(\tilde{x}_1\right)+\lambda_n f\left(x_n\right)\\
&\leq \left(1-\lambda_n\right)\left(\begin{array}{c}\frac{\lambda_1}{1-\lambda_n} f\left(x_1\right)+\frac{\lambda_2}{1-\lambda_n} f\left(x_2\right)\\ +\cdots+\frac{\lambda_{n-1}}{1-\lambda_n} f\left(x_{n-1}\right)\end{array}\right)+\lambda_n f\left(x_n\right)\\
&= \lambda_1 f\left(x_1\right)+\lambda_2 f\left(x_2\right)+\cdots+\lambda_{n-1} f\left(x_{n-1}\right)+\lambda_n f\left(x_n\right).
\end{aligned}$$

Problem 1.31 *[A**] Show that the Cobb-Douglas functional form*

$$Q=F(L)=L_1^{\alpha_1} L_2^{\alpha_2}\cdots L_n^{\alpha_n}$$

with $\alpha_i>0$ for $i=1,2,\ldots,n$ is strictly concave if $\sum_{i=1}^n \alpha_i<1$ (decreasing returns to scale) and is concave if $\sum_{i=1}^n \alpha_i=1$ (constant returns to scale).

Answer: Define

$$\alpha \equiv \alpha_1+\alpha_2+\cdots+\alpha_n.$$

Our goal is to show that $F(L)$ is strictly concave if $\alpha<1$, and concave if $\alpha=1$. To this end we have

$$\begin{aligned}
\frac{\partial^2 F(L)}{\partial L_i^2} &= \frac{\alpha_i\left(\alpha_i-1\right) F(L)}{L_i^2}=\frac{\alpha_i^2 F(L)}{L_i^2}-\frac{\alpha_i F(L)}{L_i^2}\\
\frac{\partial F(L)}{\partial L_i \partial L_j} &= \frac{\alpha_i \alpha_j F(L)}{L_i L_j} \text{ for } i \neq j
\end{aligned}$$

so that the Hessian can be written as

$$H(L)=F(L)\left(a(L)\, a(L)^T-A(L)\right)$$

where

$$a(L)=\begin{bmatrix}\frac{\alpha_1}{L_1}\\ \frac{\alpha_2}{L_2}\\ \vdots\\ \frac{\alpha_n}{L_n}\end{bmatrix}, A(L)=\begin{bmatrix}\frac{\alpha_1}{L_1^2} & 0 & \cdots & 0\\ 0 & \frac{\alpha_2}{L_2^2} & 0 & \vdots\\ \vdots & \ddots & \ddots & 0\\ 0 & \cdots & 0 & \frac{\alpha_n}{L_n^2}\end{bmatrix}.$$

The quadratic form then is

$$\begin{aligned}
x^T H(L)\, x &= F(L)\left(x^T a(L)\, a(L)^T x-x^T A(L)\, x\right)\\
&= F(L)\left(\left(a(L)^T x\right)^2-x^T A(L)\, x\right)\\
&= F(L)\left(\begin{array}{c}\left(\frac{\alpha_1 x_1}{L_1}+\frac{\alpha_2 x_2}{L_2}+\cdots+\frac{\alpha_n x_n}{L_n}\right)^2\\ -\left(\frac{\alpha_1 x_1^2}{L_1^2}+\frac{\alpha_2 x_2^2}{L_2^2}+\cdots+\frac{\alpha_n x_n^2}{L_n^2}\right)\end{array}\right).
\end{aligned}$$

Now

$$\left(\frac{\alpha_1 x_1}{L_1}+\frac{\alpha_2 x_2}{L_2}+\cdots+\frac{\alpha_n x_n}{L_n}\right)^2 = \left(\alpha\left(\begin{array}{c}\frac{\frac{\alpha_1}{\alpha}x_1}{L_1}+\frac{\frac{\alpha_2}{\alpha}x_2}{L_2}\\+\cdots+\frac{\frac{\alpha_n}{\alpha}x_n}{L_n}\end{array}\right)\right)^2 = \alpha^2\left(\begin{array}{c}\lambda_1\frac{x_1}{L_1}+\lambda_2\frac{\frac{\alpha_2}{\alpha}x_2}{L_2}\\+\cdots+\lambda_n\frac{\frac{\alpha_n}{\alpha}x_n}{L_n}\end{array}\right)^2$$

and

$$\frac{\alpha_1 x_1^2}{L_1^2}+\frac{\alpha_2 x_2^2}{L_2^2}+\cdots+\frac{\alpha_n x_n^2}{L_n^2}=\alpha\left(\begin{array}{c}\lambda_1\left(\frac{x_1}{L_1}\right)^2+\lambda_2\left(\frac{x_2}{L_2}\right)^2\\+\cdots+\lambda_n\left(\frac{x_n}{L_n}\right)^2\end{array}\right)$$

where $\lambda_i=\frac{\alpha_i}{\alpha}$ and

$$\begin{aligned}\lambda_1+\lambda_2+\cdots+\lambda_n &= \frac{\alpha_1}{\alpha}+\frac{\alpha_2}{\alpha}+\cdots+\frac{\alpha_n}{\alpha}\\ &= \frac{\alpha_1+\alpha_2+\cdots+\alpha_n}{\alpha}=1.\end{aligned}$$

Since x^2 is a convex function we have

$$\alpha^2\left(\lambda_1\frac{x_1}{L_1}+\lambda_2\frac{\frac{\alpha_2}{\alpha}x_2}{L_2}+\cdots+\lambda_n\frac{\frac{\alpha_n}{\alpha}x_n}{L_n}\right)^2\le\alpha^2\left(\begin{array}{c}\lambda_1\left(\frac{x_1}{L_1}\right)^2+\lambda_2\left(\frac{x_2}{L_2}\right)^2\\+\cdots+\lambda_n\left(\frac{x_n}{L_n}\right)^2\end{array}\right)$$

so that

$$x^T H(L)\,x\le F(L)\,\alpha(\alpha-1)\left(\begin{array}{c}\lambda_1\left(\frac{x_1}{L_1}\right)^2+\lambda_2\left(\frac{x_2}{L_2}\right)^2\\+\cdots+\lambda_n\left(\frac{x_n}{L_n}\right)^2\end{array}\right).$$

Thus if $0<\alpha<1$ we conclude that $x^T H(L)\,x<0$ for $x\neq 0$ and so $H(L)$ is negative definite for all L, and so $F(L)$ is strictly concave. If $\alpha=1$ then $x^T H(L)\,x\le 0$ and so $H(L)$ is negative semi-definite for all L, and so $F(L)$ is concave.

Problem 1.32 *[A**] If $f(x)=x^T Ax$ where A is a positive definite matrix, show that $f(x)$ is strictly convex 1) using calculus and 2) without using calculus.*

Answer: The Hessian of $f(x)=x^T Ax$ is $H(x)=2A$. Since A is positive definite it follows that $H(x)$ is positive definite for all x and hence that $f(x)$ is strictly convex.

To show strict convexity without calculus we need to show that for $0<\lambda<1$ and $x_1\neq x_2$ that

$$f(\lambda x_1+(1-\lambda)x_2)<\lambda f(x_1)+(1-\lambda)f(x_2)$$

or

$$\left(\lambda x_1 + (1-\lambda)\, x_2\right)^T A \left(\lambda x_1 + (1-\lambda)\, x_2\right) < \lambda x_1^T A x_1 + (1-\lambda)\, x_2^T A x_2.$$

Consider subtracting the left side of the inequality and simplifying to show

$$\left(\lambda x_1 + (1-\lambda)\, x_2\right)^T A \left(\lambda x_1 + (1-\lambda)\, x_2\right) - \lambda x_1^T A x_1 + (1-\lambda)\, x_2^T A x_2$$

equals

$$-\lambda (1-\lambda) (x_1 - x_2)^T A (x_1 - x_2)$$

so

$$\begin{aligned}\left(\lambda x_1 + (1-\lambda)\, x_2\right)^T A \left(\lambda x_1 + (1-\lambda)\, x_2\right) &= \lambda x_1^T A x_1 + (1-\lambda)\, x_2^T A x_2 \\ &\quad -\lambda (1-\lambda) (x_1 - x_2)^T A (x_1 - x_2).\end{aligned}$$

Since $0 < \lambda < 1$, $x_1 - x_2 \neq 0$, and A is positive definite it follows that

$$\lambda (1-\lambda) (x_1 - x_2)^T A (x_1 - x_2) > 0$$

so that

$$\left(\lambda x_1 + (1-\lambda)\, x_2\right)^T A \left(\lambda x_1 + (1-\lambda)\, x_2\right) < \lambda x_1^T A x_1 + (1-\lambda)\, x_2^T A x_2.$$

Chapter 2

Total Differentials

2.1 Linear Models

Problem 2.1 *[D***] Consider the following implicit function*

$$g(y, x_1, x_2) = 3y - 6x_1 - 5x_2 = 0.$$

Solve for the reduced form and find $\frac{\partial y}{\partial x_1}$ *and* $\frac{\partial y}{\partial x_2}$ *from the reduced form. Now find* $\frac{\partial y}{\partial x_1}$ *and* $\frac{\partial y}{\partial x_2}$ *from the total differential.*

Answer: We have

$$g(y, x_1, x_2) = 3y - 6x_1 - 5x_2 = 0 \Longrightarrow y = \frac{6}{3}x_1 + \frac{5}{3}x_2$$

and so

$$\begin{aligned} y &= f(x_1, x_2) = 2x_1 + \frac{5}{3}x_2 \\ &\Longrightarrow \frac{\partial y}{\partial x_1} = 2, \frac{\partial y}{\partial x_2} = \frac{5}{3}. \end{aligned}$$

The total differential is

$$3dy - 6dx_1 - 5dx_2 = 0.$$

To find $\frac{\partial y}{\partial x_1}$ we 1) set $dx_2 = 0$ and 2) replace the remaining $d's$ with $\partial's$ as

$$\begin{aligned} 3\partial y - 6\partial x_1 &= 0 \\ &\Longrightarrow \frac{\partial y}{\partial x_1} = 2. \end{aligned}$$

To find $\frac{\partial y}{\partial x_2}$ we 1) set $dx_1 = 0$ and 2) replace the remaining $d's$ with $\partial's$ as

$$\begin{aligned} 3\partial y - 5\partial x_2 &= 0 \\ &\Longrightarrow \frac{\partial y}{\partial x_2} = \frac{5}{3}. \end{aligned}$$

Problem 2.2 *[D***] Consider the following system of equations*

$$\begin{aligned} 3y_1 + 4y_2 - 6x_1 - 5x_2 - 5 &= 0 \\ 2y_1 + 5y_2 + 6x_1 + 5x_2 + 5 &= 0. \end{aligned}$$

Write these equations in matrix form and solve for the reduced form. Find $\frac{\partial y_1}{\partial x_1}, \frac{\partial y_2}{\partial x_1}$ *from the reduced form. Now find* $\frac{\partial y_1}{\partial x_1}, \frac{\partial y_2}{\partial x_1}$ *from the total differential using Cramer's rule.*

Answer: We have

$$\begin{bmatrix} 3 & 4 \\ 2 & 5 \end{bmatrix} \begin{bmatrix} y_1 \\ y_2 \end{bmatrix} = \begin{bmatrix} 1 \\ -1 \end{bmatrix} (6x_1 + 5x_2 + 5).$$

Since

$$\det[A] = \det \begin{bmatrix} 3 & 4 \\ 2 & 5 \end{bmatrix} = 7 \neq 0$$

it follows that A^{-1} exists and

$$\begin{aligned} \begin{bmatrix} y_1 \\ y_2 \end{bmatrix} &= \begin{bmatrix} 3 & 4 \\ 2 & 5 \end{bmatrix}^{-1} \begin{bmatrix} 1 \\ -1 \end{bmatrix} (6x_1 + 5x_2 + 5) \\ &= \begin{bmatrix} \frac{9}{7} \\ -\frac{5}{7} \end{bmatrix} (6x_1 + 5x_2 + 5) \end{aligned}$$

so that

$$\begin{aligned} y_1 &= f_1(x_1, x_2) = \frac{9}{7}(6x_1 + 5x_2 + 5) \\ &\Longrightarrow \frac{\partial y_1}{\partial x_1} = \frac{9}{7}6 = \frac{54}{7} \\ y_2 &= f_2(x_1, x_2) = -\frac{5}{7}(6x_1 + 5x_2 + 5) \\ &\Longrightarrow \frac{\partial y_2}{\partial x_1} = -\frac{30}{7}. \end{aligned}$$

Alternatively we have the total differential

$$\begin{aligned} 3dy_1 + 4dy_2 - 6dx_1 - 5dx_2 &= 0 \\ 2dy_1 + 5dy_2 + 6dx_1 + 5dx_2 &= 0 \end{aligned}$$

which in matrix notation becomes

$$\begin{bmatrix} 3 & 4 \\ 2 & 5 \end{bmatrix} \begin{bmatrix} dy_1 \\ dy_2 \end{bmatrix} = \begin{bmatrix} 6dx_1 + 5dx_2 \\ -6dx_1 - 5dx_2 \end{bmatrix}.$$

To find $\frac{\partial y_1}{\partial x_1}, \frac{\partial y_2}{\partial x_1}$ we 1) set $dx_2 = 0$ and 2) replace the remaining $d's$ with $\partial's$ as

$$\begin{aligned} \begin{bmatrix} 3 & 4 \\ 2 & 5 \end{bmatrix} \begin{bmatrix} \partial y_1 \\ \partial y_2 \end{bmatrix} &= \begin{bmatrix} 6 \\ -6 \end{bmatrix} \partial x_1 \\ &\Longrightarrow \begin{bmatrix} 3 & 4 \\ 2 & 5 \end{bmatrix} \begin{bmatrix} \frac{\partial y_1}{\partial x_1} \\ \frac{\partial y_2}{\partial x_1} \end{bmatrix} = \begin{bmatrix} 6 \\ -6 \end{bmatrix}. \end{aligned}$$

Using Cramer's rule

$$\frac{\partial y_1}{\partial x_1} = \frac{\det\begin{bmatrix} 6 & 4 \\ -6 & 5 \end{bmatrix}}{\det\begin{bmatrix} 3 & 4 \\ 2 & 5 \end{bmatrix}} = \frac{54}{7}$$

$$\frac{\partial y_2}{\partial x_1} = \frac{\det\begin{bmatrix} 3 & 6 \\ 2 & -6 \end{bmatrix}}{\det\begin{bmatrix} 3 & 4 \\ 2 & 5 \end{bmatrix}} = -\frac{30}{7}.$$

Problem 2.3 *[C**] Consider the following structural model*

$$\begin{aligned} y_1 &= y_2^2 x_1^3 x_2^4 e^6 \\ y_2 &= y_1 x_1^2 x_2^{-3} e^5 \end{aligned}$$

where $y_1 > 0, y_2 > 0$ are the endogenous variables and $x_1 > 0$ and $x_2 > 0$ are the exogenous variables. Using the $\ln(\ )$ function rewrite the structural model as two linear implicit functions in terms of

$$\ln(y_1), \ln(y_2), \ln(x_1), \ln(x_2).$$

Solve the implicit functions for $\ln(y_1), \ln(y_2)$ and use this to find the reduced form $y_1 = f_1(x_1, x_2)$ and $y_2 = f_1(x_1, x_2)$. From the reduced form show that $\frac{\partial y_1}{\partial x_2} > 0$. Now redo the calculation but use total differentials.

Answer: We have

$$\begin{aligned} y_1 &= y_2^2 x_1^3 x_2^4 e^6 \Longrightarrow \ln(y_1) = 2\ln(y_2) + 3\ln(x_1) + 4\ln(x_2) + 6 \\ y_2 &= y_1 x_1^2 x_2^{-3} e^5 \Longrightarrow \ln(y_2) = \ln(y_1) + 2\ln(x_1) - 3\ln(x_2) + 5 \end{aligned}$$

or

$$\begin{aligned} \ln(y_1) &= 2\ln(y_2) + 3\ln(x_1) + 4\ln(x_2) + 6 \\ &\Longrightarrow \ln(y_1) - 2\ln(y_2) - 3\ln(x_1) - 4\ln(x_2) - 6 = 0 \\ \ln(y_2) &= \ln(y_1) + 2\ln(x_1) - 3\ln(x_2) + 5 \\ &\Longrightarrow -\ln(y_1) + \ln(y_2) - 2\ln(x_1) + 3\ln(x_2) - 5 = 0. \end{aligned}$$

Thus

$$\begin{aligned} g_1(y_1, y_2, x_1, x_2) &= \ln(y_1) - 2\ln(y_2) - 3\ln(x_1) - 4\ln(x_2) - 6 = 0 \\ g_2(y_1, y_2, x_1, x_2) &= -\ln(y_1) + \ln(y_2) - 2\ln(x_1) + 3\ln(x_2) - 5 = 0. \end{aligned}$$

Using matrix notation

$$\underbrace{\begin{bmatrix} 1 & -2 \\ -1 & 1 \end{bmatrix}}_{A}\begin{bmatrix} \ln(y_1) \\ \ln(y_2) \end{bmatrix} = \begin{bmatrix} 3\ln(x_1) + 4\ln(x_2) + 6 \\ 2\ln(x_1) - 3\ln(x_2) + 5 \end{bmatrix}.$$

Here

$$A^{-1} = \frac{1}{\det[A]} adj\,[A] = \frac{1}{-1}\begin{bmatrix} 1 & 2 \\ 1 & 1 \end{bmatrix} = \begin{bmatrix} -1 & -2 \\ -1 & -1 \end{bmatrix}.$$

We then have

$$\begin{aligned} \begin{bmatrix} \ln(y_1) \\ \ln(y_2) \end{bmatrix} &= \begin{bmatrix} 1 & -2 \\ -1 & 1 \end{bmatrix}^{-1} \begin{bmatrix} 3\ln(x_1) + 4\ln(x_2) + 6 \\ 2\ln(x_1) - 3\ln(x_2) + 5 \end{bmatrix} \\ &= \begin{bmatrix} -1 & -2 \\ -1 & -1 \end{bmatrix} \begin{bmatrix} 3\ln(x_1) + 4\ln(x_2) + 6 \\ 2\ln(x_1) - 3\ln(x_2) + 5 \end{bmatrix} \\ &= \begin{bmatrix} -7\ln(x_1) + 2\ln(x_2) - 16 \\ -5\ln(x_1) - \ln(x_2) - 11 \end{bmatrix} \end{aligned}$$

so that

$$\begin{aligned} \ln(y_1) &= -7\ln(x_1) + 2\ln(x_2) - 16 \Longrightarrow y_1 = e^{-16}x_1^{-7}x_2^2 \\ \ln(y_2) &= -5\ln(x_1) - \ln(x_2) - 11 \Longrightarrow y_2 = e^{-11}x_1^{-5}x_2^{-1} \end{aligned}$$

and

$$y_1 = e^{-16}x_1^{-7}x_2^2 \Longrightarrow \frac{\partial y_1}{\partial x_2} = 2x_1^{-7}x_2e^{-16} = 2\frac{y_1}{x_2} > 0.$$

The implicit functions in terms of y_1, y_2, x_1, x_2 are

$$\begin{aligned} g_1(y_1, y_2, x_1, x_2) &= \ln(y_1) - 2\ln(y_2) - 3\ln(x_1) - 4\ln(x_2) - 6 = 0 \\ g_2(y_1, y_2, x_1, x_2) &= -\ln(y_1) + \ln(y_2) - 2\ln(x_1) + 3\ln(x_2) - 5 = 0. \end{aligned}$$

The total differential is

$$\begin{aligned} \frac{1}{y_1}dy_1 - \frac{2}{y_2}dy_2 - \frac{3}{x_1}dx_1 - \frac{4}{x_2}dx_2 &= 0 \\ -\frac{1}{y_1}dy_1 + \frac{1}{y_2}dy_2 - \frac{2}{x_1}dx_1 + \frac{3}{x_2}dx_2 &= 0 \end{aligned}$$

or in matrix notation

$$\begin{bmatrix} \frac{1}{y_1} & -\frac{2}{y_2} \\ -\frac{1}{y_1} & \frac{1}{y_2} \end{bmatrix} \begin{bmatrix} dy_1 \\ dy_2 \end{bmatrix} = \begin{bmatrix} \frac{3}{x_1}dx_1 + \frac{4}{x_2}dx_2 \\ \frac{2}{x_1}dx_1 - \frac{3}{x_2}dx_2 \end{bmatrix}.$$

To find $\frac{\partial y_1}{\partial x_2}$ we 1) set $dx_1 = 0$ and 2) change the remaining $d's$ to $\partial's$ as

$$\begin{bmatrix} \frac{1}{y_1} & -\frac{2}{y_2} \\ -\frac{1}{y_1} & \frac{1}{y_2} \end{bmatrix} \begin{bmatrix} \frac{\partial y_1}{\partial x_2} \\ \frac{\partial y_2}{\partial x_2} \end{bmatrix} = \begin{bmatrix} \frac{4}{x_2} \\ -\frac{3}{x_2} \end{bmatrix}$$

so that

$$\frac{\partial y_1}{\partial x_2} = \frac{\det\begin{bmatrix} \frac{4}{x_2} & -\frac{2}{y_2} \\ -\frac{3}{x_2} & \frac{1}{y_2} \end{bmatrix}}{\det\begin{bmatrix} \frac{1}{y_1} & -\frac{2}{y_2} \\ -\frac{1}{y_1} & \frac{1}{y_2} \end{bmatrix}} = \frac{-\frac{2}{x_2y_2}}{-\frac{1}{y_1y_2}} = 2\frac{y_1}{x_2} > 0.$$

Problem 2.4 *[D***] Consider the implicit functions*

$$\begin{aligned} g_1(y_1,y_2,x_1,x_2) &= (y_1)^{-\frac{1}{2}}(y_2)^{\frac{1}{4}} - 2x_1 = 0 \\ g_2(y_1,y_2,x_1,x_2) &= (y_1)^{\frac{1}{2}}(y_2)^{-\frac{3}{4}} - 4x_2 = 0 \end{aligned}$$

where y_1, y_2 are the endogenous variables and x_1, x_2 are the exogenous variables. Use the $\ln(\,)$ *function to convert this into a system of linear equations and use this to find the reduced form. From the reduced form find $\frac{\partial y_1}{\partial x_1}, \frac{\partial y_1}{\partial x_2}, \frac{\partial y_2}{\partial x_1}, \frac{\partial y_2}{\partial x_2}$, show that $\frac{\partial y_1}{\partial x_1} < 0$ and that $\frac{\partial y_2}{\partial x_1} = \frac{\partial y_1}{\partial x_2}$. Calculate the total differential for this system of equations. Show that the conditions for the explicit functions to exist are satisfied. Use the total differential to calculate $\frac{\partial y_1}{\partial x_1}$ and $\frac{\partial y_2}{\partial x_1}$ and show that $\frac{\partial y_1}{\partial x_1} < 0$ and that $\frac{\partial y_2}{\partial x_1} = \frac{\partial y_1}{\partial x_2}$.*

Answer: We have

$$\begin{aligned} (y_1)^{-\frac{1}{2}}(y_2)^{\frac{1}{4}} - 2x_1 &= 0 \Longrightarrow (y_1)^{-\frac{1}{2}}(y_2)^{\frac{1}{4}} = 2x_1 \\ &\Longrightarrow -\frac{1}{2}\ln(y_1) + \frac{1}{4}\ln(y_2) - \ln(x_1) - \ln(2) = 0 \\ (y_1)^{\frac{1}{2}}(y_2)^{-\frac{3}{4}} - 4x_2 &= 0 \Longrightarrow (y_1)^{\frac{1}{2}}(y_2)^{-\frac{3}{4}} = 4x_2 \\ &\Longrightarrow \frac{1}{2}\ln(y_1) - \frac{3}{4}\ln(y_2) - \ln(x_2) - \ln(4) = 0 \end{aligned}$$

so that in matrix notation we have

$$\underbrace{\begin{bmatrix} -\frac{1}{2} & \frac{1}{4} \\ \frac{1}{2} & -\frac{3}{4} \end{bmatrix}}_{A} \begin{bmatrix} \ln(y_1) \\ \ln(y_2) \end{bmatrix} = \begin{bmatrix} \ln(x_1) + \ln(2) \\ \ln(x_2) + \ln(4) \end{bmatrix}.$$

We have

$$\begin{aligned} \det[A] &= \det\begin{bmatrix} -\frac{1}{2} & \frac{1}{4} \\ \frac{1}{2} & -\frac{3}{4} \end{bmatrix} = \frac{1}{4} \\ &\Longrightarrow A^{-1} = \frac{1}{\frac{1}{4}}\begin{bmatrix} -\frac{3}{4} & -\frac{1}{4} \\ -\frac{1}{2} & -\frac{1}{2} \end{bmatrix} = \begin{bmatrix} -3 & -1 \\ -2 & -2 \end{bmatrix}. \end{aligned}$$

Solving we have

$$\begin{aligned} \begin{bmatrix} \ln(y_1) \\ \ln(y_2) \end{bmatrix} &= \begin{bmatrix} -\frac{1}{2} & \frac{1}{4} \\ \frac{1}{2} & -\frac{3}{4} \end{bmatrix}^{-1} \begin{bmatrix} \ln(x_1) + \ln(2) \\ \ln(x_2) + \ln(4) \end{bmatrix} \\ &= \begin{bmatrix} -3 & -1 \\ -2 & -2 \end{bmatrix} \begin{bmatrix} \ln(x_1) + \ln(2) \\ \ln(x_2) + \ln(4) \end{bmatrix} \\ &= \begin{bmatrix} -3\ln(x_1) - \ln(x_2) - 3\ln(2) - \ln(4) \\ -2\ln(x_1) - 2\ln(x_2) - 2\ln(2) - 2\ln(4) \end{bmatrix} \end{aligned}$$

or

$$\begin{aligned}
\ln(y_1) &= -3\ln(x_1) - \ln(x_2) - 3\ln(2) - \ln(4) \\
&\implies y_1 = \frac{1}{32} x_1^{-3} x_2^{-1} \\
\ln(y_2) &= -2\ln(x_1) - 2\ln(x_2) - 2\ln(2) - 2\ln(4) \\
&\implies y_2 = \frac{1}{64} x_1^{-2} x_2^{-2}.
\end{aligned}$$

From this we then have

$$\begin{aligned}
\frac{\partial y_1}{\partial x_1} &= -\frac{3}{32} x_1^{-4} x_2^{-1} < 0, \frac{\partial y_1}{\partial x_2} = -\frac{1}{32} x_1^{-3} x_2^{-2} < 0 \\
\frac{\partial y_2}{\partial x_1} &= -\frac{1}{32} x_1^{-3} x_2^{-2} = \frac{\partial y_1}{\partial x_2}, \frac{\partial y_2}{\partial x_2} = -\frac{1}{32} x_1^{-2} x_2^{-3} < 0.
\end{aligned}$$

The total differential is

$$\begin{aligned}
\left(-\frac{1}{2}(y_1)^{-\frac{3}{2}}(y_2)^{\frac{1}{4}}\right) dy_1 + \left(\frac{1}{4}(y_1)^{-\frac{1}{2}}(y_2)^{-\frac{3}{4}}\right) dy_2 - 2dx_1 &= 0 \\
\left(\frac{1}{2}(y_1)^{-\frac{1}{2}}(y_2)^{-\frac{3}{4}}\right) dy_1 + \left(-\frac{3}{4}(y_1)^{\frac{1}{2}}(y_2)^{-\frac{7}{4}}\right) dy_2 - 4dx_2 &= 0
\end{aligned}$$

or in matrix notation

$$\underbrace{\begin{bmatrix} -\frac{1}{2}(y_1)^{-\frac{3}{2}}(y_2)^{\frac{1}{4}} & \frac{1}{4}(y_1)^{-\frac{1}{2}}(y_2)^{-\frac{3}{4}} \\ \frac{1}{2}(y_1)^{-\frac{1}{2}}(y_2)^{-\frac{3}{4}} & -\frac{3}{4}(y_1)^{\frac{1}{2}}(y_2)^{-\frac{7}{4}} \end{bmatrix}}_{A} \begin{bmatrix} dy_1 \\ dy_2 \end{bmatrix} = \begin{bmatrix} 2dx_1 \\ 4dx_2 \end{bmatrix}.$$

We have

$$\begin{aligned}
\det[A] &= \det \begin{bmatrix} -\frac{1}{2}(y_1)^{-\frac{3}{2}}(y_2)^{\frac{1}{4}} & \frac{1}{4}(y_1)^{-\frac{1}{2}}(y_2)^{-\frac{3}{4}} \\ \frac{1}{2}(y_1)^{-\frac{1}{2}}(y_2)^{-\frac{3}{4}} & -\frac{3}{4}(y_1)^{\frac{1}{2}}(y_2)^{-\frac{7}{4}} \end{bmatrix}. \\
&= \frac{3}{8} y_1^{-1} y_2^{-\frac{6}{4}} - \frac{1}{8} y_1^{-1} y_2^{-\frac{6}{4}} = \frac{2}{8} y_1^{-1} y_2^{-\frac{6}{4}} > 0
\end{aligned}$$

To find $\frac{\partial y_1}{\partial x_1}, \frac{\partial y_2}{\partial x_1}$ we 1) set $dx_2 = 0$ and 2) change the remaining $d's$ to $\partial' s$ as

$$\begin{bmatrix} -\frac{1}{2}(y_1)^{-\frac{3}{2}}(y_2)^{\frac{1}{4}} & \frac{1}{4}(y_1)^{-\frac{1}{2}}(y_2)^{-\frac{3}{4}} \\ \frac{1}{2}(y_1)^{-\frac{1}{2}}(y_2)^{-\frac{3}{4}} & -\frac{3}{4}(y_1)^{\frac{1}{2}}(y_2)^{-\frac{7}{4}} \end{bmatrix} \begin{bmatrix} \frac{\partial y_1}{\partial x_1} \\ \frac{\partial y_2}{\partial x_1} \end{bmatrix} = \begin{bmatrix} 2 \\ 0 \end{bmatrix}$$

so that using Cramer's rule

$$\frac{\partial y_1}{\partial x_1} = \frac{\det\begin{bmatrix} 2 & \frac{1}{4}(y_1)^{-\frac{1}{2}}(y_2)^{-\frac{3}{4}} \\ 0 & -\frac{3}{4}(y_1)^{\frac{1}{2}}(y_2)^{-\frac{7}{4}} \end{bmatrix}}{\underbrace{\det[A]}_{+}} = \frac{-\frac{6}{4}(y_1)^{\frac{1}{2}}(y_2)^{-\frac{7}{4}}}{\underbrace{\det[A]}_{+}} < 0$$

$$\frac{\partial y_2}{\partial x_1} = \frac{\det\begin{bmatrix} -\frac{1}{2}(y_1)^{-\frac{3}{2}}(y_2)^{\frac{1}{4}} & 2 \\ \frac{1}{2}(y_1)^{-\frac{1}{2}}(y_2)^{-\frac{3}{4}} & 0 \end{bmatrix}}{\underbrace{\det[A]}_{+}} = \frac{-(y_1)^{-\frac{1}{2}}(y_2)^{-\frac{3}{4}}}{\underbrace{\det[A]}_{+}} < 0.$$

To find $\frac{\partial y_1}{\partial x_2}, \frac{\partial y_2}{\partial x_2}$ we 1) set $dx_1 = 0$ and 2) change the remaining $d's$ to $\partial's$ as

$$\begin{bmatrix} -\frac{1}{2}(y_1)^{-\frac{3}{2}}(y_2)^{\frac{1}{4}} & \frac{1}{4}(y_1)^{-\frac{1}{2}}(y_2)^{-\frac{3}{4}} \\ \frac{1}{2}(y_1)^{-\frac{1}{2}}(y_2)^{-\frac{3}{4}} & -\frac{3}{4}(y_1)^{\frac{1}{2}}(y_2)^{-\frac{7}{4}} \end{bmatrix}\begin{bmatrix} \frac{\partial y_1}{\partial x_2} \\ \frac{\partial y_2}{\partial x_2} \end{bmatrix} = \begin{bmatrix} 0 \\ 4 \end{bmatrix}$$

so that using Cramer's rule

$$\frac{\partial y_1}{\partial x_2} = \frac{\det\begin{bmatrix} 0 & \frac{1}{4}(y_1)^{-\frac{1}{2}}(y_2)^{-\frac{3}{4}} \\ 4 & -\frac{3}{4}(y_1)^{\frac{1}{2}}(y_2)^{-\frac{7}{4}} \end{bmatrix}}{\underbrace{\det[A]}_{+}} = -\frac{(y_1)^{-\frac{1}{2}}(y_2)^{-\frac{3}{4}}}{\underbrace{\det[A]}_{+}} < 0$$

$$\frac{\partial y_2}{\partial x_2} = \frac{\det\begin{bmatrix} -\frac{1}{2}(y_1)^{-\frac{3}{2}}(y_2)^{\frac{1}{4}} & 0 \\ \frac{1}{2}(y_1)^{-\frac{1}{2}}(y_2)^{-\frac{3}{4}} & 4 \end{bmatrix}}{\underbrace{\det[A]}_{+}} = \frac{-2(y_1)^{-\frac{3}{2}}(y_2)^{\frac{1}{4}}}{\underbrace{\det[A]}_{+}} < 0.$$

Problem 2.5 *[C***] Consider the following system of implicit linear functions*

$$\begin{aligned} 3y_1 + 2y_2 + 4y_3 - 6x_1 + 2x_2 - 4 &= 0 \\ 2y_1 - 5y_2 + 2y_3 + x_1 - x_2 + 3 &= 0 \\ y_1 + 2y_2 - y_3 + 2x_1 - 3x_2 - 2 &= 0. \end{aligned}$$

If this system is written as $Ay + Bx + c = 0$, what are $A, y, B, x,$ and c? Show that A is non-singular by calculating its determinant. If we write the reduced form as $y = Dx + E$, calculate $D = -A^{-1}B$ and $E = -A^{-1}c$. From the matrix D determine the multipliers

$$\frac{\partial y_i}{\partial x_j} \text{ for } i = 1, 2, 3 \text{ and } j = 1, 2.$$

What do the signs of each of these multipliers tell you? Write out the total differential as $Ady = -Bdx$ and solve for the multipliers $\frac{\partial y_1}{\partial x_2}, \frac{\partial y_2}{\partial x_2}, \frac{\partial y_3}{\partial x_2}$ using Cramer's rule.

Answer: In matrix notation we have

$$\begin{bmatrix} 3 & 2 & 4 \\ 2 & -5 & 2 \\ 1 & 2 & -1 \end{bmatrix}\begin{bmatrix} y_1 \\ y_2 \\ y_3 \end{bmatrix}+\begin{bmatrix} -6 & 2 \\ 1 & -1 \\ 2 & -3 \end{bmatrix}\begin{bmatrix} x_1 \\ x_2 \end{bmatrix}+\begin{bmatrix} -4 \\ 3 \\ -2 \end{bmatrix}=\begin{bmatrix} 0 \\ 0 \\ 0 \end{bmatrix}.$$

The matrix A is non-singular since

$$\det\begin{bmatrix} 3 & 2 & 4 \\ 2 & -5 & 2 \\ 1 & 2 & -1 \end{bmatrix}=47\neq 0.$$

We have

$$\begin{aligned} D &= -\begin{bmatrix} 3 & 2 & 4 \\ 2 & -5 & 2 \\ 1 & 2 & -1 \end{bmatrix}^{-1}\begin{bmatrix} -6 & 2 \\ 1 & -1 \\ 2 & -3 \end{bmatrix}=\begin{bmatrix} -\frac{52}{47} & \frac{80}{47} \\ \frac{27}{47} & -\frac{9}{47} \\ \frac{96}{47} & -\frac{79}{47} \end{bmatrix} \\ E &= -\begin{bmatrix} 3 & 2 & 4 \\ 2 & -5 & 2 \\ 1 & 2 & -1 \end{bmatrix}^{-1}\begin{bmatrix} -4 \\ 3 \\ -2 \end{bmatrix}=\begin{bmatrix} \frac{22}{47} \\ \frac{41}{47} \\ \frac{10}{47} \end{bmatrix} \end{aligned}$$

so the reduced form is

$$\begin{bmatrix} y_1 \\ y_2 \\ y_3 \end{bmatrix}=\begin{bmatrix} -\frac{52}{47} & \frac{80}{47} \\ \frac{27}{47} & -\frac{9}{47} \\ \frac{96}{47} & -\frac{79}{47} \end{bmatrix}\begin{bmatrix} x_1 \\ x_2 \end{bmatrix}+\begin{bmatrix} \frac{22}{47} \\ \frac{41}{47} \\ \frac{10}{47} \end{bmatrix}$$

or

$$\begin{aligned} y_1 &= f_1(x_1,x_2)=-\frac{52}{47}x_1+\frac{80}{47}x_2+\frac{22}{47} \\ y_2 &= f_2(x_1,x_2)=\frac{27}{47}x_1-\frac{9}{47}x_2+\frac{41}{47} \\ y_3 &= f_3(x_1,x_2)=\frac{96}{47}x_1-\frac{79}{47}x_2+\frac{10}{47}. \end{aligned}$$

It follows then that

$$\begin{bmatrix} \frac{\partial y_1}{\partial x_1} & \frac{\partial y_1}{\partial x_2} \\ \frac{\partial y_2}{\partial x_1} & \frac{\partial y_2}{\partial x_2} \\ \frac{\partial y_3}{\partial x_1} & \frac{\partial y_3}{\partial x_2} \end{bmatrix}=D=\begin{bmatrix} -\frac{52}{47} & \frac{80}{47} \\ \frac{27}{47} & -\frac{9}{47} \\ \frac{96}{47} & -\frac{79}{47} \end{bmatrix}.$$

For example $\frac{\partial y_1}{\partial x_2}=\frac{80}{47}>0$.

The total differential in matrix notation is

$$\begin{bmatrix} 3 & 2 & 4 \\ 2 & -5 & 2 \\ 1 & 2 & -1 \end{bmatrix}\begin{bmatrix} dy_1 \\ dy_2 \\ dy_3 \end{bmatrix}=-\begin{bmatrix} -6 & 2 \\ 1 & -1 \\ 2 & -3 \end{bmatrix}\begin{bmatrix} dx_1 \\ dx_2 \end{bmatrix}=\begin{bmatrix} 6dx_1-2dx_2 \\ -dx_1+dx_2 \\ -2dx_1+3dx_2 \end{bmatrix}.$$

To find $\frac{\partial y_1}{\partial x_2}, \frac{\partial y_2}{\partial x_2}, \frac{\partial y_3}{\partial x_2}$ we 1) set $dx_1 = 0$ and 2) replace the remaining $d's$ with $\partial's$ as

$$\begin{bmatrix} 3 & 2 & 4 \\ 2 & -5 & 2 \\ 1 & 2 & -1 \end{bmatrix} \begin{bmatrix} \frac{\partial y_1}{\partial x_2} \\ \frac{\partial y_2}{\partial x_2} \\ \frac{\partial y_3}{\partial x_2} \end{bmatrix} = \begin{bmatrix} -2 \\ 1 \\ 3 \end{bmatrix}$$

so that using Cramer's rule we have

$$\frac{\partial y_1}{\partial x_2} = \frac{\det \begin{bmatrix} -2 & 2 & 4 \\ 1 & -5 & 2 \\ 3 & 2 & -1 \end{bmatrix}}{47} = \frac{80}{47}, \frac{\partial y_2}{\partial x_2} = \frac{\det \begin{bmatrix} 3 & -2 & 4 \\ 2 & 1 & 2 \\ 1 & 3 & -1 \end{bmatrix}}{47} = -\frac{9}{47}$$

$$\frac{\partial y_3}{\partial x_2} = \frac{\det \begin{bmatrix} 3 & 2 & -2 \\ 2 & -5 & 1 \\ 1 & 2 & 3 \end{bmatrix}}{47} = -\frac{79}{47}.$$

Problem 2.6 *[C**] Consider the following set of linear equations*

$$\begin{aligned} 5y_1 - 2y_2 + 3y_3 + 4x_1 + 2x_2 + x_3 &= 0 \\ -5y_2 + 2y_3 + 4x_1 - x_2 &= 0 \\ 4y_3 - x_1 - x_2 - x_3 &= 0. \end{aligned}$$

If this system is written as $Ay + Bx + c = 0$, what are A, y, B, x, c and 0? Show that A is non-singular by calculating its determinant. Show that A is upper triangular. Recursive calculations are methods whereby one problem is solved using the solution to the previous problem. (There is a close connection between triangular matrices and recursive calculations.) Show in particular that it is possible to solve for the reduced form by first solving for y_3, then solving for y_2 using the solution for y_3, and then solving for y_1 using the solution for y_2 and y_3. If we write the reduced form as $y = Dx + E$, calculate $D = -A^{-1}B$ and $E = -A^{-1}c$. From the matrix D determine the multipliers $\frac{\partial y_i}{\partial x_j}$ for $i = 1, 2, 3$ and $j = 1, 2$. Write out the total differential as $Ady = -Bdx$ and solve for the multipliers $\frac{\partial y_1}{\partial x_3}, \frac{\partial y_2}{\partial x_3}, \frac{\partial y_3}{\partial x_3}$ using Cramer's rule.

Answer: Here we have $c = 0$. In matrix notation we have

$$\begin{bmatrix} 5 & -2 & 3 \\ 0 & 5 & 2 \\ 0 & 0 & 4 \end{bmatrix} \begin{bmatrix} y_1 \\ y_2 \\ y_3 \end{bmatrix} + \begin{bmatrix} 4 & 2 & 1 \\ 4 & -1 & 0 \\ -1 & -1 & -1 \end{bmatrix} \begin{bmatrix} x_1 \\ x_2 \\ x_3 \end{bmatrix} = \begin{bmatrix} 0 \\ 0 \\ 0 \end{bmatrix}.$$

The matrix A is upper triangular (all elements below the diagonal are 0) and

$$\det \begin{bmatrix} 5 & -2 & 3 \\ 0 & -5 & 2 \\ 0 & 0 & 4 \end{bmatrix} = 5\,(-5)\,4 = -100 \neq 0.$$

Note that the third equation has no y_2 or y_1. We can therefore easily calculate y_3 as

$$4y_3 - x_1 - x_2 - x_3 = 0 \Longrightarrow y_3 = \frac{1}{4}x_1 + \frac{1}{4}x_2 + \frac{1}{4}x_3.$$

Now to calculate y_2 we use our solution for y_3 and the second equation, (which does not contain y_1) as

$$\begin{aligned} -5y_2 + 2y_3 + 4x_1 - x_2 &= 0 \Longrightarrow 5y_2 = 2y_3 + 4x_1 - x_2 \\ &\Longrightarrow \quad 5y_2 = 2\underbrace{\left(\frac{1}{4}x_1 + \frac{1}{4}x_2 + \frac{1}{4}x_3\right)}_{y_3} + 4x_1 - x_2 \\ &\Longrightarrow \quad y_2 = \frac{9}{10}x_1 - \frac{1}{10}x_2 + \frac{1}{10}x_3. \end{aligned}$$

Now that we have y_2 and y_3 we can calculate y_1 from the first equation as

$$5y_1 - 2y_2 + 3y_3 + 4x_1 + 2x_2 + x_3 = 0$$

so that

$$\begin{aligned} 5y_1 &= 2y_2 - 3y_3 - 4x_1 - 2x_2 - x_3 \\ &\Longrightarrow \quad 5y_1 = 2\left(\frac{9}{10}x_1 - \frac{1}{10}x_2 + \frac{1}{10}x_3\right) - 3\left(\frac{1}{4}x_1 + \frac{1}{4}x_2 + \frac{1}{4}x_3\right) - 4x_1 - 2x_2 - x_3 \\ &\Longrightarrow \quad y_1 = -\frac{59}{100}x_1 - \frac{59}{100}x_2 - \frac{31}{100}x_3. \end{aligned}$$

The inverse of an upper triangular matrix is also upper triangular and hence easier to calculate. We have

$$\begin{bmatrix} 5 & -2 & 3 \\ 0 & -5 & 2 \\ 0 & 0 & 4 \end{bmatrix}^{-1} = \frac{-1}{100}\begin{bmatrix} -20 & 8 & 11 \\ 0 & 20 & -10 \\ 0 & 0 & -25 \end{bmatrix}.$$

We have

$$\begin{aligned} D &= -\begin{bmatrix} 5 & -2 & 3 \\ 0 & -5 & 2 \\ 0 & 0 & 4 \end{bmatrix}^{-1}\begin{bmatrix} 4 & 2 & 1 \\ 4 & -1 & 0 \\ -1 & -1 & -1 \end{bmatrix} \\ &= \frac{1}{100}\begin{bmatrix} -20 & 8 & 11 \\ 0 & 20 & -10 \\ 0 & 0 & -25 \end{bmatrix}\begin{bmatrix} 4 & 2 & 1 \\ 4 & -1 & 0 \\ -1 & -1 & -1 \end{bmatrix} = \begin{bmatrix} -\frac{59}{100} & -\frac{59}{100} & -\frac{31}{100} \\ \frac{9}{10} & -\frac{1}{10} & \frac{1}{10} \\ \frac{1}{4} & \frac{1}{4} & \frac{1}{4} \end{bmatrix}. \end{aligned}$$

Since $c = 0$ it follows that $E = -D^{-1}c = 0$. The reduced form is therefore

$$\begin{bmatrix} y_1 \\ y_2 \\ y_3 \end{bmatrix} = \begin{bmatrix} -\frac{59}{100} & -\frac{59}{100} & -\frac{31}{100} \\ \frac{9}{10} & -\frac{1}{10} & \frac{1}{10} \\ \frac{1}{4} & \frac{1}{4} & \frac{1}{4} \end{bmatrix}\begin{bmatrix} x_1 \\ x_2 \\ x_3 \end{bmatrix} + \begin{bmatrix} 0 \\ 0 \\ 0 \end{bmatrix}$$

or a reduced form

$$\begin{aligned} y_1 &= f_1(x_1,x_2,x_3) = -\frac{59}{100}x_1 - \frac{59}{100}x_2 - \frac{31}{100}x_3 \\ y_2 &= f_2(x_1,x_2,x_3) = \frac{9}{10}x_1 - \frac{1}{10}x_2 + \frac{1}{10}x_3 \\ y_3 &= f_3(x_1,x_2,x_3) = \frac{1}{4}x_1 + \frac{1}{4}x_2 + \frac{1}{4}x_3. \end{aligned}$$

It follows then that

$$\begin{bmatrix} \frac{\partial y_1}{\partial x_1} & \frac{\partial y_1}{\partial x_2} & \frac{\partial y_1}{\partial x_3} \\ \frac{\partial y_2}{\partial x_1} & \frac{\partial y_2}{\partial x_2} & \frac{\partial y_2}{\partial x_3} \\ \frac{\partial y_3}{\partial x_1} & \frac{\partial y_3}{\partial x_2} & \frac{\partial y_3}{\partial x_3} \end{bmatrix} = D = \begin{bmatrix} -\frac{59}{100} & -\frac{59}{100} & -\frac{31}{100} \\ \frac{9}{10} & -\frac{1}{10} & \frac{1}{10} \\ \frac{1}{4} & \frac{1}{4} & \frac{1}{4} \end{bmatrix}.$$

For example $\frac{\partial y_2}{\partial x_3} = \frac{1}{10} > 0$ implies that there is a positive relationship between x_3 and y_2.

We have

$$\begin{aligned} \begin{bmatrix} 5 & -2 & 3 \\ 0 & -5 & 2 \\ 0 & 0 & 4 \end{bmatrix} \begin{bmatrix} dy_1 \\ dy_2 \\ dy_3 \end{bmatrix} &= -\begin{bmatrix} 4 & 2 & 1 \\ 4 & -1 & 0 \\ -1 & -1 & -1 \end{bmatrix} \begin{bmatrix} dx_1 \\ dx_2 \\ dx_3 \end{bmatrix} \\ &= \begin{bmatrix} -4dx_1 - 2dx_2 - dx_3 \\ -4dx_1 + dx_2 \\ dx_1 + dx_2 + dx_3 \end{bmatrix}. \end{aligned}$$

To find $\frac{\partial y_1}{\partial x_3}, \frac{\partial y_2}{\partial x_3}, \frac{\partial y_3}{\partial x_3}$ we 1) set $dx_1 = dx_2 = 0$ and 2) replace the remaining $d's$ with $\partial's$ to obtain

$$\begin{bmatrix} 5 & -2 & 3 \\ 0 & -5 & 2 \\ 0 & 0 & 4 \end{bmatrix} \begin{bmatrix} \frac{\partial y_1}{\partial x_3} \\ \frac{\partial y_2}{\partial x_3} \\ \frac{\partial y_3}{\partial x_3} \end{bmatrix} = \begin{bmatrix} -1 \\ 0 \\ 1 \end{bmatrix}$$

so that

$$\begin{aligned} \frac{\partial y_1}{\partial x_3} &= \frac{\det \begin{bmatrix} -1 & -2 & 3 \\ 0 & -5 & 2 \\ 1 & 0 & 4 \end{bmatrix}}{-100} = -\frac{31}{100}, \frac{\partial y_2}{\partial x_3} = \frac{\det \begin{bmatrix} 5 & -1 & 3 \\ 0 & 0 & 2 \\ 0 & 1 & 4 \end{bmatrix}}{-100} = \frac{1}{10} \\ \frac{\partial y_3}{\partial x_3} &= \frac{\det \begin{bmatrix} 5 & -2 & -1 \\ 0 & -5 & 0 \\ 0 & 0 & 1 \end{bmatrix}}{-100} = \frac{1}{4}. \end{aligned}$$

Problem 2.7 *[A**] Suppose y is an $n \times 1$ vector of endogenous variables and x is an $m \times 1$ vector of exogenous variables with linear implicit functions written in matrix form as*

$$g(y,x) = Ay + Bx + c = 0$$

where A is $n \times n$, B is $n \times m$ and c is $n \times 1$. Calculate the reduced form

$$y = f(x) = Dx + e.$$

What is $\frac{\partial y_i}{\partial x_j}$?

Answer: We have

$$Ay + Bx + c = 0 \Longrightarrow Ay = -Bx - c \Longrightarrow y = -A^{-1}Bx - A^{-1}c \Longrightarrow y = Dx + e$$

where $D = -A^{-1}B$ and $e = -A^{-1}c$. Written explicitly for y_i with d_{ij} the i, j^{th} element of D and e_i the ith element of e we have

$$y_i = d_{i1}x_1 + d_{i2}x_2 + \cdots + d_{im}x_m + e_i$$

so that $\frac{\partial y_i}{\partial x_j} = d_{ij}$.

2.2 Non-Linear Models

Problem 2.8 *[D***] Write the equation $\ln(y) = -\frac{1}{2}x^2$ as an explicit function and as an implicit function. Calculate the total differential and from this calculate $\frac{dy}{dx}$.*

Answer: The explicit and implicit functions are

$$y = f(x) = e^{-\frac{1}{2}x^2} \text{ and } g(y, x) = \ln(y) + \frac{1}{2}x^2 = 0.$$

The direct calculation of the derivative yields

$$\frac{dy}{dx} = -xe^{-\frac{1}{2}x^2} = -xy.$$

Using the total differential we have

$$\frac{1}{y}dy + xdx = 0 \Longrightarrow \frac{dy}{dx} = -xy.$$

Problem 2.9 *[D***] Given the implicit function*

$$g(y, x) = \sqrt{y}\ln(y) - 2e^{-x} = 0$$

where $x > 0$ and $y > 1$, calculate the total differential and show that $\frac{dy}{dx} < 0$. Can you find the reduced form $y = f(x)$ directly?

Answer: The total differential is

$$\left(\frac{1}{\sqrt{y}} + \frac{\ln(y)}{2\sqrt{y}}\right) dy + 2e^{-x}dx = 0$$

so that

$$\frac{dy}{dx} = \frac{-2e^{-x}}{\left(\frac{1}{\sqrt{y}} + \frac{\ln(y)}{2\sqrt{y}}\right)} = \frac{-2e^{-x}}{\frac{1}{\sqrt{y}}\left(1 + \frac{\ln(y)}{2}\right)} = \frac{-2\sqrt{y}e^{-x}}{\left(1 + \frac{\ln(y)}{2}\right)} < 0$$

since the numerator is negative and the denominator is positive since

$$y > 1 \Longrightarrow 1 + \frac{\ln(y)}{2} > 0.$$

It is not possible to directly calculate the reduced form since there is no way of simultaneously liberating y from $\ln(y)$ and $\sqrt{y}$ in $\sqrt{y}\ln(y)$.

Problem 2.10 *[C***] Consider the following set of implicit functions*

$$\begin{aligned} g_1(y_1, y_2, x_1, x_2) &= \ln(y_1) - y_2 + 4e^{-x_1} - 2x_2 = 0 \\ g_2(y_1, y_2, x_1, x_2) &= -y_1 + \ln(y_2) - 2e^{x_1} + 3x_2 = 0. \end{aligned}$$

Calculate the total differential. Why is it necessary to assume that $y_1y_2 \neq 1$? Using Cramer's rule and assuming that $y_1 > 1, y_2 > 1$ find $\frac{\partial y_1}{\partial x_1}$ and $\frac{\partial y_2}{\partial x_1}$ and determine their signs.

Answer: The total differential is

$$\begin{aligned} \frac{1}{y_1}dy_1 - dy_2 - 4e^{-x_1}dx_1 - 2dx_2 &= 0 \\ -dy_1 + \frac{1}{y_2}dy_2 - 2e^{x_1}dx_1 + 3dx_2 &= 0 \end{aligned}$$

or in matrix notation

$$\underbrace{\begin{bmatrix} \frac{1}{y_1} & -1 \\ -1 & \frac{1}{y_2} \end{bmatrix}}_{A} \begin{bmatrix} dy_1 \\ dy_2 \end{bmatrix} = \begin{bmatrix} 4e^{-x_1}dx_1 + 2dx_2 \\ 2e^{x_1}dx_1 - 3dx_2 \end{bmatrix}.$$

We have

$$\det[A] = \frac{1}{y_1y_2} - 1 \neq 0 \Longrightarrow y_1y_2 \neq 1.$$

Assuming $y_1 > 1$ and $y_2 > 1$ we have

$$\det[A] = \frac{1}{y_1y_2} - 1 < 0.$$

To find $\frac{\partial y_1}{\partial x_1}, \frac{\partial y_2}{\partial x_1}$ we 1) set $dx_2 = 0$ and 2) changing the remaining $d's$ to $\partial's$ as

$$\begin{aligned} \frac{1}{y_1}\partial y_1 - \partial y_2 - 4e^{-x_1}\partial x_1 &= 0 \\ -\partial y_1 + \frac{1}{y_2}\partial y_2 - 2e^{x_1}\partial x_1 &= 0 \end{aligned}$$

or in matrix notation

$$\underbrace{\begin{bmatrix} \frac{1}{y_1} & -1 \\ -1 & \frac{1}{y_2} \end{bmatrix}}_{A} \begin{bmatrix} \frac{\partial y_1}{\partial x_1} \\ \frac{\partial y_2}{\partial x_1} \end{bmatrix} = \begin{bmatrix} 4e^{-x_1} \\ 2e^{x_1} \end{bmatrix}.$$

This yields

$$\frac{\partial y_1}{\partial x_1} = \frac{\begin{bmatrix} 4e^{-x_1} & -1 \\ 2e^{x_1} & \frac{1}{y_2} \end{bmatrix}}{\det[A]} = \frac{\overbrace{\frac{4e^{-x_1}}{y_2} + 2e^{x_1}}^{+}}{\underbrace{\det[A]}_{-}} < 0$$

$$\frac{\partial y_1}{\partial x_2} = \frac{\begin{bmatrix} \frac{1}{y_1} & 4e^{-x_1} \\ -1 & 2e^{x_1} \end{bmatrix}}{\det[A]} = \frac{\overbrace{\frac{2e^{x_1}}{y_1} + 4e^{-x_1}}^{+}}{\underbrace{\det[A]}_{-}} < 0.$$

Problem 2.11 *[D***] Consider a model with implicit function*

$$g(y, x_1, x_2) = y^{\frac{1}{2}} \ln(y) - e^{x_1 - 2x_2} = 0$$

where $x_1 > 0, x_2 > 0$ are exogenous and $y > 0$ is endogenous. Prove that $y > 1$. What form does the explicit function take? Calculate the total differential and show that the explicit function exists. From the total differential find $\frac{\partial y}{\partial x_2}$ and determine if $\frac{\partial y}{\partial x_2}$ is positive or negative.

Answer: We have

$$y^{\frac{1}{2}} \ln(y) - e^{x_1 - 2x_2} = 0 \Longrightarrow y^{\frac{1}{2}} \ln(y) = e^{x_1 - 2x_2} > 0$$

since $e^x > 0$. Since $y^{\frac{1}{2}} > 0$ it follows that $\ln(y) > 0 \Longrightarrow y > 1$.

The explicit function would take the form $y = f(x_1, x_2)$. This exists since $\frac{\partial g(y, x_1, x_2)}{\partial y} \neq 0$ as

$$\frac{\partial g(y, x_1, x_2)}{\partial y} = y^{\frac{1}{2}} \frac{1}{y} + \frac{1}{2} y^{-\frac{1}{2}} \ln(y) = y^{-\frac{1}{2}} \left(1 + \frac{1}{2} \ln(y)\right) > 0.$$

The total differential is

$$y^{-\frac{1}{2}} \left(1 + \frac{1}{2} \ln(y)\right) dy - e^{x_1 - 2x_2} dx_1 + 2e^{x_1 - 2x_2} dx_2 = 0.$$

To calculate $\frac{\partial y}{\partial x_2}$ we 1) set $dx_1 = 0$ and 2) replace the remaining $d's$ with $\partial's$ to obtain

$$y^{-\frac{1}{2}} \left(1 + \frac{1}{2} \ln(y)\right) \partial y + 2e^{x_1 - 2x_2} \partial x_2 = 0 \Longrightarrow \frac{\partial y}{\partial x_2} = -\frac{2e^{x_1 - 2x_2}}{y^{-\frac{1}{2}} \left(1 + \frac{1}{2} \ln(y)\right)} < 0$$

since $2e^{x_1 - 2x_2} > 0$ and $y^{-\frac{1}{2}} \left(1 + \frac{1}{2} \ln(y)\right) > 0$.

Problem 2.12 *[D***] Consider a model with implicit function*

$$g(y, x_1, x_2) = y \ln(y) - e^{-2x_1+3x_2} = 0$$

where $x_1 > 0, x_2 > 0$ are exogenous and $y > 0$ is endogenous. Prove that in fact $y > 1$. What is the form of the explicit function? Calculate the total differential and show that the explicit function exists. From the total differential find $\frac{\partial y}{\partial x_1}, \frac{\partial y}{\partial x_2}$ and determine whether they are positive or negative.

Answer: We have

$$y \ln(y) - e^{-2x_1+3x_2} = 0 \Longrightarrow y \ln(y) = e^{-2x_1+3x_2} > 0$$

since $e^x > 0$. Since $y > 0$ it follows that $\ln(y) > 0 \Longrightarrow y > 1$.

The explicit function would take the from $y = f(x_1, x_2)$ and this exists since $\frac{\partial g(y,x_1,x_2)}{\partial y} \neq 0$ as

$$\frac{\partial g(y, x_1, x_2)}{\partial y} = 1 + \ln(y) > 1.$$

The total differential is

$$(1 + \ln(y))\, dy + 2e^{-2x_1+3x_2} dx_1 - 3e^{-2x_1+3x_2} dx_2 = 0.$$

To calculate $\frac{\partial y}{\partial x_1}$ we 1) set $dx_2 = 0$ and 2) replace the remaining $d's$ with $\partial's$ to obtain

$$(1 + \ln(y))\, \partial y + 2e^{-2x_1+3x_2} \partial x_1 = 0 \Longrightarrow \frac{\partial y}{\partial x_1} = -\frac{2e^{-2x_1+3x_2}}{1 + \ln(y)} < 0$$

since $1 + \ln(y) > 0$ follows from $y > 1$.

To calculate $\frac{\partial y}{\partial x_2}$ we 1) set $dx_1 = 0$ and 2) replace the remaining $d's$ with $\partial's$ to obtain

$$(1 + \ln(y))\, \partial y - 3e^{-2x_1+3x_2} \partial x_2 = 0 \Longrightarrow \frac{\partial y}{\partial x_2} = \frac{3e^{-2x_1+3x_2}}{1 + \ln(y)} > 0$$

since again $1 + \ln(y) > 0$ follows from $y > 1$.

Problem 2.13 *[D***] Consider the implicit function*

$$g(y, x) = \ln(y) + 3x^2 = 0$$

for $-\infty < x < \infty$ and $y > 0$. Find the explicit function $y = f(x)$ and use this to show that $\frac{dy}{dx} = -6xy$ and that $\frac{dy}{dx} > 0$. Now calculate the total differential, use this to show that the explicit function exists and find $\frac{dy}{dx}$.

Answer: We have

$$\ln(y) + 3x^2 = 0 \Longrightarrow y = f(x) = e^{-3x^2}.$$

Now

$$\frac{dy}{dx} = f'(x) = -6x\underbrace{e^{-3x^2}}_{y} = -6xy$$

and so $f'(x) < 0$ for $x > 0$ and $f'(x) > 0$ for $x < 0$ contrary to what the question asserts.

The total differential is

$$\frac{1}{y}dy + 6xdx = 0 \Longrightarrow \frac{dy}{dx} = -6xy.$$

The explicit function exists since $\frac{\partial g(y,x)}{\partial y} = \frac{1}{y} > 0$.

Problem 2.14 *[D***] Consider the implicit function*

$$g(y, x_1, x_2) = y\ln(y) - x_2 e^{-x_1 y} = 0$$

where $x_1 > 0, x_2 > 0$. *Prove that* $y > 1$. *What is the form of the explicit function? Calculate the total differential and show that the explicit function exists. From the total differential show that* $\frac{\partial y}{\partial x_1} < 0$ *and* $\frac{\partial y}{\partial x_2} > 0$.

Answer: We require $y > 0$ so that $\ln(y)$ is defined. It follows that $y > 1$ since

$$y\ln(y) - x_2 e^{-x_1 y} = 0 \Longrightarrow y\ln(y) = x_2 e^{-x_1 y} > 0$$

and $y\ln(y) > 0$ and

$$y > 0 \Longrightarrow \ln(y) > 0 \Longrightarrow y > 1.$$

The total differential is

$$\left(1 + \ln(y) + x_1 x_2 e^{-x_1 y}\right) dy + y x_2 e^{-x_1 y} dx_1 - e^{-x_1 y} dx_2 = 0.$$

Since $\ln(y) > 0$ we have

$$\frac{\partial g(y, x_1, x_2)}{\partial y} = 1 + \ln(y) + x_1 x_2 e^{-x_1 y} > 0$$

and so the explicit function exists which takes the form $y = f(x_1, x_2)$.

To find $\frac{\partial y}{\partial x_1}$ we 1) set $dx_2 = 0$ and 2) change the remaining $d's$ to $\partial' s$ as

$$\left(1 + \ln(y) + x_1 x_2 e^{-x_1 y}\right) \partial y + y x_2 e^{-x_1 y} \partial x_1 = 0$$

so that

$$\frac{\partial y}{\partial x_1} = -\frac{y x_2 e^{-x_1 y}}{1 + \ln(y) + x_1 x_2 e^{-x_1 y}} < 0$$

since $y > 1 \Longrightarrow \ln(y) > 0$ and so the denominator is positive.

To find $\frac{\partial y}{\partial x_2}$ we 1) set $dx_1 = 0$ and 2) change the remaining $d's$ to $\partial' s$ as

$$\left(1 + \ln(y) + x_1 x_2 e^{-x_1 y}\right) \partial y - e^{-x_1 y} \partial x_2 = 0$$

so that

$$\frac{\partial y}{\partial x_2} = \frac{e^{-x_1 y}}{1 + \ln(y) + x_1 x_2 e^{-x_1 y}} > 0.$$

Problem 2.15 *[C***] Consider the following system of implicit functions*

$$\begin{aligned} g_1(y_1,y_2,y_3,x_1,x_2,x_3) &= 2\ln(x_3)-\ln(y_2)-\ln(y_3)=0 \\ g_2(y_1,y_2,y_3,x_1,x_2,x_3) &= 2x_1-\frac{y_1}{y_2}=0 \\ g_3(y_1,y_2,y_3,x_1,x_2,x_3) &= 2x_2-\frac{y_1}{y_3}=0 \end{aligned}$$

where $y_1, y_2,$ and y_3 are the endogenous variables (assumed to be positive) and x_1, x_2 and x_3 are the exogenous variables (also assumed to be positive). Calculate the total differential and write it in matrix notation. Are the conditions of the implicit function theorem satisfied? From the total differential calculate $\frac{\partial y_1}{\partial x_3}$ and $\frac{\partial y_2}{\partial x_2}$ and determine their signs. Solve the three implicit functions above for the explicit functions.

Answer: The total differential is

$$\begin{aligned} -\frac{1}{y_2}dy_2-\frac{1}{y_3}dy_3+\frac{2}{x_3}dx_3 &= 0 \\ -\frac{1}{y_2}dy_1+\frac{y_1}{y_2^2}dy_2+2dx_1 &= 0 \\ -\frac{1}{y_3}dy_1+\frac{y_1}{y_3^2}dy_3+2dx_2 &= 0 \end{aligned}$$

or in matrix notation

$$\begin{bmatrix} 0 & -\frac{1}{y_2} & -\frac{1}{y_3} \\ -\frac{1}{y_2} & \frac{y_1}{y_2^2} & 0 \\ -\frac{1}{y_3} & 0 & \frac{y_1}{y_3^2} \end{bmatrix}\begin{bmatrix} dy_1 \\ dy_2 \\ dy_3 \end{bmatrix}=\begin{bmatrix} -\frac{2}{x_3}dx_3 \\ -2dx_1 \\ -2dx_2 \end{bmatrix}.$$

We have

$$\det[A]=\det\begin{bmatrix} 0 & -\frac{1}{y_2} & -\frac{1}{y_3} \\ -\frac{1}{y_2} & \frac{y_1}{y_2^2} & 0 \\ -\frac{1}{y_3} & 0 & \frac{y_1}{y_3^2} \end{bmatrix}=-\frac{2}{y_2^2}\frac{y_1}{y_3^2}<0.$$

To find $\frac{\partial y_1}{\partial x_3}$ we 1) set $dx_1=dx_2=0$ and 2) replacing the remaining $d's$ with $\partial's$ to obtain

$$\begin{bmatrix} 0 & -\frac{1}{y_2} & -\frac{1}{y_3} \\ -\frac{1}{y_2} & \frac{y_1}{y_2^2} & 0 \\ -\frac{1}{y_3} & 0 & \frac{y_1}{y_3^2} \end{bmatrix}\begin{bmatrix} \frac{\partial y_1}{\partial x_3} \\ \frac{\partial y_2}{\partial x_3} \\ \frac{\partial y_3}{\partial x_3} \end{bmatrix}=\begin{bmatrix} -\frac{2}{x_3} \\ 0 \\ 0 \end{bmatrix}.$$

From Cramer's rule

$$\frac{\partial y_1}{\partial x_3}=\frac{\begin{bmatrix} -\frac{2}{x_3} & -\frac{1}{y_2} & -\frac{1}{y_3} \\ 0 & \frac{y_1}{y_2^2} & 0 \\ 0 & 0 & \frac{y_1}{y_3^2} \end{bmatrix}}{\det[A]}=\frac{-\frac{2}{x_3}\frac{y_1}{y_2^2}\frac{y_1}{y_3^2}}{\det[A]}>0.$$

To find $\frac{\partial y_2}{\partial x_2}$ we 1) set $dx_1 = dx_3 = 0$ and 2) replace the remaining $d's$ with $\partial' s$ to obtain

$$\begin{bmatrix} 0 & -\frac{1}{y_2} & -\frac{1}{y_3} \\ -\frac{1}{y_2} & \frac{y_1}{y_2^2} & 0 \\ -\frac{1}{y_3} & 0 & \frac{y_1}{y_3^2} \end{bmatrix} \begin{bmatrix} \frac{\partial y_1}{\partial x_2} \\ \frac{\partial y_2}{\partial x_2} \\ \frac{\partial y_3}{\partial x_2} \end{bmatrix} = \begin{bmatrix} 0 \\ 0 \\ -2 \end{bmatrix}.$$

From Cramer's rule

$$\frac{\partial y_2}{\partial x_2} = \frac{\det \begin{bmatrix} 0 & 0 & -\frac{1}{y_3} \\ -\frac{1}{y_2} & 0 & 0 \\ -\frac{1}{y_3} & -2 & \frac{y_1}{y_3^2} \end{bmatrix}}{\det [A]} = \frac{\frac{-2}{y_2 y_3}}{\det [A]} > 0.$$

From the 3 implicit functions we have

$$\begin{aligned} 2\ln(x_3) - \ln(y_2) - \ln(y_3) &= 0 \Longrightarrow y_2 y_3 = x_3^2 \\ 2x_1 - \frac{y_1}{y_2} &= 0 \Longrightarrow y_1 = 2x_1 y_2 \\ 2x_2 - \frac{y_1}{y_3} &= 0 \Longrightarrow y_1 = 2x_2 y_3 \end{aligned}$$

so that dividing the second and third results we have

$$\frac{x_1 y_2}{x_2 y_3} = 1 \Longrightarrow y_2 = \frac{x_2}{x_1} y_3$$

and placing this in the first result

$$y_2 y_3 = x_3^2 \text{ and } y_2 = \frac{x_2}{x_1} y_3 \Longrightarrow \frac{x_2}{x_1} y_3^2 = x_3^2 \Longrightarrow y_3 = f_3(x_1, x_2, x_3) = x_1^{\frac{1}{2}} x_2^{-\frac{1}{2}} x_3$$

and so

$$y_2 = \frac{x_2}{x_1} y_3 \text{ and } y_3 = x_1^{\frac{1}{2}} x_2^{-\frac{1}{2}} x_3 \Longrightarrow y_2 = \frac{x_2}{x_1} x_1^{\frac{1}{2}} x_2^{-\frac{1}{2}} x_3 \Longrightarrow y_2 = f_2(x_1, x_2, x_3) = x_1^{-\frac{1}{2}} x_2^{\frac{1}{2}} x_3.$$

Finally

$$y_1 = 2x_2 y_3 \text{ and } y_3 = x_1^{\frac{1}{2}} x_2^{-\frac{1}{2}} x_3 \Longrightarrow y_1 = 2x_2 x_1^{\frac{1}{2}} x_2^{-\frac{1}{2}} x_3 \Longrightarrow y_1 = f_1(x_1, x_2, x_3) = 2x_1^{\frac{1}{2}} x_2^{\frac{1}{2}} x_3.$$

Problem 2.16 *[C***] Consider the following system of implicit functions*

$$\begin{aligned} g_1(y_1, y_2, x_1, x_2, x_3) &= y_1^{2x_1} - 3y_2^{x_2} + x_3 = 0 \\ g_2(y_1, y_2, x_1, x_2, x_3) &= x_1 \ln(y_1) + x_2 \ln(y_2) = 0 \end{aligned}$$

where all variables (i.e., all $y's$ and $x's$) are assumed to be positive. Calculate the total differential of this system and write it in matrix form. Show that the conditions for the implicit function theorem will be satisfied. Solve for $\frac{\partial y_1}{\partial x_3}$ using Cramer's rule and determine its sign if possible.

Answer: We have the total differential

$$\begin{aligned} 2x_1 y_1^{2x_1-1} dy_1 - 3x_2 y_2^{x_2-1} dy_2 + 2\ln(y_1) y_1^{2x_1} dx_1 - 3\ln(y_2) y_2^{x_2} dx_2 + dx_3 &= 0 \\ \frac{x_1}{y_1} dy_1 + \frac{x_2}{y_2} dy_2 + \ln(y_1) dx_1 + \ln(y_2) dx_2 &= 0. \end{aligned}$$

In matrix form this becomes

$$\underbrace{\begin{bmatrix} 2x_1 y_1^{2x_1-1} & -3x_2 y_2^{x_2-1} \\ \frac{x_1}{y_1} & \frac{x_2}{y_2} \end{bmatrix}}_{A} \begin{bmatrix} dy_1 \\ dy_2 \end{bmatrix} = \begin{bmatrix} -2\ln(y_1) y_1^{2x_1} dx_1 + 3\ln(y_2) y_2^{x_2} dx_2 - dx_3 \\ -\ln(y_1) dx_1 - \ln(y_2) dx_2 \end{bmatrix}.$$

We have

$$\det[A] = \det \begin{bmatrix} 2x_1 y_1^{2x_1-1} & -3x_2 y_2^{x_2-1} \\ \frac{x_1}{y_1} & \frac{x_2}{y_2} \end{bmatrix} = \frac{x_1 x_2 \left(2y_1^{2x_1-1} y_1 + 3y_2^{x_2-1} y_2\right)}{y_2 y_1} > 0$$

since all variables (all $y's$ and $x's$) are assumed to be positive.

To find $\frac{\partial y_1}{\partial x_3}$ we 1) set $dx_1 = dx_2 = 0$ and 2) replace the remaining $d's$ with $\partial's$ to obtain

$$\begin{bmatrix} 2x_1 y_1^{2x_1-1} & -3x_2 y_2^{x_2-1} \\ \frac{x_1}{y_1} & \frac{x_2}{y_2} \end{bmatrix} \begin{bmatrix} \frac{\partial y_1}{\partial x_3} \\ \frac{\partial y_2}{\partial x_3} \end{bmatrix} = \begin{bmatrix} -1 \\ 0 \end{bmatrix}.$$

Thus using Cramer's rule we have

$$\frac{\partial y_1}{\partial x_3} = \frac{\det \begin{bmatrix} -1 & -3x_2 y_2^{x_2-1} \\ 0 & \frac{x_2}{y_2} \end{bmatrix}}{\det[A]} = \frac{-\frac{x_2}{y_2}}{x_1 x_2 \frac{2y_1^{2x_1-1} y_1 + 3y_2^{x_2-1} y_2}{y_2 y_1}} = -\frac{y_1}{x_1 \left(2y_1^{2x_1} + 3y_2^{x_2}\right)} < 0.$$

Problem 2.17 *[C***] Consider the implicit function*

$$g(y, x_1, x_2) = y\ln(y) - x_2 e^{-x_1 y} = 0$$

where $x_1 > 0$ and $x_2 > 0$. Prove that $y > 1$. What is the form of the explicit function? Calculate the total differential and show that the explicit function exists. From the total differential show that $\frac{\partial y}{\partial x_1} < 0$ and $\frac{\partial y}{\partial x_2} > 0$.

Answer: We require $y > 0$ so that $\ln(y)$ is defined. It follows that $y > 1$ since

$$y\ln(y) - x_2 e^{-x_1 y} = 0 \Longrightarrow y\ln(y) = x_2 e^{-x_1 y} > 0$$

and $y\ln(y) > 0$ and $y > 0 \Longrightarrow \ln(y) > 0 \Longrightarrow y > 1$.

The total differential is

$$\left(1 + \ln(y) + x_1 x_2 e^{-x_1 y}\right) dy + y x_2 e^{-x_1 y} dx_1 - e^{-x_1 y} dx_2 = 0.$$

Since $\ln(y) > 0$ we have

$$\frac{\partial g(y, x_1, x_2)}{\partial y} = \left(1 + \ln(y) + x_1 x_2 e^{-x_1 y}\right) > 0.$$

To find $\frac{\partial y}{\partial x_1}$ we 1) set $dx_2 = 0$ and 2) converting the remaining $d's$ to $\partial's$ as

$$\left(1 + \ln(y) + x_1 x_2 e^{-x_1 y}\right) \partial y + y x_2 e^{-x_1 y} \partial x_1 = 0$$

so that

$$\frac{\partial y}{\partial x_1} = -\frac{y x_2 e^{-x_1 y}}{1 + \ln(y) + x_1 x_2 e^{-x_1 y}} < 0$$

since $y > 1 \Longrightarrow \ln(y) > 0$ and so the denominator is positive.

To find $\frac{\partial y}{\partial x_2}$ we 1) set $dx_1 = 0$ and 2) converting the remaining $d's$ to $\partial's$ as

$$\left(1 + \ln(y) + x_1 x_2 e^{-x_1 y}\right) \partial y - e^{-x_1 y} \partial x_2 = 0$$

so that

$$\frac{\partial y}{\partial x_2} = \frac{e^{-x_1 y}}{1 + \ln(y) + x_1 x_2 e^{-x_1 y}} > 0.$$

Problem 2.18 *[C***] Consider the following set of implicit functions*

$$\begin{aligned} g_1(y_1, y_2, x_1, x_2) &= \ln(y_1) - y_2 + 4e^{-x_1} - 2x_2 = 0 \\ g_2(y_1, y_2, x_1, x_2) &= -y_1 + \ln(y_2) - 2e^{x_1} + 3x_2 = 0. \end{aligned}$$

Calculate the total differential. Why is it necessary to assume that $y_1 y_2 \neq 1$*? Using Cramer's rule and assuming that* $y_1 > 1, y_2 > 1$*, find* $\frac{\partial y_1}{\partial x_1}, \frac{\partial y_2}{\partial x_1}$*, and determine their signs.*

Answer: The total differential is

$$\begin{aligned} \frac{1}{y_1} dy_1 - dy_2 - 4e^{-x_1} dx_1 - 2dx_2 &= 0 \\ -dy_1 + \frac{1}{y_2} dy_2 - 2e^{x_1} dx_1 + 3dx_2 &= 0 \end{aligned}$$

or in matrix notation

$$\underbrace{\begin{bmatrix} \frac{1}{y_1} & -1 \\ -1 & \frac{1}{y_2} \end{bmatrix}}_{A} \begin{bmatrix} dy_1 \\ dy_2 \end{bmatrix} = \begin{bmatrix} 4e^{-x_1} dx_1 + 2dx_2 \\ 2e^{x_1} dx_1 - 3dx_2 \end{bmatrix}.$$

We require that $\det[A] \neq 0$ or

$$\det[A] = \frac{1}{y_1 y_2} - 1 \neq 0 \Longrightarrow y_1 y_2 \neq 1.$$

Assuming $y_1 > 1$ and $y_2 > 1$ we have

$$\det[A] = \frac{1}{y_1 y_2} - 1 < 0.$$

To find $\frac{\partial y_1}{\partial x_1}, \frac{\partial y_2}{\partial x_1}$ we 1) set $dx_2 = 0$ and 2) change the remaining $d's$ to $\partial's$ as

$$\begin{aligned} \frac{1}{y_1}\partial y_1 - \partial y_2 - 4e^{-x_1}\partial x_1 &= 0 \\ -\partial y_1 + \frac{1}{y_2}\partial y_2 - 2e^{x_1}\partial x_1 &= 0 \end{aligned}$$

or in matrix notation

$$\underbrace{\begin{bmatrix} \frac{1}{y_1} & -1 \\ -1 & \frac{1}{y_2} \end{bmatrix}}_{A} \begin{bmatrix} \frac{\partial y_1}{\partial x_1} \\ \frac{\partial y_2}{\partial x_1} \end{bmatrix} = \begin{bmatrix} 4e^{-x_1} \\ 2e^{x_1} \end{bmatrix}.$$

Using Cramer's rule we have

$$\frac{\partial y_1}{\partial x_1} = \frac{\begin{bmatrix} 4e^{-x_1} & -1 \\ 2e^{x_1} & \frac{1}{y_2} \end{bmatrix}}{\det[A]} = \frac{\overbrace{\frac{4e^{-x_1}}{y_2} + 2e^{x_1}}^{+}}{\underbrace{\det[A]}_{-}} < 0, \frac{\partial y_2}{\partial x_1} = \frac{\begin{bmatrix} \frac{1}{y_1} & 4e^{-x_1} \\ -1 & 2e^{x_1} \end{bmatrix}}{\det[A]} = \frac{\overbrace{\frac{2e^{x_1}}{y_1} + 4e^{-x_1}}^{+}}{\underbrace{\det[A]}_{-}} < 0.$$

Problem 2.19 *[D***] Consider a model with implicit function*

$$g(y, x_1, x_2) = y^{\frac{1}{2}} \ln(y) - e^{x_1 - 2x_2} = 0$$

where $x_1 > 0$ and $x_2 > 0$ are exogenous and $y > 0$ is endogenous. Prove that $y > 1$. What form does the explicit function take? Calculate the total differential and show that the explicit function exists. From the total differential find $\frac{\partial y}{\partial x_2}$ and determine if $\frac{\partial y}{\partial x_2}$ is positive or negative.

Answer: We have

$$y^{\frac{1}{2}} \ln(y) - e^{x_1 - 2x_2} = 0 \Longrightarrow y^{\frac{1}{2}} \ln(y) = e^{x_1 - 2x_2} > 0$$

since $e^x > 0$. Since $y^{\frac{1}{2}} > 0$ it follows that $\ln(y) > 0 \Longrightarrow y > 1$.

For the explicit function to exist we require $\frac{\partial g(y, x_1, x_2)}{\partial y} \neq 0$ as

$$\frac{\partial g(y, x_1, x_2)}{\partial y} = y^{\frac{1}{2}}\frac{1}{y} + \frac{1}{2}y^{-\frac{1}{2}}\ln(y) = y^{-\frac{1}{2}}\left(1 + \frac{1}{2}\ln(y)\right) > 0.$$

The total differential is

$$y^{-\frac{1}{2}}\left(1 + \frac{1}{2}\ln(y)\right)dy - e^{x_1 - 2x_2}dx_1 + 2e^{x_1 - 2x_2}dx_2 = 0.$$

To calculate $\frac{\partial y}{\partial x_2}$ we 1) set $dx_1 = 0$ and 2) replace the remaining $d's$ with $\partial's$ to obtain

$$y^{-\frac{1}{2}}\left(1 + \frac{1}{2}\ln(y)\right)\partial y + 2e^{x_1 - 2x_2}\partial x_2 = 0 \Longrightarrow \frac{\partial y}{\partial x_2} = -\frac{2e^{x_1 - 2x_2}}{y^{-\frac{1}{2}}\left(1 + \frac{1}{2}\ln(y)\right)} < 0$$

since $2e^{x_1 - 2x_2} > 0$ and $y^{-\frac{1}{2}}\left(1 + \frac{1}{2}\ln(y)\right) > 0$.

2.3 Supply and Demand

Problem 2.20 *Using total differentials on the supply and demand model*

$$Q = D(P + T_H) \text{ and } Q = S(P - T_F)$$

it was shown in the textbook that

$$\begin{aligned} \frac{\partial P}{\partial T_H} &= \frac{-D_P}{-S_P + D_P}, \frac{\partial Q}{\partial T_H} = \frac{-D_P S_P}{-S_P + D_P} \\ \frac{\partial P}{\partial T_F} &= \frac{-S_P}{-S_P + D_P}, \frac{\partial Q}{\partial T_F} = \frac{-D_P S_P}{-S_P + D_P} \end{aligned}$$

where $D_P \equiv D'(P + T_H) < 0$ *and* $S_P \equiv S'(P - T_F) > 0$. *Consider a linear version of the same supply and demand model*

$$\begin{aligned} Q_D &= a - b(P + T_H) \text{ for } b > 0 \\ Q_S &= c + d(P - T_F) \text{ for } d > 0. \end{aligned}$$

*[D**] Solve for the reduced form* P, Q, *and use this to calculate* $\frac{\partial P}{\partial T_H}, \frac{\partial Q}{\partial T_H}, \frac{\partial P}{\partial T_F}, \frac{\partial Q}{\partial T_F}$. *Show that your results are consistent with those obtained using total differentials.*

Answer: From the linear model we have

$$\begin{aligned} D_P &\equiv D'(P + T_H) = \frac{d}{dP}(a - b(P + T_H)) = -b \\ S_P &\equiv S'(P - T_F) = \frac{d}{dP}(c + d(P - T_F)) = d. \end{aligned}$$

We solve for the reduced form as

$$c + d(P - T_F) = Q_S = Q_D = a - b(P + T_H) \Longrightarrow P = \frac{a - c}{d + b} - \frac{b}{d + b}T_H + \frac{d}{d + b}T_F.$$

Putting P in the supply curve then yields

$$\begin{aligned} Q &= c + d(P - T_F) = c + d\left(\frac{a - c}{d + b} - \frac{b}{d + b}T_H + \frac{d}{d + b}T_F - T_F\right) \\ &= c + d\frac{a - c}{d + b} + \left(-\frac{b}{d + b}T_H + \frac{d}{d + b}T_F - \frac{d + b}{d + b}T_F\right) \\ &= \frac{ad + bc}{d + b} - \frac{bd}{d + b}(T_H + T_F). \end{aligned}$$

Using $D_P = -b, S_P = d$ it follows that

$$\begin{aligned} \frac{\partial P}{\partial T_H} &= -\frac{b}{d + b} = \frac{-(-b)}{(-d) + (-b)} = \frac{-D_P}{-S_P + D_P} \\ \frac{\partial Q}{\partial T_H} &= -\frac{bd}{d + b} = \frac{-(-b)(d)}{(-d) + (-b)} = \frac{-D_P S_P}{-S_P + D_P} \\ \frac{\partial Q}{\partial T_F} &= -\frac{bd}{d + b} = \frac{-(-b)(d)}{(-d) + (-b)} = \frac{-D_P S_P}{-S_P + D_P} \\ \frac{\partial P}{\partial T_F} &= \frac{d}{d + b} = \frac{-(d)}{(-d) + (-b)} = \frac{-S_P}{-S_P + D_P}. \end{aligned}$$

Problem 2.21 *Consider a demand and supply model where households pay a tax T for every unit they buy. Demand and supply are therefore given by*

$$Q^d = D(P+T) \text{ and } Q^s = S(P).$$

Letting $Q^s = Q^d = Q$ we then have

$$Q = D(P+T) \text{ and } Q = S(P)$$

and

$$D_P \equiv D'(P+T) < 0,\ S_P \equiv S'(P) > 0$$

*[D***] Write the structural model as a system of implicit functions indicating what are the endogenous and exogenous variables. Calculate the total differential. From the total differential calculate $\frac{dQ}{dT}$ and $\frac{dP}{dT}$ and show that both are negative. [C***] If $P_H = P + T$ is the net price that households pay, show that $\frac{dP_H}{dT} > 0$.*

Answer: We have

$$\begin{aligned} g_1(Q,P,T) &= Q - D(P+T) = 0 \\ g_2(Q,P,T) &= Q - S(P) = 0. \end{aligned}$$

The total differential then is

$$\begin{aligned} dQ - D'(P+T)\,dP - D'(P+T)\,dT &= 0 \\ dQ - S'(P)\,dP &= 0 \end{aligned}$$

or

$$\begin{aligned} dQ - D_P dP - D_P dT &= 0 \\ dQ - S_P dP &= 0 \end{aligned}$$

which in matrix notation becomes

$$\begin{bmatrix} 1 & -D_P \\ 1 & -S_P \end{bmatrix} \begin{bmatrix} dQ \\ dP \end{bmatrix} = \begin{bmatrix} D_P dT \\ 0 \end{bmatrix} = \begin{bmatrix} D_P \\ 0 \end{bmatrix} dT.$$

Dividing both sides by dT results in

$$\begin{bmatrix} 1 & -D_P \\ 1 & -S_P \end{bmatrix} \begin{bmatrix} \frac{dQ}{dT} \\ \frac{dP}{dT} \end{bmatrix} = \begin{bmatrix} D_P \\ 0 \end{bmatrix}.$$

We have

$$A \equiv \begin{bmatrix} 1 & -D_P \\ 1 & -S_P \end{bmatrix} \Longrightarrow \det[A] = -S_P + D_P < 0$$

so that

$$\frac{dQ}{dT} = \frac{\det\begin{bmatrix} D_P & -D_P \\ 0 & -S_P \end{bmatrix}}{\det[A]} = \frac{-\overbrace{S_P}^{+}\overbrace{D_P}^{-}}{\underbrace{\det[A]}_{-}} < 0$$

$$\frac{dP}{dT} = \frac{\det\begin{bmatrix} 1 & D_P \\ 1 & 0 \end{bmatrix}}{\det[A]} = \frac{-\overbrace{D_P}^{-}}{\underbrace{\det[A]}_{-}} < 0.$$

Now since $P_H = P + T$ we have

$$\frac{dP_H}{dT} = \frac{dP}{dT} + 1 = \frac{-D_P}{-S_P + D_P} + 1 = \frac{-S_P}{-S_P + D_P} > 0.$$

Problem 2.22 *Suppose that demand is given by $Q^d = D(P)$ where $D'(P) < 0$, supply is given by $Q^s = S(P)$ where $S'(P) > 0$, and that in equilibrium $Q^d = Q^s = Q$. The government decides to impose a tax on this good such that if the household purchases one dollar of the good it must pay t dollars to the government where $t > 0$. This means that the effective price the household must pay is $P \times (1+t)$. Hence in equilibrium the following two implicit functions will hold*

$$Q - D(P \times (1+t)) = 0 \text{ and } Q - S(P) = 0.$$

These two implicit functions then determine the explicit functions $P(t)$ and $Q(t)$, that is, the price and quantity as a function of the tax t. To simplify the notation define

$$D_P \equiv D'(P \times (1+t)) < 0,\ S_P \equiv S'(P) > 0$$

*[C***] Take the total differential of the two implicit functions and show that the conditions of the implicit function theorem are satisfied; that is, the appropriate matrix is non-singular. Calculate $\frac{dP}{dt}$ and $\frac{dQ}{dt}$ and determine their signs. What do these signs tell you about the impact of the tax on the market? The price that households pay is $P_H = P \times (1+t)$. [B***] Show that an increase in the tax increases P_H; that is $\frac{dP_H}{dt} > 0$.*

Answer: The total differential then is

$$\begin{aligned} dQ - (1+t)D'(P \times (1+t))\,dP - D'(P \times (1+t)) \times P dt &= 0 \\ dQ - S'(P)\,dP &= 0 \end{aligned}$$

or

$$\begin{aligned} dQ - (1+t)D_P dP - D_P P dt &= 0 \\ dQ - S_P dP &= 0. \end{aligned}$$

In matrix notation this becomes

$$\begin{bmatrix} 1 & -(1+t)D_P \\ 1 & -S_P \end{bmatrix} \begin{bmatrix} dQ \\ dP \end{bmatrix} = \begin{bmatrix} D_P P dt \\ 0 \end{bmatrix} = \begin{bmatrix} D_P P \\ 0 \end{bmatrix} dt$$

so that dividing both sides by dt yields

$$\begin{bmatrix} 1 & -(1+t)D_P \\ 1 & -S_P \end{bmatrix} \begin{bmatrix} \frac{dQ}{dt} \\ \frac{dP}{dt} \end{bmatrix} = \begin{bmatrix} D_P P \\ 0 \end{bmatrix}.$$

We have

$$A = \begin{bmatrix} 1 & -(1+t)D_P \\ 1 & -S_P \end{bmatrix} \Longrightarrow \det[A] = -S_P + (1+t)D_P < 0$$

so that

$$\frac{dQ}{dt} = \frac{\det \begin{bmatrix} D_P P & -(1+t)D_P \\ 0 & -S_P \end{bmatrix}}{\det[A]} = \frac{-\overbrace{S_P}^{+}\overbrace{D_P P}^{-}}{\underbrace{\det[A]}_{-}} < 0$$

$$\frac{dP}{dt} = \frac{\det \begin{bmatrix} 1 & D_P P \\ 1 & 0 \end{bmatrix}}{\det[A]} = \frac{\overbrace{-D_P P}^{-}}{\underbrace{\det[A]}_{-}} < 0.$$

Now since $P_H = P \times (1+t)$ we have

$$\begin{aligned} \frac{dP_H}{dt} &= \frac{dP}{dt} \times (1+t) + P = \frac{-D_P P}{-S_P + D_P(1+t)}(1+t) + P \\ &= P\left(\frac{-D_P(1+t)}{-S_P + D_P(1+t)} + 1\right) = P\left(\frac{-S_P}{-S_P + D_P(1+t)}\right) > 0. \end{aligned}$$

Problem 2.23 *Consider a demand and supply model where households pay a tax of T_H and firms a tax of T_F. This results in a demand and supply model of the form*

$$\begin{aligned} Q &= D(P + T_H), \text{ 'demand curve'} \\ Q &= S(P - T_F), \text{ 'supply curve'}. \end{aligned}$$

Assume that

$$D_P \equiv D'(P + T_H) < 0, \; S_P \equiv S'(P - T_F) > 0.$$

*[D***] Determine the sign of $\frac{\partial P}{\partial T_H}$ and show that an increase in T_H increases the net price $P_H = P + T_H$ that households pay. Determine the sign of $\frac{\partial P}{\partial T_F}$ and show that an increase in T_F decreases the net price $P_F = P - T_F$ that firm's receive. [B**] Given that the tax revenue that the government collects*

is $R = (T_H + T_F) \times Q$*, show that the government will be indifferent between increasing* T_H *and* T_F *in that* $\frac{\partial R}{\partial T_H} = \frac{\partial R}{\partial T_F}$*. Normally one thinks that households would prefer governments to raise taxes on firms by using* T_F *while firm's would prefer that governments tax households by using* T_H*. Prove that households, firms, and government care only about the total tax* $T = T_H + T_F$*.*

Answer: We have

$$\begin{aligned} g_1(Q, P, T_H, T_F) &= Q - D(P + T_H) = 0 \\ g_2(Q, P, T_H, T_F) &= Q - S(P - T_F) = 0. \end{aligned}$$

The total differential then is

$$\begin{aligned} dQ - D_P dP - D_P dT_H &= 0 \\ dQ - S_P dP + S_P dT_F &= 0 \end{aligned}$$

or in matrix notation

$$\underbrace{\begin{bmatrix} 1 & -D_P \\ 1 & -S_P \end{bmatrix}}_{A} \begin{bmatrix} dQ \\ dP \end{bmatrix} = \begin{bmatrix} D_P dT_H \\ -S_P dT_F \end{bmatrix}.$$

We have

$$\det[A] = -S_P + D_P < 0.$$

To find $\frac{\partial Q}{\partial T_H}, \frac{\partial P}{\partial T_H}$ we 1) set $dT_F = 0$ and 2) replacing the remaining $d's$ with $\partial's$ as

$$\begin{bmatrix} 1 & -D_P \\ 1 & -S_P \end{bmatrix} \begin{bmatrix} \partial Q \\ \partial P \end{bmatrix} = \begin{bmatrix} D_P \partial T_H \\ 0 \end{bmatrix} = \begin{bmatrix} D_P \\ 0 \end{bmatrix} \partial T_H$$

and so dividing both sides by ∂T_H yields

$$\begin{bmatrix} 1 & -D_P \\ 1 & -S_P \end{bmatrix} \begin{bmatrix} \frac{\partial Q}{\partial T_H} \\ \frac{\partial P}{\partial T_H} \end{bmatrix} = \begin{bmatrix} D_P \\ 0 \end{bmatrix}.$$

From Cramer's rule we then have

$$\begin{aligned} \frac{\partial Q}{\partial T_H} &= \frac{\det \begin{bmatrix} D_P & -D_P \\ 0 & -S_P \end{bmatrix}}{\det[A]} = \frac{-\overbrace{S_P}^{+}\overbrace{D_P}^{-}}{\underbrace{\det[A]}_{-}} < 0 \\ \frac{\partial P}{\partial T_H} &= \frac{\det \begin{bmatrix} 1 & D_P \\ 1 & 0 \end{bmatrix}}{\det[A]} = \frac{-\overbrace{D_P}^{-}}{\underbrace{\det[A]}_{-}} < 0. \end{aligned}$$

Now since $P_H = P + T_H$ we have

$$\frac{\partial P_H}{\partial T_H} = \frac{\partial P}{\partial T_H} + 1 = \frac{-D_P}{-S_P + D_P} + 1 = \frac{-S_P}{-S_P + D_P} > 0.$$

To find $\frac{\partial Q}{\partial TF}, \frac{\partial P}{\partial T_F}$ we 1) set $dT_H = 0$ and 2) replacing the remaining $d's$ with $\partial's$ as

$$\begin{bmatrix} 1 & -D_P \\ 1 & -S_P \end{bmatrix} \begin{bmatrix} \frac{\partial Q}{\partial T_F} \\ \frac{\partial P}{\partial T_F} \end{bmatrix} = \begin{bmatrix} 0 \\ -S_P \end{bmatrix}$$

so that using Cramer's rule

$$\frac{\partial Q}{\partial T_F} = \frac{\det \begin{bmatrix} 0 & -D_P \\ -S_P & -S_P \end{bmatrix}}{\det[A]} = \frac{-\overbrace{S_P}^{+}\overbrace{D_P}^{-}}{\underbrace{\det[A]}_{-}} < 0$$

$$\frac{\partial P}{\partial T_F} = \frac{\det \begin{bmatrix} 1 & 0 \\ 1 & -S_P \end{bmatrix}}{\det[A]} = \frac{-\overbrace{S_P}^{+}}{\underbrace{\det[A]}_{-}} > 0.$$

Now from $R = (T_H + T_F) Q$ we have

$$\frac{\partial R}{\partial T_H} = Q + (T_H + T_F) \frac{\partial Q}{\partial T_H} = \frac{\partial R}{\partial T_F} = Q + (T_H + T_F) \frac{\partial Q}{\partial T_F}$$

since $\frac{\partial Q}{\partial T_H} = \frac{\partial Q}{\partial T_F}$ by the above calculation.

The equilibrium price P satisfies

$$Q = D(P + T_H) \text{ and } Q = S(P - T_F)$$

Now

$$P_F = P - T_F = P + T_H - (T_H + T_F) = P_H - (T_H + T_F)$$

so

$$Q = D(P_H) \text{ and } Q = S(P_F) = S(P_H - (T_H + T_F)).$$

Thus P_H is a function only of $T_H + T_F$ is defined by

$$D(P_H) = S(P_H - (T_H + T_F)) \Longrightarrow P_H = f_1(T_H + T_F).$$

Similarly since

$$P_H = P + T_H = P - T_F + (T_H + T_F) = P_F + (T_H + T_F)$$

it follows that

$$D(P_F + T_H + T_F) = S(P_F) \Longrightarrow P_F = f_2(T_H + T_F).$$

Finally Q is a function only of $T_H + T_F$ since $Q = D(f_1(T_H + T_F))$.

Problem 2.24 *Consider a demand and supply model where for each unit they buy, consumers must pay tax at a rate of t_H and producers a tax at a rate of t_F. This results in a demand and supply model of the form*

$$Q = D(P \times (1 + t_H)), \quad Q = S(P \times (1 - t_F)).$$

Assume

$$D_P \equiv D'(P \times (1+t_H)) < 0, \; S_P \equiv S'(P \times (1-t_F)) > 0.$$

*[C***] Determine the sign of* $\frac{\partial P}{\partial t_H}, \frac{\partial Q}{\partial t_H}, \frac{\partial P}{\partial t_F}$ *and* $\frac{\partial Q}{\partial t_F}$. *[B***] The tax revenue that the government collects is* $R = (t_H + t_F) PQ$. *Find* $\frac{\partial R}{\partial t_H}$ *and* $\frac{\partial R}{\partial t_F}$.

Answer: We have

$$\begin{aligned} g_1(Q, P, t_H, t_F) &= Q - D(P \times (1+t_H)) = 0 \\ g_2(Q, P, t_H, t_F) &= Q - S(P \times (1-t_F)) = 0. \end{aligned}$$

The total differential then is

$$\begin{aligned} dQ - D_P(1+t_H)\,dP - D_P P dt_H &= 0 \\ dQ - S_P(1-t_F)\,dP + S_P P dt_F &= 0 \end{aligned}$$

or in matrix notation

$$\begin{bmatrix} 1 & -D_P(1+t_H) \\ 1 & -S_P(1-t_F) \end{bmatrix} \begin{bmatrix} dQ \\ dP \end{bmatrix} = \begin{bmatrix} D_P P dt_H \\ -S_P P dt_F \end{bmatrix}.$$

We have

$$\det[A] = \det \begin{bmatrix} 1 & -D_P(1+t_H) \\ 1 & -S_P(1-t_F) \end{bmatrix} = -S_P(1-t_F) + D_P(1+t_H) < 0.$$

To find $\frac{\partial Q}{\partial t_H}, \frac{\partial P}{\partial t_H}$ we 1) set $dt_F = 0$ in the total differential and 2) replace the remaining $d's$ with $\partial's$ as

$$\begin{bmatrix} 1 & -D_P(1+t_H) \\ 1 & -S_P(1-t_F) \end{bmatrix} \begin{bmatrix} \frac{\partial Q}{\partial t_H} \\ \frac{\partial P}{\partial t_H} \end{bmatrix} = \begin{bmatrix} D_P P \\ 0 \end{bmatrix}.$$

Using Cramer's rule it follows that

$$\frac{\partial Q}{\partial t_H} = \frac{\det \begin{bmatrix} D_P P & -D_P(1+t_H) \\ 0 & -S_P(1-t_F) \end{bmatrix}}{\det[A]} = \frac{\overbrace{-D_P P}^{-}\overbrace{S_P(1-t_F)}^{+}}{\underbrace{\det[A]}_{-}} < 0$$

$$\frac{\partial P}{\partial t_H} = \frac{\det \begin{bmatrix} 1 & D_P P \\ 1 & 0 \end{bmatrix}}{\det[A]} = \frac{\overbrace{-D_P P}^{-}}{\underbrace{\det[A]}_{-}} < 0.$$

To find $\frac{\partial Q}{\partial t_F}, \frac{\partial P}{\partial t_F}$ we 1) set $dt_H = 0$ and 2) replace the remaining $d's$ with $\partial's$ as

$$\begin{bmatrix} 1 & -D_P(1+t_H) \\ 1 & -S_P(1-t_F) \end{bmatrix} \begin{bmatrix} \frac{\partial Q}{\partial t_F} \\ \frac{\partial P}{\partial t_F} \end{bmatrix} = \begin{bmatrix} 0 \\ -S_P P \end{bmatrix}$$

so that

$$\frac{\partial Q}{\partial t_F} = \frac{\det\begin{bmatrix} 0 & -D_P(1+t_H) \\ -S_P P & -S_P(1-t_F) \end{bmatrix}}{\det[A]} = \frac{\overbrace{-S_P P}^{+}\overbrace{D_P(1+t_H)}^{-}}{\underbrace{\det[A]}_{-}} < 0$$

$$\frac{\partial P}{\partial t_F} = \frac{\det\begin{bmatrix} 1 & 0 \\ 1 & -S_P P \end{bmatrix}}{\det[A]} = \frac{\overbrace{-S_P P}^{+}}{\underbrace{\det[A]}_{-}} > 0.$$

Now from $R = (t_H + t_F) PQ$ we have

$$\begin{aligned} \frac{\partial R}{\partial t_H} &= PQ + (t_H + t_F)\left(\frac{\partial Q}{\partial t_H}P + \frac{\partial P}{\partial t_H}Q\right) \\ &= PQ + (t_H + t_F)\left(\frac{-D_P P^2 S_P(1-t_F) - D_P PQ}{-S_P(1-t_F) + D_P(1+t_H)}\right) \\ \frac{\partial R}{\partial t_F} &= PQ + (t_H + t_F)\left(\frac{\partial Q}{\partial t_F}P + \frac{\partial P}{\partial t_F}Q\right) \\ &= PQ + (t_H + t_F)\left(\frac{-S_P P^2 D_P(1+t_H) - S_P PQ}{\det[A]}\right). \end{aligned}$$

2.4 The *IS/LM* Model

Problem 2.25 *Consider a linear version of the IS/LM model*

$$\begin{aligned} c^d &= c(y) = c_o + c_1 \times y \textit{ for } 0 < c_1 < 1 \\ i^d &= i(r) = i_o - i_1 \times r \textit{ for } i_1 > 0 \\ g^d &= g_o \\ m^d &= k(y) + l(r) \\ k(y) &= k_0 + k_1 \times y \textit{ for } k_1 > 0 \\ l(r) &= l_0 - l_1 \times r \textit{ for } l_1 > 0 \\ m^s &= m^s_o. \end{aligned}$$

*[D***] Find the IS curve and the LM curve. Find the slopes of the IS and LM curves, show that the IS curve slopes downwards, and that the LM curve slopes upwards. Compare your answers with those derived from total differentials in the textbook. Calculate the equilibrium y as a function of the exogenous variables and use this to calculate* $\frac{\partial y}{\partial g_0}, \frac{\partial y}{\partial m^s_0}$. *Compare these with those derived from total differentials in the textbook.*

Answer: The *IS* curve is found from

$$y = c(y) + i(r) + g_0 = c_o + c_1 \times y + i_o - i_1 \times r + g_0.$$

Solving for r we have

$$\begin{aligned} i_1 \times r &= c_o + c_1 \times y + i_o - i_1 \times r + g_0 - y = c_o + i_o + g_0 - (1 - c_1) \times y \\ &\Longrightarrow r = \frac{c_o + i_o + g_0}{i_1} - \frac{1 - c_1}{i_1} \times y. \end{aligned}$$

The *IS* curve slopes downwards $\frac{\partial r}{\partial y} = -\frac{1-c_1}{i_1} < 0$ since $i_1 > 0$ and

$$0 < c_1 < 1 \Longrightarrow 1 - c_1 > 0.$$

Using total differentials (see the textbook) the predicted slope of the *IS* curve is $\frac{\partial r}{\partial y} = \frac{1-c_y}{i'(r)}$ which agrees with $\frac{\partial r}{\partial y} = -\frac{1-c_1}{i_1}$ since

$$\begin{aligned} c(y) &= c_o + c_1 \times y \Longrightarrow c_y \equiv c'(y) \Longrightarrow \frac{d}{dy}(c_o + c_1 \times y) = c_1 \\ i(r) &= i_o - i_1 \times r \Longrightarrow i'(r) = \frac{d}{dr}(i_o - i_1 \times r) = -i_1. \end{aligned}$$

The *LM* curve is found from

$$m^s = m_o^s = m^d = k(y) + l(r) = k_0 + k_1 \times y + l_0 - l_1 \times r$$

so that solving for r we have

$$l_1 \times r = k_0 + k_1 \times y + l_0 - m_o^s \Longrightarrow r = \frac{k_0 + l_0 - m_o^s}{l_1} + \frac{k_1}{l_1} y.$$

The *LM* curve slopes upwards as

$$r = \frac{k_0 + l_0 - m_o^s}{l_1} + \frac{k_1}{l_1} y \Longrightarrow \frac{\partial r}{\partial y} = \frac{k_1}{l_1} > 0$$

since $k_1 > 0, l_1 > 0$. Using total differentials the predicted slope of the *LM* curve is $\frac{\partial r}{\partial y} = -\frac{k'(y)}{l'(r)}$. This agrees with $\frac{\partial r}{\partial y} = \frac{k_1}{l_1}$ as

$$k(y) = k_0 + k_1 \times y \Longrightarrow k'(y) = k_1 \text{ and } l(r) = l_0 - l_1 \times r \Longrightarrow l'(r) = -l_1.$$

We now find the reduced form y by equating the r from the *IS* and *LM* curves as

$$r = \frac{c_o + i_o + g_0}{i_1} - \frac{1 - c_1}{i_1} \times y \text{ and } r = \frac{k_0 + l_0 - m_o^s}{l_1} + \frac{k_1}{l_1} y$$

so that

$$\begin{aligned} \frac{c_o + i_o + g_0}{i_1} - \frac{1 - c_1}{i_1} \times y &= r = \frac{k_0 + l_0 - m_o^s}{l_1} + \frac{k_1}{l_1} y \\ &\Longrightarrow \left(\frac{1 - c_1}{i_1} + \frac{k_1}{l_1} \right) \times y = \frac{c_o + i_o + g_0}{i_1} - \frac{k_0 + l_0 - m_o^s}{l_1} \\ &\Longrightarrow y = \frac{\frac{c_o + i_o + g_0}{i_1} - \frac{k_0 + l_0 - m_o^s}{l_1}}{\frac{1 - c_1}{i_1} + \frac{k_1}{l_1}} = \frac{\frac{c_o + i_o + g_0}{i_1} - \frac{k_0 + l_0 - m_o^s}{l_1}}{\frac{(1 - c_1) l_1 + i_1 k_1}{i_1 l_1}}. \end{aligned}$$

Simplifying somewhat we have

$$y=\frac{\frac{c_o+i_o+g_0}{i_1}-\frac{k_0+l_0-m_o^s}{l_1}}{\frac{1-c_1}{i_1}+\frac{k_1}{l_1}}=\frac{\frac{c_o+i_o+g_0}{i_1}-\frac{k_0+l_0-m_o^s}{l_1}}{\frac{(1-c_1)l_1+i_1k_1}{i_1l_1}}=\frac{(c_o+i_o+g_0)\,l_1-i_1\,(k_0+l_0-m_o^s)}{(1-c_1)\,l_1+i_1k_1}.$$

Thus

$$\begin{aligned}\frac{\partial y}{\partial g_0}&=\frac{l_1}{(1-c_1)\,l_1+i_1k_1}=\frac{l_1}{(1-c_1)\,l_1+i_1k_1}\times\frac{\frac{1}{l_1}}{\frac{1}{l_1}}=\frac{1}{1-c_1+i_1\frac{k_1}{l_1}}\\ \frac{\partial y}{\partial m_0^s}&=\frac{i_1}{(1-c_1)\,l_1+i_1k_1}.\end{aligned}$$

This agrees with the expressions found using total differentials

$$\frac{\partial y}{\partial g_0}=\frac{1}{1-c'(y)-i'(r)\frac{k'(y)}{l'(r)}},\ \frac{\partial y}{\partial m_0^s}=\frac{i'(r)}{-(1-c'(y))\,l'(r)-i'(r)\,k'(y)}$$

since

$$c'(y)=c_1, i'(r)=-i_1, k'(y)=k_1, l'(r)=-l_1.$$

Problem 2.26 *Consider the IS/LM model where instead of having consumption as $c(y)$ we introduce taxes $t=\tau y$ for $0<\tau<1$ so that disposable income is $y\times(1-\tau)$ and we have $c(y\times(1-\tau))$. To simplify the notation you may define $c_y=c'(y\times(1-\tau))$ and you may assume that $0<c_y<1$. Thus the IS/LM model becomes*

$$\begin{aligned}IS&: \; y=c(y\times(1-\tau))+i(r)+g_0, \text{ where } i'(r)<0\\ LM&: \; m_0^s=k(y)+l(r), \text{ where: } k'(y)>0 \text{ and } l'(r)<0\end{aligned}$$

*and where y and r are endogenous and g_0 and m_0^s are exogenous. [D***] Using total differentials show that*

$$\frac{\partial y}{\partial m_0^s}>0, \frac{\partial r}{\partial m_0^s}<0$$

and $\frac{\partial i}{\partial m_0^s}>0$.

Answer: Written as implicit functions we have

$$\begin{aligned}g_1(y,r,g_0,m_0^s)&=y-c(y\times(1-\tau))-i(r)-g_0=0\\ g_2(y,r,g_0,m_0^s)&=k(y)+l(r)-m_0^s=0\end{aligned}$$

with total differential

$$\begin{aligned}(1-c_y\times(1-\tau))\,dy-i'(r)\,dr-dg_0&=0\\ k'(y)\,dy+l'(r)\,dr-dm_0^s&=0.\end{aligned}$$

The only tricky part here is calculating

$$\frac{\partial g_1}{\partial y} = \frac{\partial}{\partial y}\left(y - c\left(y - t\left(y\right)\right)\right) = 1 - c'\left(y \times (1-\tau)\right)(1-\tau) = 1 - c_y \times (1-\tau)$$

using the chain rule to differentiate $c\left(y \times (1-\tau)\right)$ with respect to y. Note that since $0 < c_y < 1$ and $0 < \tau < 1$ it follows that

$$0 < 1 - c_y \times (1-\tau) < 1.$$

In matrix notation the total differential becomes

$$\underbrace{\begin{bmatrix} 1 - c_y \times (1-\tau) & -i'(r) \\ k'(y) & l'(r) \end{bmatrix}}_{A} \begin{bmatrix} dy \\ dr \end{bmatrix} = \begin{bmatrix} dg_0 \\ dm_0^s \end{bmatrix}.$$

To find $\frac{\partial y}{\partial m_0^s}, \frac{\partial r}{\partial m_0^s}$ we 1) set $dg_0 = 0$ and 2) change the remaining $d's$ to $\partial's$ to obtain

$$\begin{aligned} (1 - c_y \times (1-\tau))\,\partial y - i'(r)\,\partial r &= 0 \\ k'(y)\,\partial y + l'(r)\,\partial r - \partial m_0^s &= 0 \end{aligned}$$

or in matrix notation

$$\underbrace{\begin{bmatrix} 1 - c_y \times (1-\tau) & -i'(r) \\ k'(y) & l'(r) \end{bmatrix}}_{A} \begin{bmatrix} \frac{\partial y}{\partial m_0^s} \\ \frac{\partial r}{\partial m_0^s} \end{bmatrix} = \begin{bmatrix} 0 \\ 1 \end{bmatrix}$$

and where

$$\det[A] = \det\begin{bmatrix} 1 - c_y \times (1-\tau) & -i'(r) \\ k'(y) & l'(r) \end{bmatrix} = \overbrace{(1 - c_y \times (1-\tau))}^{+}\overbrace{l'(r)}^{-} + \overbrace{i'(r)}^{-}\overbrace{k'(y)}^{+} < 0.$$

Using Cramer's rule and $\det[A] < 0$ we have

$$\begin{aligned} \frac{\partial y}{\partial m_0^s} &= \frac{\det\begin{bmatrix} 0 & -i'(r) \\ 1 & l'(r) \end{bmatrix}}{\det[A]} = \frac{\overbrace{i'(r)}^{-}}{\det[A]} > 0 \\ \frac{\partial r}{\partial m_0^s} &= \frac{\det\begin{bmatrix} 1 - c_y \times (1-\tau) & 0 \\ k'(y) & 1 \end{bmatrix}}{\det[A]} = \frac{\overbrace{1 - c_y \times (1-\tau)}^{+}}{\det[A]} < 0. \end{aligned}$$

To calculate $\frac{\partial i}{\partial m_0^s}$ we have using the chain rule that

$$\frac{\partial i}{\partial m_0^s} = \overbrace{i'(r)}^{-}\overbrace{\frac{\partial r}{\partial m_0^s}}^{-} > 0.$$

Problem 2.27 *Consider the following IS/LM model*

$$\begin{aligned}
c^d &= c(y-t)\ ,\ \textit{where } c_y \equiv c'(y-t) \textit{ and } 0 < c_y < 1 \\
i^d &= i(r),\ \textit{where } i_r \equiv i'(r) < 0 \\
g^d &= g_o (\textit{government expenditure is exogenous}) \\
t &= t(y),\ \textit{where } t_y \equiv t'(y) \textit{ with } 0 < t_y < 1 \\
m^d &= l(y,r),\ \textit{where } l_y \equiv \frac{\partial l(y,r)}{\partial y} > 0 \textit{ and } l_r \equiv \frac{\partial l(y,r)}{\partial r} < 0
\end{aligned}$$

*[C**] Show that for this model the IS curve is downward sloping and the LM curve is upward sloping. [D***] Show that $\frac{\partial y}{\partial g_o} > 0$ and $\frac{\partial r}{\partial g_o} > 0$. Show that $\frac{\partial y}{\partial m_o^s} > 0$ and $\frac{\partial r}{\partial m_o^s} < 0$. [B***] Show that increasing the money supply increases investment. [A***] The government's deficit is given by*

$$d = g_0 - t(y).$$

An increase in government expenditure g_0 increases y and hence increases taxes $t(y)$. Could increasing g_0 then increase tax revenue so much that it would actually reduce the deficit?

Answer: The *IS* curve is defined by

$$y = y^d \equiv c^d + i^d + g^d = c(y - t(y)) + i(r) + g_o$$

leading to the implicit function

$$g_1(y, r, g_o, m_o^s) = y - c(y - t(y)) - i(r) - g_o = 0$$

with total differential

$$(1 - c_y (1 - t_y))\, dy - i_r dr - dg_o = 0.$$

Setting $dg_0 = 0$ and solving for the slope of the *IS* curve $\frac{dr}{dy}$ we obtain

$$\frac{dr}{dy} = \frac{1 - c_y (1 - t_y)}{i_r} < 0$$

since $i_r < 0$ and $0 < 1 - c_y (1 - t_y) < 1$.

The *LM* curve is defined by

$$m_0^s = m^d = l(y, r)$$

leading to the implicit function

$$g_2(y, r, g_o, m_o^s) = l(y, r) - m_0^s = 0$$

with total differential

$$l_y dy + l_r dr - dm_0^s = 0.$$

Setting $dm_o^s = 0$ and solving for the slope of the LM curve $\frac{dr}{dy}$ we obtain

$$\frac{dr}{dy} = -\frac{l_y}{l_r} > 0$$

since $l_y > 0$ and $l_r < 0$.

The two total differentials in matrix notation are

$$\begin{bmatrix} 1 - c_y(1 - t_y) & -i_r \\ l_y & l_r \end{bmatrix} \begin{bmatrix} dy \\ dr \end{bmatrix} = \begin{bmatrix} dg_o \\ dm_0^s \end{bmatrix}$$

where

$$\det[A] = \det \begin{bmatrix} 1 - c_y(1 - t_y) & -i_r \\ l_y & l_r \end{bmatrix} = (1 - c_y(1 - t_y)) l_r + i_r l_y < 0$$

since

$$0 < 1 - c_y(1 - t_y) < 1, \; l_r < 0, \text{ and } i_r l_y < 0.$$

To find $\frac{\partial y}{\partial g_o}, \frac{\partial r}{\partial g_o}$ we 1) setting $dm_0^s = 0$ and 2) change the remaining $d's$ to ∂'s to obtain

$$\begin{bmatrix} 1 - c_y(1 - t_y) & -i_r \\ l_y & l_r \end{bmatrix} \begin{bmatrix} \frac{\partial y}{\partial g_o} \\ \frac{\partial r}{\partial g_o} \end{bmatrix} = \begin{bmatrix} 1 \\ 0 \end{bmatrix}.$$

Thus

$$\frac{\partial y}{\partial g_o} = \frac{\det \begin{bmatrix} 1 & -i'(r) \\ 0 & l_r \end{bmatrix}}{\det[A]} = \frac{l_r}{\det[A]} = \frac{1}{(1 - c_y(1 - t_y)) + i_r \frac{l_y}{l_r}} > 0.$$

Similarly

$$\frac{\partial r}{\partial g_o} = \frac{\det \begin{bmatrix} 1 - c_y(1 - t_y) & 1 \\ l_y & 0 \end{bmatrix}}{\det[A]} = \frac{-l_y}{\det[A]} > 0$$

since $l_y > 0$.

To find $\frac{\partial y}{\partial m_o^s}, \frac{\partial r}{\partial m_o^s}$ we 1) setting $dg_0 = 0$ and 2) change the remaining $d's$ to ∂'s to obtain

$$\begin{bmatrix} 1 - c_y(1 - t_y) & -i_r \\ l_y & l_r \end{bmatrix} \begin{bmatrix} \frac{\partial y}{\partial m_o^s} \\ \frac{\partial r}{\partial m_o^s} \end{bmatrix} = \begin{bmatrix} 0 \\ 1 \end{bmatrix}$$

so that

$$\begin{aligned} \frac{\partial y}{\partial m_o^s} &= \frac{\det \begin{bmatrix} 0 & -i_r \\ 1 & l_r \end{bmatrix}}{\det[A]} = \frac{i_r}{\det[A]} > 0 \\ \frac{\partial r}{\partial m_o^s} &= \frac{\det \begin{bmatrix} 1 - c_y(1 - t_y) & 0 \\ l_y & 1 \end{bmatrix}}{\det[A]} = \frac{1 - c_y(1 - t_y)}{\det[A]} < 0. \end{aligned}$$

We have

$$i = i(r) \Longrightarrow \frac{\partial i}{\partial m_o^s} = i'(r)\frac{\partial r}{\partial m_o^s} = i_r \frac{\partial r}{\partial m_o^s} > 0$$

since $i_r < 0$ and from the above problem $\frac{\partial r}{\partial m_o^s} < 0$. Thus increasing the money supply increases investment.

The government's deficit is given by $d = g_0 - t(y)$. We have

$$\frac{\partial d}{\partial g_o} = \frac{\partial}{\partial g_o}(g_0 - t(y)) = 1 - t_y \frac{\partial y}{\partial g_o}.$$

We have calculated $\frac{\partial y}{\partial g_o} > 0$ above so it appears the result is ambiguous since $t_y \frac{\partial y}{\partial g_o} > 0$. However

$$\begin{aligned}
\frac{\partial d}{\partial g_o} &= 1 - t_y \frac{\partial y}{\partial g_o} = 1 - t_y \frac{1}{(1 - c_y(1 - t_y)) + i_r \frac{l_y}{l_r}} = \frac{(1 - c_y(1 - t_y)) + i_r \frac{l_y}{l_r} - t_y}{(1 - c_y(1 - t_y)) + i_r \frac{l_y}{l_r}} \\
&= \frac{((1 - c_y) + c_y t_y) + i_r \frac{l_y}{l_r} - t_y}{(1 - c_y(1 - t_y)) + i_r \frac{l_y}{l_r}} = \frac{(1 - c_y) + i_r \frac{l_y}{l_r} - (1 - c_y) t_y}{(1 - c_y(1 - t_y)) + i_r \frac{l_y}{l_r}} \\
&= \frac{\overbrace{(1 - c_y)}^{+}\overbrace{(1 - t_y)}^{+} + \overbrace{i_r \frac{l_y}{l_r}}^{+}}{(1 - c_y(1 - t_y)) + i_r \frac{l_y}{l_r}} > 0
\end{aligned}$$

and so increasing government expenditure does increase the deficit.

Problem 2.28 *Consider the following IS/LM model*

$$\begin{aligned}
c^d &= c(y, r), \text{ where } 0 < c_y \equiv \frac{\partial c(y, r)}{\partial y} < 1 \text{ and } c_r \equiv \frac{\partial c(y, r)}{\partial r} < 0 \\
i^d &= i_o \text{ (i.e. investment is exogenous)} \\
g^d &= g_o \text{ (i.e. government expenditure is exogenous)} \\
m^d &= l(y, r), \text{ where } l_y \equiv \frac{\partial l(y, r)}{\partial y} > 0 \text{ and } l_r \equiv \frac{\partial l(y, r)}{\partial r} < 0.
\end{aligned}$$

*[D**] Show that the IS curve is downward sloping. [A*] Show that the conditions of the implicit function theorem are satisfied. [D***] Show that*

$$\frac{\partial y}{\partial g_o} > 0 \quad \frac{\partial y}{\partial m_o^s} > 0 \text{ and } \frac{\partial r}{\partial m_o^s} < 0.$$

*[B**] Show that an increase in the money supply increases consumption. Does it follow that increasing government expenditure increases consumption?*

Answer: We have

$$g_1(y, r, i_o, g_o, m_o^s) = y - c(y, r) - i_0 - g_o = 0$$

so that the total differential is

$$(1-c_y)\,dy - c_r dr - di_0 - dg_0 = 0.$$

Setting $di_0 = dg_0 = 0$ and solving for the slope of the *IS* curve $\frac{dr}{dy}$ we obtain

$$\frac{dr}{dy} = \frac{(1-c_y)}{c_r} < 0$$

since $0 < (1-c_y) < 1$ and $c_r < 0$.

From the money market we have

$$g_2(y, r, i_o, g_o, m_o^s) = l(y,r) - m_o^s = 0$$

so that the total differential is

$$l_y dy + l_r dr - dm_0^s = 0.$$

Altogether we have two total differentials as

$$\begin{aligned} (1-c_y)\,dy - c_r dr - di_0 - dg_0 &= 0 \\ l_y dy + l_r dr - dm_0^s &= 0 \end{aligned}$$

or in matrix notation

$$\underbrace{\begin{bmatrix} 1-c_y & -c_r \\ l_y & l_r \end{bmatrix}}_{A} \begin{bmatrix} dy \\ dr \end{bmatrix} = \begin{bmatrix} di_0 + dg_0 \\ dm_0^s \end{bmatrix}.$$

We have

$$\det[A] = \det\begin{bmatrix} 1-c_y & -c_r \\ l_y & l_r \end{bmatrix} = (1-c_y)\,l_r + c_r l_y < 0.$$

To find $\frac{\partial y}{\partial g_o}$ we 1) set $di_0 = dm_0^s = 0$ and 2) replace the remaining $d's$ with $\partial's$ to obtain

$$\begin{bmatrix} 1-c_y & -c_r \\ l_y & l_r \end{bmatrix} \begin{bmatrix} \frac{\partial y}{\partial g_o} \\ \frac{\partial r}{\partial g_o} \end{bmatrix} = \begin{bmatrix} 1 \\ 0 \end{bmatrix}$$

so that

$$\frac{\partial y}{\partial g_o} = \frac{\det\begin{bmatrix} 1 & -c_r \\ 0 & l_r \end{bmatrix}}{\det[A]} = \frac{l_r}{\det[A]} > 0$$

since $l_r < 0$ and $\det[A] < 0$.

To find $\frac{\partial y}{\partial m_o^s}, \frac{\partial r}{\partial m_o^s}$ we 1) set $di_0 = dg_0 = 0$ and 2) replace the remaining $d's$ with $\partial's$ to obtain

$$\begin{bmatrix} 1-c_y & -c_r \\ l_y & l_r \end{bmatrix} \begin{bmatrix} \frac{\partial y}{\partial m_o^s} \\ \frac{\partial r}{\partial m_o^s} \end{bmatrix} = \begin{bmatrix} 0 \\ 1 \end{bmatrix}$$

so that

$$\frac{\partial y}{\partial m_o^s} = \frac{\det \begin{bmatrix} 0 & -c_r \\ 1 & l_r \end{bmatrix}}{\det [A]} = \frac{c_r}{\det [A]} > 0$$

since $c_r < 0$. As well

$$\frac{\partial r}{\partial m_o^s} = \frac{\det \begin{bmatrix} 1 - c_y & 0 \\ l_y & 1 \end{bmatrix}}{\det [A]} = \frac{1 - c_y}{\det [A]} < 0$$

since $1 - c_y > 0$.

We have

$$c = c(y, r) \Longrightarrow \frac{\partial c(y,r)}{\partial m_o^s} = c_y \frac{\partial y}{\partial m_o^s} + c_r \frac{\partial r}{\partial m_o^s} > 0$$

since $\frac{\partial y}{\partial m_o^s} > 0, c_y > 0, c_r < 0$, and $\frac{\partial r}{\partial m_o^s} < 0$.

For government expenditure we have

$$c = c(y, r) \Longrightarrow \frac{\partial c(y,r)}{\partial g_o} = c_y \frac{\partial y}{\partial g_o} + c_r \frac{\partial r}{\partial g_o}.$$

We cannot here show that $\frac{\partial c(y,r)}{\partial g_o} > 0$ since while the first term is positive the second is negative since $c_r < 0$ and $\frac{\partial r}{\partial g_0} > 0$.

Problem 2.29 *Consider the following IS/LM model where*

$$\begin{aligned} c^d &= c(r)\ ,\ \text{where } c_r \equiv c'(r) < 0 \\ i^d &= i(y),\ \text{where } 0 < i_y \equiv i'(y) < 1 \\ g^d &= g_o (\text{government expenditure is exogenous}) \\ m^d &= l(y, r),\ \text{where } l_y \equiv \frac{\partial l(y,r)}{\partial y} > 0 \text{ and } l_r \equiv \frac{\partial l(y,r)}{\partial r} < 0. \end{aligned}$$

*[B**] Show that the IS curve is downward sloping and the LM curve is upward sloping. Show that the conditions of the implicit function theorem are satisfied. [D***] Use this to show that* $\frac{\partial y}{\partial g_o} > 0$ *and* $\frac{\partial r}{\partial g_o} > 0$, *and* $\frac{\partial c}{\partial g_0} < 0$. *Show that* $\frac{\partial y}{\partial m_0^s} > 0$ *and* $\frac{\partial r}{\partial m_0^s} < 0$.

Answer: We have

$$g_1(y, r, g_o, m_o^s) = y - c(r) - i(y) - g_o = 0$$

so that the total differential is

$$(1 - i_y)\, dy - c_r dr - dg_0 = 0.$$

Setting $dg_0 = 0$ and solving for the slope of the IS curve $\frac{dr}{dy}$ we obtain

$$\frac{dr}{dy} = \frac{1 - i_y}{c_r} < 0$$

since $1 - i_y > 0$ and $c_r < 0$.

The LM curve is defined by the implicit function

$$g_2(y, r, g_o, m_o^s) = l(y, r) - m_0^s = 0$$

with total differential

$$l_y dy + l_r dr - dm_0^s = 0.$$

Setting $dm_o^s = 0$ and solving for the slope of the LM curve we obtain

$$\frac{dr}{dy} = -\frac{l_y}{l_r} > 0$$

sinee $l_y > 0$ and $l_r < 0$.

Writing the total differential in matrix notation we have

$$\underbrace{\begin{bmatrix} 1 - i_y & -c_r \\ l_y & l_r \end{bmatrix}}_{A} \begin{bmatrix} dy \\ dr \end{bmatrix} = \begin{bmatrix} dg_0 \\ dm_0^s \end{bmatrix}.$$

We have

$$\det[A] = \det \begin{bmatrix} 1 - i_y & -c_r \\ l_y & l_r \end{bmatrix} = (1 - i_y) l_r + c_r l_y < 0.$$

To find $\frac{\partial y}{\partial g_o}, \frac{\partial r}{\partial g_o}$ we 1) set $dm_0^s = 0$ and 2) replace the remaining $d's$ with $\partial's$ to obtain

$$\begin{bmatrix} 1 - i_y & -c_r \\ l_y & l_r \end{bmatrix} \begin{bmatrix} \frac{\partial y}{\partial g_o} \\ \frac{\partial r}{\partial g_o} \end{bmatrix} = \begin{bmatrix} 1 \\ 0 \end{bmatrix}$$

so that

$$\begin{aligned} \frac{\partial y}{\partial g_o} &= \frac{\det \begin{bmatrix} 1 & -c_r \\ 0 & l_r \end{bmatrix}}{\det[A]} = \frac{l_r}{\det[A]} > 0 \\ \frac{\partial r}{\partial g_o} &= \frac{\det \begin{bmatrix} 1 - i_y & 1 \\ l_y & 0 \end{bmatrix}}{\det[A]} = \frac{-l_y}{\det[A]} > 0 \end{aligned}$$

since $l_r < 0, l_y > 0$ and $\det[A] < 0$. To calculate $\frac{\partial c}{\partial g_0}$ we use the chain rule as

$$\frac{\partial c(r)}{\partial g_0} = \overbrace{c'(r)}^{-} \overbrace{\frac{\partial r}{\partial g_0}}^{+} <= c'(r) \frac{-l_y}{(1 - i_y) l_r + c_r l_y} < 0.$$

To find $\frac{\partial y}{\partial m_o^s}, \frac{\partial r}{\partial m_o^s}$ we 1) set $dg_0 = 0$ and 2) replace the remaining $d's$ with $\partial' s$ to obtain

$$\begin{bmatrix} 1-i_y & -c_r \\ l_y & l_r \end{bmatrix} \begin{bmatrix} \frac{\partial y}{\partial m_o^s} \\ \frac{\partial r}{\partial m_o^s} \end{bmatrix} = \begin{bmatrix} 0 \\ 1 \end{bmatrix}$$

so that

$$\frac{\partial y}{\partial m_0^s} = \frac{\det \begin{bmatrix} 0 & -c'(r) \\ 1 & l'(r) \end{bmatrix}}{\det [A]} = \frac{\overbrace{c'(r)}^{-}}{\det [A]} > 0$$

$$\frac{\partial r}{\partial m_o^s} = \frac{\det \begin{bmatrix} 1-i_y & 0 \\ l_y & 1 \end{bmatrix}}{\det \begin{bmatrix} 1-i_y & -c_r \\ l_y & l_r \end{bmatrix}} = \frac{1-i_y}{(1-i_y)\, l_r + c_r l_y} < 0.$$

2.5 Profit Maximization

In this section we work with the concave production function $Q = F(L, K)$ and use the notation

$$F_L \equiv \frac{\partial F(L^*, K^*)}{\partial L}, F_K \equiv \frac{\partial F(L^*, K^*)}{\partial K}$$
$$F_{LL} \equiv \frac{\partial^2 F(L^*, K^*)}{\partial L^2}, F_{LK} \equiv \frac{\partial^2 F(L^*, K^*)}{\partial L \partial K}, F_{KK} \equiv \frac{\partial^2 F(L^*, K^*)}{\partial K^2}$$

with the Hessian of $Q = F(L, K)$ at L^*, K^* given by

$$H = \begin{bmatrix} F_{LL} & F_{LK} \\ F_{LK} & F_{KK} \end{bmatrix}.$$

Since $F(L, K)$ is concave it follows that

$$F_{LL} < 0, F_{KK} < 0 \text{ and } \det[H] = F_{LL}F_{KK} - F_{LK}^2 > 0.$$

Problem 2.30 *Suppose the production function is*

$$Q = F(L, K) = 3L^{\frac{1}{3}} + 3K^{\frac{1}{3}}.$$

*[D***] Find the first-order conditions for maximizing profits and [B***] solve for L^*, K^*, Q^*. From your solution show that*

$$\frac{\partial L^*}{\partial W} < 0, \frac{\partial K^*}{\partial R} < 0, \frac{\partial Q^*}{\partial P} > 0.$$

Now show that

$$\frac{\partial L^*}{\partial W} < 0, \frac{\partial K^*}{\partial R} < 0, \frac{\partial Q^*}{\partial P} > 0$$

*using total differentials. [B**] Show that the expressions for $\frac{\partial L^*}{\partial W}, \frac{\partial K^*}{\partial R}, \frac{\partial Q^*}{\partial P}$ obtained from the two methods are equivalent.*

Answer: The first-order conditions yield

$$\begin{aligned} P(L^*)^{-\frac{2}{3}} - W &= 0 \Longrightarrow L^* = W^{-\frac{3}{2}}P^{\frac{3}{2}} \\ P(K^*)^{-\frac{2}{3}} - R &= 0 \Longrightarrow K^* = R^{-\frac{3}{2}}P^{\frac{3}{2}}. \end{aligned}$$

The supply curve then is

$$Q^* = 3(L^*)^{\frac{1}{3}} + 3(K^*)^{\frac{1}{3}} = 3\left(W^{-\frac{3}{2}}P^{\frac{3}{2}}\right)^{\frac{1}{3}} + 3\left(R^{-\frac{3}{2}}P^{\frac{3}{2}}\right)^{\frac{1}{3}} = 3P^{\frac{1}{2}}\left(W^{-\frac{1}{2}} + R^{-\frac{1}{2}}\right).$$

Thus

$$\begin{aligned} \frac{\partial L^*}{\partial W} &= \frac{\partial}{\partial W}\left(W^{-\frac{3}{2}}P^{\frac{3}{2}}\right) = -\frac{3}{2}W^{-\frac{5}{2}}P^{\frac{3}{2}} < 0 \\ \frac{\partial K^*}{\partial R} &= \frac{\partial}{\partial R}\left(R^{-\frac{3}{2}}P^{\frac{3}{2}}\right) = -\frac{3}{2}R^{-\frac{5}{2}}P^{\frac{3}{2}} < 0 \\ \frac{\partial Q^*}{\partial P} &= \frac{\partial}{\partial P}\left(3P^{\frac{1}{2}}\left(W^{-\frac{1}{2}} + R^{-\frac{1}{2}}\right)\right) = \frac{3}{2}P^{-\frac{1}{2}}\left(W^{-\frac{1}{2}} + R^{-\frac{1}{2}}\right) > 0. \end{aligned}$$

Now let's use total differentials. The first-order conditions yield two implicit functions

$$\begin{aligned} g_1(L^*, K^*, P, W, R) &= P(L^*)^{-\frac{2}{3}} - W = 0 \\ g_2(L^*, K^*, P, W, R) &= P(K^*)^{-\frac{2}{3}} - R = 0 \end{aligned}$$

with total differential

$$\begin{aligned} -\frac{2}{3}P(L^*)^{-\frac{5}{3}}dL^* + (L^*)^{-\frac{2}{3}}dP - dW &= 0 \\ -\frac{2}{3}P(K^*)^{-\frac{5}{3}}dK^* + (K^*)^{-\frac{2}{3}}dP - dR &= 0. \end{aligned}$$

To find $\frac{\partial L^*}{\partial W}$ we 1) set $dP = 0$ in the first total differential (we do not need the second total differential since dL^* does not appear there) and 2) replace the remaining $d's$ to $\partial's$ to obtain

$$-\frac{2}{3}P(L^*)^{-\frac{5}{3}}\partial L^* - \partial W = 0 \Longrightarrow \frac{\partial L^*}{\partial W} = -\frac{1}{\frac{2}{3}P(L^*)^{-\frac{5}{3}}} < 0.$$

Since $L^* = W^{-\frac{3}{2}}P^{\frac{3}{2}}$ this reduces to

$$\begin{aligned} \frac{\partial L^*}{\partial W} &= -\frac{1}{\frac{2}{3}P(L^*)^{-\frac{5}{3}}} = -\frac{3}{2}\frac{1}{P\left(W^{-\frac{3}{2}}P^{\frac{3}{2}}\right)^{-\frac{5}{3}}} = -\frac{3}{2}\frac{1}{PW^{\frac{5}{2}}P^{-\frac{5}{2}}} = -\frac{3}{2}\frac{1}{W^{\frac{5}{2}}P^{-\frac{3}{2}}} \\ &= -\frac{3}{2}W^{-\frac{5}{2}}P^{\frac{3}{2}} \end{aligned}$$

which is the same expression found by directly differentiating $L^* = W^{-\frac{3}{2}}P^{\frac{3}{2}}$ with respect to W.

To find $\frac{\partial K^*}{\partial R}$ we 1) set $dP = 0$ in the second total differential and 2) replace the remaining $d's$ to $\partial's$ to obtain

$$-\frac{2}{3}P(K^*)^{-\frac{5}{3}}\partial K^* - \partial R = 0 \Longrightarrow \frac{\partial K^*}{\partial R} = -\frac{1}{\frac{2}{3}P(K^*)^{-\frac{5}{3}}} < 0.$$

Since $K^* = R^{-\frac{3}{2}}P^{\frac{3}{2}}$ this reduces to

$$\frac{\partial K^*}{\partial R} = -\frac{1}{\frac{2}{3}P(K^*)^{-\frac{5}{3}}} = -\frac{3}{2}\frac{1}{P\left(R^{-\frac{3}{2}}P^{\frac{3}{2}}\right)^{-\frac{5}{3}}} = -\frac{3}{2}R^{-\frac{5}{2}}P^{\frac{3}{2}}$$

which is the same expression found by directly differentiating $K^* = R^{-\frac{3}{2}}P^{\frac{3}{2}}$ with respect to R.

To find $\frac{\partial L^*}{\partial P}$ we 1) set $dW = 0$ in the first total differential and 2) replace the remaining $d's$ to $\partial's$ to obtain

$$-\frac{2}{3}P(L^*)^{-\frac{5}{3}}\partial L^* + (L^*)^{-\frac{2}{3}}\partial P = 0 \Longrightarrow \frac{\partial L^*}{\partial P} = \frac{(L^*)^{-\frac{2}{3}}}{\frac{2}{3}P(L^*)^{-\frac{5}{3}}} = \frac{3}{2}\frac{L^*}{P} > 0.$$

To find $\frac{\partial K^*}{\partial P}$ set $dR = 0$ in the second total differential and replace the remaining $d's$ to $\partial's$ to obtain

$$-\frac{2}{3}P(K^*)^{-\frac{5}{3}}\partial L^* + (K^*)^{-\frac{2}{3}}\partial P = 0 \Longrightarrow \frac{\partial K^*}{\partial P} = \frac{3}{2}\frac{K^*}{P} > 0.$$

To find $\frac{\partial Q^*}{\partial P}$ use

$$Q^* = F(L^*, K^*) = 3(L^*)^{\frac{1}{3}} + 3(K^*)^{\frac{1}{3}}$$

so that

$$\frac{\partial Q^*}{\partial P} = (L^*)^{-\frac{2}{3}}\frac{\partial L^*}{\partial P} + (K^*)^{-\frac{2}{3}}\frac{\partial K^*}{\partial P} > 0.$$

This reduces to

$$\begin{aligned}\frac{\partial Q^*}{\partial P} &= (L^*)^{-\frac{2}{3}}\left(\frac{3}{2}\frac{L^*}{P}\right) + (K^*)^{-\frac{2}{3}}\left(\frac{3}{2}\frac{K^*}{P}\right) = \frac{1}{2P}3\left((L^*)^{\frac{1}{3}} + (K^*)^{\frac{1}{3}}\right)\\ &= \frac{Q^*}{2P} = \frac{3P^{\frac{1}{2}}\left(W^{-\frac{1}{2}} + R^{-\frac{1}{2}}\right)}{2P} = \frac{3}{2}P^{-\frac{1}{2}}\left(W^{-\frac{1}{2}} + R^{-\frac{1}{2}}\right)\end{aligned}$$

which is equivalent to identical to the expression found by directly differentiating Q^* with respect to P.

Problem 2.31 *Suppose real profits are given by*

$$\pi_R(L, K) = F(L, K) - wL - rK$$

with Cobb-Douglas production function

$$Q = F(L, K) = 3L^{\frac{1}{3}}K^{\frac{1}{3}}.$$

*[D***] Find the first-order conditions for maximizing real profits. [B***] Rewrite the first-order conditions in terms of* $\ln(L^*), \ln(K^*)$ *and solve using Cramer's rule. Find* L^*, K^* *and verify that* $\frac{\partial L^*}{\partial w} < 0$ *and* $\frac{\partial K^*}{\partial r} < 0$. *Verify that* $\frac{\partial L^*}{\partial r} = \frac{\partial K^*}{\partial w}$.

Answer: The first-order conditions yield

$$\begin{aligned}
\frac{\partial \pi_R(L^*,K^*,w,r)}{\partial L} &= 0 \\
&\Longrightarrow (L^*)^{-\frac{2}{3}}(K^*)^{\frac{1}{3}} - w = 0 \\
&\Longrightarrow -\frac{2}{3}\ln(L^*) + \frac{1}{3}\ln(K^*) = \ln(w) \\
\frac{\partial \pi_R(L^*,K^*,w,r)}{\partial K} &= 0 \\
&\Longrightarrow (L^*)^{\frac{1}{3}}(K^*)^{-\frac{2}{3}} - r = 0 \\
&\Longrightarrow \frac{1}{3}\ln(L^*) - \frac{2}{3}\ln(K^*) = \ln(r).
\end{aligned}$$

In matrix notation we have

$$\begin{bmatrix} -\frac{2}{3} & \frac{1}{3} \\ \frac{1}{3} & -\frac{2}{3} \end{bmatrix}\begin{bmatrix} \ln(L^*) \\ \ln(K^*) \end{bmatrix} = \begin{bmatrix} \ln(w) \\ \ln(r) \end{bmatrix} \text{ and } \det\begin{bmatrix} -\frac{2}{3} & \frac{1}{3} \\ \frac{1}{3} & -\frac{2}{3} \end{bmatrix} = \frac{1}{3}.$$

Using Cramer's rule it follows that

$$\begin{aligned}
\ln(L^*) &= \frac{\det\begin{bmatrix} \ln(w) & \frac{1}{3} \\ \ln(r) & -\frac{2}{3} \end{bmatrix}}{\frac{1}{3}} = \frac{-\frac{2}{3}\ln(w) - \frac{1}{3}\ln(r)}{\frac{1}{3}} = -2\ln(w) - \ln(r) \\
\ln(K^*) &= \frac{\det\begin{bmatrix} -\frac{2}{3} & \ln(w) \\ \frac{1}{3} & \ln(r) \end{bmatrix}}{\frac{1}{3}} = \frac{-\frac{2}{3}\ln(r) - \frac{1}{3}\ln(w)}{\frac{1}{3}} = -\ln(w) - 2\ln(r).
\end{aligned}$$

It then follows that

$$\begin{aligned}
L^*(w,r) &= e^{\ln(L^*)} = e^{-2\ln(w)-\ln(r)} = w^{-2}r^{-1} \\
K^*(w,r) &= e^{\ln(K^*)} = e^{-\ln(w)-2\ln(r)} = w^{-1}r^{-2}.
\end{aligned}$$

Thus

$$\frac{\partial L^*(w,r)}{\partial w} = -2w^{-3}r^{-1} < 0 \text{ and } \frac{\partial K^*(w,r)}{\partial r} = -2w^{-1}r^{-3} < 0.$$

As well we have $\frac{\partial L^*}{\partial r} = \frac{\partial K^*}{\partial w}$ as

$$\frac{\partial L^*(w,r)}{\partial r} = -w^{-2}r^{-2} = \frac{\partial K^*(w,r)}{\partial w}.$$

Problem 2.32 *Suppose naive real profits are given by*

$$\pi_R(L,K) = F(L,K) - wL - rK$$

with Cobb-Douglas production function

$$Q = F(L,K) = \frac{1}{\alpha}L^\alpha K^\alpha$$

*where $\alpha < \frac{1}{2}$. [B***] Find the first-order conditions for maximizing real profits and the profit maximizing levels of labour and capital L^* and K^*. Rewrite the first-order conditions in terms of $\ln(L^*), \ln(K^*)$ and solve using Cramer's rule. Find L^*, K^* and verify that $\frac{\partial L^*}{\partial w} < 0$, $\frac{\partial K^*}{\partial r} < 0$. Verify that $\frac{\partial L^*}{\partial r} = \frac{\partial K^*}{\partial w}$.*

Answer: The first-order conditions yield

$$\begin{aligned}
\frac{\partial \pi_R(L^*, K^*)}{\partial L} &= 0 \Longrightarrow (L^*)^{\alpha-1}(K^*)^{\alpha} - w = 0 \Longrightarrow (\alpha-1)\ln(L^*) + \alpha\ln(K^*) = \ln(w) \\
\frac{\partial \pi_R(L^*, K^*)}{\partial K} &= 0 \Longrightarrow (L^*)^{\alpha}(K^*)^{\alpha-1} - r = 0 \Longrightarrow \alpha\ln(L^*) + (\alpha-1)\ln(K^*) = \ln(r)
\end{aligned}$$

so that in matrix notation

$$\begin{bmatrix} \alpha-1 & \alpha \\ \alpha & \alpha-1 \end{bmatrix} \begin{bmatrix} \ln(L^*) \\ \ln(K^*) \end{bmatrix} = \begin{bmatrix} \ln(w) \\ \ln(r) \end{bmatrix}$$

and

$$\det \begin{bmatrix} \alpha-1 & \alpha \\ \alpha & \alpha-1 \end{bmatrix} = (\alpha-1)^2 - \alpha^2 = 1 - 2\alpha > 0$$

since $\alpha < \frac{1}{2}$. Using Cramer's rule we have

$$\begin{aligned}
\ln(L^*) &= \frac{\det \begin{bmatrix} \ln(w) & \alpha \\ \ln(r) & \alpha-1 \end{bmatrix}}{1-2\alpha} = \frac{(\alpha-1)\ln(w) - \alpha\ln(r)}{1-2\alpha} \\
&= -\frac{1-\alpha}{1-2\alpha}\ln(w) - \frac{\alpha}{1-2\alpha}\ln(r) \\
\ln(K^*) &= \frac{\det \begin{bmatrix} \alpha-1 & \ln(w) \\ \alpha & \ln(r) \end{bmatrix}}{1-2\alpha} = \frac{(\alpha-1)\ln(r) - \alpha\ln(w)}{1-2\alpha} \\
&= \frac{\alpha}{1-2\alpha}\ln(w) - \frac{1-\alpha}{1-2\alpha}\ln(r).
\end{aligned}$$

It then follows that

$$\begin{aligned}
L^*(w,r) &= e^{\ln(L^*)} = e^{-\frac{1-\alpha}{1-2\alpha}\ln(w) - \frac{\alpha}{1-2\alpha}\ln(r)} = w^{-\frac{1-\alpha}{1-2\alpha}} r^{-\frac{\alpha}{1-2\alpha}} \\
K^*(w,r) &= e^{\ln(K^*)} = e^{\frac{\alpha}{1-2\alpha}\ln(2) - \frac{1-\alpha}{1-2\alpha}\ln(r)} = w^{-\frac{\alpha}{1-2\alpha}} r^{-\frac{1-\alpha}{1-2\alpha}}.
\end{aligned}$$

Thus $\frac{\partial L^*}{\partial w} < 0$ as

$$\frac{\partial L^*(w,r)}{\partial w} = -\frac{1-\alpha}{1-2\alpha} w^{-\frac{1-\alpha}{1-2\alpha}-1} r^{-\frac{\alpha}{1-2\alpha}} < 0$$

since

$$0 < \alpha < \frac{1}{2} \Longrightarrow -\frac{1-\alpha}{1-2\alpha} < 0.$$

Similarly

$$\frac{\partial K^*(w,r)}{\partial r} = -\frac{1-\alpha}{1-2\alpha} w^{-\frac{1-\alpha}{1-2\alpha}} r^{-\frac{\alpha}{1-2\alpha}-1} < 0.$$

We have $\frac{\partial L^*}{\partial r} = \frac{\partial K^*}{\partial w}$ as

$$\begin{aligned}
\frac{\partial L^*}{\partial r} &= -\frac{\alpha}{1-2\alpha} w^{-\frac{1-\alpha}{1-2\alpha}} r^{-\frac{\alpha}{1-2\alpha}-1} = -\frac{\alpha}{1-2\alpha} w^{-\frac{1-\alpha}{1-2\alpha}} r^{-\frac{1-\alpha}{1-2\alpha}} \\
\frac{\partial K^*}{\partial w} &= -\frac{\alpha}{1-2\alpha} w^{-\frac{\alpha}{1-2\alpha}-1} r^{-\frac{1-\alpha}{1-2\alpha}} = -\frac{\alpha}{1-2\alpha} w^{-\frac{1-\alpha}{1-2\alpha}} r^{-\frac{1-\alpha}{1-2\alpha}}.
\end{aligned}$$

Problem 2.33 *[A***] Suppose real profits are given by*

$$\pi_R(L,K) = F(L,K) - wL - rK$$

with concave production function $Q = F(L,K)$. Write the first-order conditions for maximization as implicit functions and calculate the total differential. Show that $\frac{\partial L^*}{\partial w} < 0$, $\frac{\partial K^*}{\partial r} < 0$, and $\frac{\partial L^*}{\partial r} = \frac{\partial K^*}{\partial w}$.

Answer: The first order conditions are

$$\frac{\partial F(L^*,K^*)}{\partial L} - w = 0 \text{ and } \frac{\partial F(L^*,K^*)}{\partial K} - r = 0$$

so that

$$\begin{aligned}
g_1(L^*,K^*,w,r) &= \frac{\partial F(L^*,K^*)}{\partial L} - w = 0 \\
g_2(L^*,K^*,w,r) &= \frac{\partial F(L^*,K^*)}{\partial K} - r = 0.
\end{aligned}$$

The total differential is then

$$\begin{aligned}
\frac{\partial^2 F(L^*,K^*)}{\partial L^2} dL^* + \frac{\partial^2 F(L^*,K^*)}{\partial L \partial K} dK^* - dw &= 0 \\
\frac{\partial^2 F(L^*,K^*)}{\partial L \partial K} dL^* + \frac{\partial^2 F(L^*,K^*)}{\partial K^2} dK^* - dr &= 0
\end{aligned}$$

or

$$\begin{aligned}
F_{LL} dL^* + F_{LK} dK^* - dw &= 0 \\
F_{LK} dL^* + F_{KK} dK^* - dr &= 0
\end{aligned}$$

or in matrix notation

$$\underbrace{\begin{bmatrix} F_{LL} & F_{LK} \\ F_{LK} & F_{KK} \end{bmatrix}}_{H} \begin{bmatrix} dL^* \\ dK^* \end{bmatrix} = \begin{bmatrix} dw \\ dr \end{bmatrix}.$$

The matrix H given by

$$H = \begin{bmatrix} F_{LL} & F_{LK} \\ F_{LK} & F_{KK} \end{bmatrix}$$

is negative definite since $F(L,K)$ is concave, and so $F_{LL} < 0, F_{KK} < 0$, and $\det[H] > 0$.

To find $\frac{\partial L^*}{\partial w}, \frac{\partial K^*}{\partial w}$ we 1) set $dr = 0$ and 2) change the remaining $d's$ to $\partial's$ as

$$\begin{aligned} F_{LL}\partial L^* + F_{LK}\partial K^* - \partial w &= 0 \\ F_{LK}\partial L^* + F_{KK}\partial K^* - 0 &= 0 \end{aligned}$$

or in matrix notation

$$\begin{aligned} \begin{bmatrix} F_{LL} & F_{LK} \\ F_{LK} & F_{KK} \end{bmatrix} \begin{bmatrix} \partial L^* \\ \partial K^* \end{bmatrix} &= \begin{bmatrix} \partial w \\ 0 \end{bmatrix} \\ &\implies \begin{bmatrix} F_{LL} & F_{LK} \\ F_{LK} & F_{KK} \end{bmatrix} \begin{bmatrix} \partial L^* \\ \partial K^* \end{bmatrix} = \begin{bmatrix} 1 \\ 0 \end{bmatrix} \partial w \\ &\implies \begin{bmatrix} F_{LL} & F_{LK} \\ F_{LK} & F_{KK} \end{bmatrix} \begin{bmatrix} \frac{\partial L^*}{\partial w} \\ \frac{\partial K^*}{\partial w} \end{bmatrix} = \begin{bmatrix} 1 \\ 0 \end{bmatrix}. \end{aligned}$$

Thus

$$\begin{aligned} \frac{\partial L^*}{\partial w} &= \frac{\det \begin{bmatrix} 1 & F_{LK} \\ 0 & F_{KK} \end{bmatrix}}{\det[H]} = \frac{F_{KK}}{\det[H]} < 0 \\ \frac{\partial K^*}{\partial w} &= \frac{\det \begin{bmatrix} F_{LL} & 1 \\ F_{LK} & 0 \end{bmatrix}}{\det[H]} = \frac{-F_{LK}}{\det[H]}. \end{aligned}$$

To find $\frac{\partial L^*}{\partial r}, \frac{\partial K^*}{\partial r}$ we 1) set $dw = 0$ and 2) change the remaining $d's$ to $\partial's$ as

$$\begin{bmatrix} F_{LL} & F_{LK} \\ F_{LK} & F_{KK} \end{bmatrix} \begin{bmatrix} \frac{\partial L^*}{\partial r} \\ \frac{\partial K^*}{\partial r} \end{bmatrix} = \begin{bmatrix} 0 \\ 1 \end{bmatrix}$$

so that

$$\begin{aligned} \frac{\partial K^*}{\partial r} &= \frac{\det \begin{bmatrix} F_{LL} & 0 \\ F_{LK} & 1 \end{bmatrix}}{\det[H]} = \frac{F_{LL}}{\det[H]} < 0 \\ \frac{\partial L^*}{\partial r} &= \frac{\det \begin{bmatrix} 0 & F_{LK} \\ 1 & F_{KK} \end{bmatrix}}{\det[H]} = \frac{-F_{LK}}{\det[H]}. \end{aligned}$$

Thus

$$\frac{\partial L^*}{\partial r} = \frac{\partial K^*}{\partial w} = \frac{-F_{LK}}{\det[H]}.$$

Problem 2.34 *[A ***] Consider the nominal profit maximization problem where*

$$\pi(L,K) = PF(L,K) \quad WL \quad RK.$$

Show that the firm's supply curve

$$Q^*(P,W,R) = F(L^*(P,W,R), K^*(P,W,R))$$

is upward sloping (i.e., $\frac{\partial Q^*(P,W,R)}{\partial P} > 0$ *) using total differentials. Show that* $\frac{\partial Q^*(P,W,R)}{\partial W} = -\frac{\partial L^*(P,W,R)}{\partial P}$ *using total differentials.*

Answer: The first-order conditions for profit maximization yield

$$\begin{aligned}\frac{\partial \pi (L^*, K^*)}{\partial L} &= 0 \Longrightarrow g_1 (L^*, K^*, P, W, R) = P\frac{\partial F (L^*, K^*)}{\partial L} - W = 0 \\ \frac{\partial \pi (L^*, K^*)}{\partial K} &= 0 \Longrightarrow g_2 (L^*, K^*, P, W, R) = P\frac{\partial F (L^*, K^*)}{\partial K} - R = 0.\end{aligned}$$

The total differential is then

$$\begin{aligned}PF_{LL}dL^* + PF_{LK}dK^* + F_L dP - dW &= 0 \\ PF_{LK}dL^* + PF_{KK}dK^* + F_K dP - dR &= 0.\end{aligned}$$

We will need $\frac{\partial L^*}{\partial P}$ and $\frac{\partial K^*}{\partial P}$. Setting $dW = dR = 0$ and changing the $d's$ to $\partial's$ we obtain

$$\begin{aligned}\begin{bmatrix} PF_{LL} & PF_{LK} \\ PF_{LK} & PF_{KK} \end{bmatrix}\begin{bmatrix} \frac{\partial L^*}{\partial P} \\ \frac{\partial K^*}{\partial P} \end{bmatrix} &= \begin{bmatrix} -F_L \\ -F_K \end{bmatrix} \\ &\Longrightarrow \begin{bmatrix} F_{LL} & F_{LK} \\ F_{LK} & F_{KK} \end{bmatrix}\begin{bmatrix} \frac{\partial L^*}{\partial P} \\ \frac{\partial K^*}{\partial P} \end{bmatrix} = -\frac{1}{P}\begin{bmatrix} F_L \\ F_K \end{bmatrix}\end{aligned}$$

so that

$$\begin{bmatrix} \frac{\partial L^*}{\partial P} \\ \frac{\partial K^*}{\partial P} \end{bmatrix} = -\frac{1}{P}\begin{bmatrix} F_{LL} & F_{LK} \\ F_{LK} & F_{KK} \end{bmatrix}^{-1}\begin{bmatrix} F_L \\ F_K \end{bmatrix}.$$

Recall that

$$H = \begin{bmatrix} F_{LL} & F_{LK} \\ F_{LK} & F_{KK} \end{bmatrix}$$

is negative definite since $F(L, K)$ is concave and thus

$$A = \begin{bmatrix} F_{LL} & F_{LK} \\ F_{LK} & F_{KK} \end{bmatrix}^{-1} = H^{-1}$$

is also negative definite (since the inverse of a negative definite matrix is negative definite).

Now from

$$Q^* (P, W, R) = F (L^* (P, W, R), K^* (P, W, R))$$

we have

$$\begin{aligned}\frac{\partial Q^*}{\partial P} &= \frac{\partial F (L^*, K^*)}{\partial L}\frac{\partial L^*}{\partial P} + \frac{\partial F (L^*, K^*)}{\partial K}\frac{\partial K^*}{\partial P} \\ &= F_L\frac{\partial L^*}{\partial P} + F_K\frac{\partial K^*}{\partial P} = \begin{bmatrix} F_L & F_K \end{bmatrix}\begin{bmatrix} \frac{\partial L^*}{\partial P} \\ \frac{\partial K^*}{\partial P} \end{bmatrix} \\ &= -\frac{1}{P}\begin{bmatrix} F_L & F_K \end{bmatrix}\begin{bmatrix} F_{LL} & F_{LK} \\ F_{LK} & F_{KK} \end{bmatrix}^{-1}\begin{bmatrix} F_L \\ F_K \end{bmatrix} \\ &= -\frac{1}{P}x^T Ax\end{aligned}$$

where

$$x = \begin{bmatrix} F_L \\ F_K \end{bmatrix} \neq 0$$

since $F_L > 0$ and $F_K > 0$. Since A is negative definite and $x \neq 0$ it follows that $x^T A x < 0$ and so

$$\frac{\partial Q^*}{\partial P} = -\frac{1}{P} x^T A x > 0$$

and the supply curve slopes upwards.

We now show that $\frac{\partial Q^*}{\partial W} = -\frac{\partial L^*}{\partial P}$. Using the total differential and calculating $\frac{\partial L^*(P,W,R)}{\partial P}$ first from

$$\underbrace{\begin{bmatrix} PF_{LL} & PF_{LK} \\ PF_{LK} & PF_{KK} \end{bmatrix}}_{\tilde{H}} \begin{bmatrix} \frac{\partial L^*}{\partial P} \\ \frac{\partial K^*}{\partial P} \end{bmatrix} = \begin{bmatrix} -F_L \\ -F_K \end{bmatrix}$$

we find that

$$\frac{\partial L^*}{\partial P} = \frac{\det \begin{bmatrix} -F_L & PF_{LK} \\ -F_K & PF_{KK} \end{bmatrix}}{\det \left[\tilde{H}\right]} = \frac{-PF_L F_{KK} + PF_K F_{LK}}{\det \left[\tilde{H}\right]}.$$

Now we calculate $\frac{\partial Q^*(P,W,R)}{\partial W}$ from $Q^* = F(L^*, K^*)$ using the chain rule as

$$\begin{aligned} \frac{\partial Q^*}{\partial W} &= \frac{\partial F(L^*, K^*)}{\partial L} \frac{\partial L^*}{\partial W} + \frac{\partial F(L^*, K^*)}{\partial K} \frac{\partial K^*}{\partial W} \\ &= F_L \frac{\partial L^*}{\partial W} + F_K \frac{\partial K^*}{\partial W} = \begin{bmatrix} F_L & F_K \end{bmatrix} \begin{bmatrix} \frac{\partial L^*}{\partial W} \\ \frac{\partial K^*}{\partial W} \end{bmatrix}. \end{aligned}$$

From

$$\underbrace{\begin{bmatrix} PF_{LL} & PF_{LK} \\ PF_{LK} & PF_{KK} \end{bmatrix}}_{\tilde{H}} \begin{bmatrix} \frac{\partial L^*}{\partial W} \\ \frac{\partial K^*}{\partial W} \end{bmatrix} = \begin{bmatrix} 1 \\ 0 \end{bmatrix}$$

we have

$$\frac{\partial L^*}{\partial W} = \frac{PF_{KK}}{\det \left[\tilde{H}\right]}, \frac{\partial K^*}{\partial W} = -\frac{PF_{LK}}{\det \left[\tilde{H}\right]}$$

so that

$$\begin{aligned}
\frac{\partial Q^*}{\partial W} &= \begin{bmatrix} F_L & F_K \end{bmatrix} \begin{bmatrix} \frac{\partial L^*}{\partial W} \\ \frac{\partial K^*}{\partial W} \end{bmatrix} \\
&= \begin{bmatrix} F_L & F_K \end{bmatrix} \begin{bmatrix} \frac{PF_{KK}}{\det\left[\tilde{H}\right]} \\ -\frac{PF_{LK}}{\det\left[\tilde{H}\right]} \end{bmatrix} \\
&= F_L \frac{PF_{KK}}{\det\left[\tilde{H}\right]} - F_K \frac{PF_{LK}}{\det\left[\tilde{H}\right]} \\
&= -\left(\frac{-PF_L F_{KK} + PF_K F_{LK}}{\det\left[\tilde{H}\right]} \right) = -\frac{\partial L^*}{\partial P}.
\end{aligned}$$

Problem 2.35 *Consider a profit maximizing firm with a production function* $Q = F(L, K) T^{-\frac{1}{2}}$ *where* T *is the (exogenous) outside temperature and* $F(L, K)$ *is concave. [C***] Starting with real profits given by*

$$\pi(L, K) = F(L, K) T^{-\frac{1}{2}} - wL - rK$$

find the first-order conditions for the profit maximizing L^*, K^*. *[B***] From the first-order conditions calculate the total differential and use this to show that* $\frac{\partial L^*}{\partial w} < 0$. *Find* $\frac{\partial K^*}{\partial T}$ *and show that* $\frac{\partial K^*}{\partial T} < 0$ *if* $F_{LK} > 0$.

Answer: The first-order conditions yield

$$\begin{aligned}
\frac{\partial \pi(L^*, K^*)}{\partial L} &= 0 \Longrightarrow g_1(L^*, K^*, w, r, T) = \frac{\partial F(L^*, K^*)}{\partial L} T^{-\frac{1}{2}} - w = 0 \\
\frac{\partial \pi(L^*, K^*)}{\partial K} &= 0 \Longrightarrow g_2(L^*, K^*, w, r, T) = \frac{\partial F(L^*, K^*)}{\partial K} T^{-\frac{1}{2}} - r = 0.
\end{aligned}$$

The total differential from the first-order conditions then is

$$\begin{aligned}
T^{-\frac{1}{2}} F_{LL} dL^* + T^{-\frac{1}{2}} F_{LK} dK^* - dw - \frac{1}{2} T^{-\frac{3}{2}} F_L dT &= 0 \\
T^{-\frac{1}{2}} F_{LK} dL^* + T^{-\frac{1}{2}} F_{KK} dK^* - dr - \frac{1}{2} T^{-\frac{3}{2}} F_K dT &= 0
\end{aligned}$$

or in matrix notation

$$\underbrace{\begin{bmatrix} T^{-\frac{1}{2}} F_{LL} & T^{-\frac{1}{2}} F_{LK} \\ T^{-\frac{1}{2}} F_{LK} & T^{-\frac{1}{2}} F_{KK} \end{bmatrix}}_{\tilde{H}} \begin{bmatrix} dL^* \\ dK^* \end{bmatrix} = \begin{bmatrix} dw + \frac{1}{2} T^{-\frac{3}{2}} F_L dT \\ dr + \frac{1}{2} T^{-\frac{3}{2}} F_K dT \end{bmatrix}$$

where $\tilde{H} = T^{-\frac{1}{2}} H$ as

$$\tilde{H} \equiv \begin{bmatrix} T^{-\frac{1}{2}} F_{LL} & T^{-\frac{1}{2}} F_{LK} \\ T^{-\frac{1}{2}} F_{LK} & T^{-\frac{1}{2}} F_{KK} \end{bmatrix} = T^{-\frac{1}{2}} \underbrace{\begin{bmatrix} F_{LL} & F_{LK} \\ F_{LK} & F_{KK} \end{bmatrix}}_{H} = T^{-\frac{1}{2}} H.$$

Since H is negative definite by the concavity of $F(L,K)$ and $T^{-\frac{1}{2}} > 0$, it follows that $\tilde{H}$ is also negative definite, and so $\det\left[\tilde{H}\right] > 0$.

To find $\frac{\partial L^*}{\partial w}$ we 1) set $dT = dr = 0$ and 2) change the remaining $d's$ to $\partial's$ so that

$$\underbrace{\begin{bmatrix} T^{-\frac{1}{2}}F_{LL} & T^{-\frac{1}{2}}F_{LK} \\ T^{-\frac{1}{2}}F_{LK} & T^{-\frac{1}{2}}F_{KK} \end{bmatrix}}_{\tilde{H}} \begin{bmatrix} \frac{\partial L^*}{\partial w} \\ \frac{\partial K^*}{\partial w} \end{bmatrix} = \begin{bmatrix} 1 \\ 0 \end{bmatrix}.$$

Using Cramer's rule we have

$$\frac{\partial L^*}{\partial w} = \frac{\det\begin{bmatrix} 1 & T^{-\frac{1}{2}}F_{LK} \\ 0 & T^{-\frac{1}{2}}F_{KK} \end{bmatrix}}{\det\left[\tilde{H}\right]} = \frac{T^{-\frac{1}{2}}\overbrace{F_{KK}}^{-}}{\underbrace{\det[H]}_{+}} < 0.$$

To find $\frac{\partial K^*}{\partial T}$ we 1) set $dw = dr = 0$ and 2) change the remaining $d's$ to $\partial's$ as

$$\begin{bmatrix} T^{-\frac{1}{2}}F_{LL} & T^{-\frac{1}{2}}F_{LK} \\ T^{-\frac{1}{2}}F_{LK} & T^{-\frac{1}{2}}F_{KK} \end{bmatrix} \begin{bmatrix} \frac{\partial L^*}{\partial T} \\ \frac{\partial K^*}{\partial T} \end{bmatrix} = \begin{bmatrix} \frac{1}{2}T^{-\frac{3}{2}}F_L \\ \frac{1}{2}T^{-\frac{3}{2}}F_K \end{bmatrix}$$

so that using Cramer's rule we have

$$\frac{\partial K^*}{\partial T} = \frac{\det\begin{bmatrix} T^{-\frac{1}{2}}F_{LL} & \frac{1}{2}T^{-\frac{3}{2}}F_L \\ T^{-\frac{1}{2}}F_{LK} & \frac{1}{2}T^{-\frac{3}{2}}F_K \end{bmatrix}}{\det\left[\tilde{H}\right]} = \frac{\frac{T^{-2}}{2}\left(F_K F_{LL} - F_L F_{LK}\right)}{\det\left[\tilde{H}\right]} < 0$$

where $F_K F_{LL} - F_L F_{LK} < 0$ since $F_{LL} < 0, F_K > 0, F_L > 0$, and (by assumption) $F_{LK} > 0$.

Problem 2.36 *Consider a profit maximizing firm with a production function $Q = F(L,K)e^{-T}$ where T is the (exogenous) outside temperature and $F(L,K)$ is concave. Starting with real profits given by*

$$\pi(L,K) = F(L,K)e^{-T} - wL - rK$$

*find the first-order conditions for the profit maximizing L^*, K^*. [B***] From the first-order conditions calculate the total differential and use this to show that $\frac{\partial L^*}{\partial w} < 0$. Find $\frac{\partial L^*}{\partial T}$ and show that $\frac{\partial L^*}{\partial T} < 0$ if $F_{LK} > 0$.*

Answer: The first-order conditions yield

$$\begin{aligned} \frac{\partial F(L^*,K^*)}{\partial L}e^{-T} - w &= 0 \Longrightarrow F_L e^{-T} = w \\ \frac{\partial F(L^*,K^*)}{\partial K}e^{-T} - r &= 0 \Longrightarrow F_K e^{-T} = r. \end{aligned}$$

The total differential from the first-order conditions is

$$\begin{aligned} e^{-T}F_{LL}dL^* + e^{-T}F_{LK}dK^* - e^{-T}F_L dT - dw &= 0 \\ e^{-T}F_{LK}dL^* + e^{-T}F_{KK}dK^* - e^{-T}F_K dT - dr &= 0 \end{aligned}$$

or in matrix notation

$$\underbrace{\begin{bmatrix} e^{-T}F_{LL} & e^{-T}F_{LK} \\ e^{-T}F_{LK} & e^{-T}F_{KK} \end{bmatrix}}_{\tilde{H}} \begin{bmatrix} dL^* \\ dK^* \end{bmatrix} = \begin{bmatrix} e^{-T}F_L dT + dw \\ e^{-T}F_K dT + dr \end{bmatrix}.$$

Here $\tilde{H} = e^{-T}H$ as

$$\tilde{H} \equiv \begin{bmatrix} e^{-T}F_{LL} & e^{-T}F_{LK} \\ e^{-T}F_{LK} & e^{-T}F_{KK} \end{bmatrix} = e^{-T} \underbrace{\begin{bmatrix} F_{LL} & F_{LK} \\ F_{LK} & F_{KK} \end{bmatrix}}_{H} = e^{-T}H.$$

Since $e^{-T} > 0$ and H is negative definite, it follows that $\tilde{H}$ is negative definite and so $\det\left[\tilde{H}\right] > 0$.

To find $\frac{\partial L^*}{\partial w}$ we 1) set $dT = dr = 0$ and 2) change the remaining $d's$ to $\partial's$ to obtain

$$\underbrace{\begin{bmatrix} e^{-T}F_{LL} & e^{-T}F_{LK} \\ e^{-T}F_{LK} & e^{-T}F_{KK} \end{bmatrix}}_{\tilde{H}} \begin{bmatrix} \frac{\partial L^*}{\partial w} \\ \frac{\partial K^*}{\partial w} \end{bmatrix} = \begin{bmatrix} 1 \\ 0 \end{bmatrix}.$$

Using Cramer's rule we have

$$\frac{\partial L^*}{\partial w} = \frac{\det \begin{bmatrix} 1 & e^{-T}F_{LK} \\ 0 & e^{-T}F_{KK} \end{bmatrix}}{\det\left[\tilde{H}\right]} = \frac{e^{-T}\overbrace{F_{KK}}^{-}}{\underbrace{\det\left[\tilde{H}\right]}_{+}} < 0.$$

To find $\frac{\partial L^*}{\partial T}$ we 1) set $dw = dr = 0$ and 2) change the remaining $d's$ to $\partial's$ as

$$\begin{bmatrix} e^{-T}F_{LL} & e^{-T}F_{LK} \\ e^{-T}F_{LK} & e^{-T}F_{KK} \end{bmatrix} \begin{bmatrix} \frac{\partial L^*}{\partial T} \\ \frac{\partial K^*}{\partial T} \end{bmatrix} = \begin{bmatrix} e^{-T}F_L \\ e^{-T}F_K \end{bmatrix}$$

or

$$\underbrace{\begin{bmatrix} F_{LL} & F_{LK} \\ F_{LK} & F_{KK} \end{bmatrix}}_{H} \begin{bmatrix} \frac{\partial L^*}{\partial T} \\ \frac{\partial K^*}{\partial T} \end{bmatrix} = \begin{bmatrix} F_L \\ F_K \end{bmatrix}$$

so that using Cramer's rule we have

$$\frac{\partial L^*}{\partial T} = \frac{\det \begin{bmatrix} F_L & F_{LK} \\ F_K & F_{KK} \end{bmatrix}}{\det\left[H\right]} = \frac{F_L F_{KK} - F_K F_{LK}}{\det\left[H\right]} < 0$$

where $F_L F_{KK} - F_K F_{LK} < 0$ since $F_{KK} < 0, F_K > 0, F_L > 0, F_{LK} > 0$.

Problem 2.37 *Consider a competitive profit maximizing firm with a concave production function $Q = F(L, K)$. Suppose that profits are*

$$\pi(L, K) = PQ - (1+e) WL - RK$$

*where $e > 0$ is the (exogenous) rate of contribution by the firm to unemployment insurance. [B***] Find the first-order conditions for the profit maximizing L^*, K^*. From the first-order conditions calculate the total differential and use this to show that $\frac{\partial L^*}{\partial W} < 0$. Show that increasing e reduces employment L^*.*

Answer: The first-order conditions yield

$$\begin{aligned}\frac{\partial \pi(L^*, K^*)}{\partial L} &= 0 \Longrightarrow P\frac{\partial F(L^*, K^*)}{\partial L} - (1+e) W = 0\\ \frac{\partial \pi(L^*, K^*)}{\partial K} &= 0 \Longrightarrow P\frac{\partial F(L^*, K^*)}{\partial K} - R = 0.\end{aligned}$$

The total differential from the first-order conditions is

$$\begin{aligned}PF_{LL}dL^* + PF_{LK}dK^* + F_L dP - (1+e) dW - W de &= 0\\ PF_{LK}dL^* + PF_{KK}dK^* + F_K dP - dR &= 0\end{aligned}$$

or in matrix notation

$$\underbrace{\begin{bmatrix} PF_{LL} & PF_{LK} \\ PF_{LK} & PF_{KK}\end{bmatrix}}_{\tilde{H}} \begin{bmatrix} dL^* \\ dK^* \end{bmatrix} = \begin{bmatrix} -F_L dP + (1+e) dW + W de \\ -F_K dP + dR \end{bmatrix}.$$

We have $\tilde{H} = P \times H$ as

$$\tilde{H} = \begin{bmatrix} PF_{LL} & PF_{LK} \\ PF_{LK} & PF_{KK}\end{bmatrix} = P \begin{bmatrix} F_{LL} & F_{LK} \\ F_{LK} & F_{KK}\end{bmatrix} = P \times H.$$

Since $P > 0$ and H is negative definite, it follows that $\tilde{H}$ is negative definite, and so $\det\left[\tilde{H}\right] > 0$.

To find $\frac{\partial L^*}{\partial W}$ we 1) set $dP = dR = de = 0$ and 2) change the remaining $d's$ to $\partial's$ as

$$\underbrace{\begin{bmatrix} PF_{LL} & PF_{LK} \\ PF_{LK} & PF_{KK}\end{bmatrix}}_{\tilde{H}} \begin{bmatrix} \frac{\partial L^*}{\partial W} \\ \frac{\partial K^*}{\partial W} \end{bmatrix} = \begin{bmatrix} 1+e \\ 0 \end{bmatrix}.$$

Using Cramer's rule

$$\frac{\partial L^*}{\partial W} = \frac{\det\begin{bmatrix} 1+e & PF_{LK} \\ 0 & PF_{KK}\end{bmatrix}}{\det\left[\tilde{H}\right]} = \frac{\overbrace{(1+e)}^{+}\overbrace{PF_{KK}}^{-}}{\underbrace{\det\left[\tilde{H}\right]}_{+}} < 0.$$

To find $\frac{\partial L^*}{\partial e}$ we 1) set $dP = dW = dR = 0$ and 2) change the remaining $d's$ to $\partial' s$ as

$$\underbrace{\begin{bmatrix} PF_{LL} & PF_{LK} \\ PF_{LK} & PF_{KK} \end{bmatrix}}_{\tilde{H}} \begin{bmatrix} \frac{\partial L^*}{\partial e} \\ \frac{\partial K^*}{\partial e} \end{bmatrix} = \begin{bmatrix} W \\ 0 \end{bmatrix}$$

so that using Cramer's rule we have

$$\frac{\partial L^*}{\partial e} = \frac{\det \begin{bmatrix} W & PF_{LK} \\ 0 & PF_{KK} \end{bmatrix}}{\det \left[\tilde{H}\right]} = \frac{\overbrace{W}^{+}\overbrace{PF_{KK}}^{-}}{\underbrace{\det \left[\tilde{H}\right]}_{+}} < 0.$$

Thus increasing e decreases employment L^*.

2.6 Additively Separable Utility Maximization

Problem 2.38 *Consider a household with utility function*

$$U(Q_1, Q_2) = 2Q_1^{\frac{1}{2}} + 2Q_2^{\frac{1}{2}}$$

*and budget constraint $Y = P_1Q_1 + P_2Q_2$. [D***] Write down the Lagrangian for the utility maximization problem, find the first-order conditions and [C***] solve. Show that Q_1 is a normal good from this solution. [B***] Now calculate the total differential and show that Q_1 is a normal good.*

Answer: We have

$$\mathcal{L}(\lambda, Q_1, Q_2) = 2Q_1^{\frac{1}{2}} + 2Q_2^{\frac{1}{2}} + \lambda(Y - P_1Q_1 - P_2Q_2)$$

so that

$$\begin{aligned} \frac{\partial \mathcal{L}(\lambda, Q_1, Q_2)}{\partial \lambda} &= Y - P_1Q_1 - P_2Q_2 \\ \frac{\partial \mathcal{L}(\lambda, Q_1, Q_2)}{\partial Q_1} &= Q_1^{-\frac{1}{2}} - \lambda P_1 \\ \frac{\partial \mathcal{L}(\lambda, Q_1, Q_2)}{\partial Q_2} &= Q_2^{-\frac{1}{2}} - \lambda P_2 \end{aligned}$$

with first-order conditions

$$\begin{aligned} Y - P_1Q_1^* - P_2Q_2^* &= 0 \\ (Q_1^*)^{-\frac{1}{2}} - \lambda^* P_1 &= 0 \\ (Q_2^*)^{-\frac{1}{2}} - \lambda^* P_2 &= 0. \end{aligned}$$

Now

$$(Q_1^*)^{-\frac{1}{2}} - \lambda^* P_1 = 0 \Longrightarrow Q_1^* = (\lambda^* P_1)^{-2} = (\lambda^*)^{-2} P_1^{-2}$$

and

$$(Q_2^*)^{-\frac{1}{2}} - \lambda^* P_2 = 0 \Longrightarrow Q_2^* = (\lambda^* P_2)^{-2} = (\lambda^*)^{-2} P_2^{-2}$$

so that combining this with the budget constraint

$$\begin{aligned} Y &= P_1 Q_1^* + P_2 Q_2^* = P_1 (\lambda^*)^{-2} P_1^{-2} + P_2 (\lambda^*)^{-2} P_2^{-2} = (\lambda^*)^{-2} \left(P_1^{-1} + P_2^{-1}\right) \\ \Longrightarrow & (\lambda^*)^{-2} = \frac{Y}{P_1^{-1} + P_2^{-1}} \Longrightarrow \lambda^* = \left(P_1^{-1} + P_2^{-1}\right)^{\frac{1}{2}} Y^{-\frac{1}{2}}. \end{aligned}$$

Therefore

$$Q_1^* = (\lambda^*)^{-2} P_1^{-2} = \frac{Y P_1^{-2}}{P_1^{-1} + P_2^{-1}}, \; Q_2^* = (\lambda^*)^{-2} P_2^{-2} = \frac{Y P_2^{-2}}{P_1^{-1} + P_2^{-1}}.$$

Thus Q_1 is a normal good as

$$\frac{\partial Q_1^*}{\partial Y} = \frac{P_1^{-2}}{P_1^{-1} + P_2^{-1}} > 0.$$

The total differential is

$$\begin{aligned} 0 d\lambda^* - P_1 dQ_1^* - P_2 dQ_2^* + dY - Q_1^* dP_1 - Q_2^* dP_2 &= 0 \\ -P_1 d\lambda^* - \frac{1}{2} (Q_1^*)^{-\frac{3}{2}} dQ_1^* + 0 dQ_2^* - \lambda^* dP_1 &= 0 \\ -P_2 d\lambda^* + 0 dQ_1^* - \frac{1}{2} (Q_2^*)^{-\frac{3}{2}} dQ_2^* - \lambda^* dP_2 &= 0 \end{aligned}$$

or

$$\begin{bmatrix} 0 & -P_1 & -P_2 \\ -P_1 & -\frac{1}{2}(Q_1^*)^{-\frac{3}{2}} & 0 \\ -P_2 & 0 & -\frac{1}{2}(Q_2^*)^{-\frac{3}{2}} \end{bmatrix} \begin{bmatrix} d\lambda^* \\ dQ_1^* \\ dQ_2^* \end{bmatrix} = \begin{bmatrix} -dY + Q_1^* dP_1 + Q_2^* dP_2 \\ \lambda^* dP_1 \\ \lambda^* dP_2 \end{bmatrix}.$$

Here

$$H = \begin{bmatrix} 0 & -P_1 & -P_2 \\ -P_1 & -\frac{1}{2}(Q_1^*)^{-\frac{3}{2}} & 0 \\ -P_2 & 0 & -\frac{1}{2}(Q_2^*)^{-\frac{3}{2}} \end{bmatrix}$$

satisfies

$$\det[H] = \frac{P_1^2}{2} (Q_2^*)^{-\frac{3}{2}} + \frac{P_2^2}{2} (Q_1^*)^{-\frac{3}{2}} > 0$$

and Q_1 is a normal good since

$$\frac{\partial Q_1^*}{\partial Y} = \frac{\det \begin{bmatrix} 0 & -1 & -P_2 \\ -P_1 & 0 & 0 \\ -P_2 & 0 & -\frac{1}{2}(Q_2^*)^{-\frac{3}{2}} \end{bmatrix}}{\det[H]} = \frac{\frac{P_1}{2}(Q_2^*)^{-\frac{3}{2}}}{\frac{P_1^2}{2}(Q_2^*)^{-\frac{3}{2}} + \frac{P_2^2}{2}(Q_1^*)^{-\frac{3}{2}}} > 0.$$

Problem 2.39 *Consider a household with concave utility function*

$$U(Q_1, Q_2) = U_1(Q_1) + U_2(Q_2)$$

and budget constraint $Y = P_1Q_1 + P_2Q_2$. *[B***] Write down the Lagrangian for utility maximization and the first-order conditions. From the first-order conditions calculate the total differential. [A***] Show that* Q_1^* *is a normal good, that is* $\frac{\partial Q_1^*}{\partial Y} > 0$. *Prove that* $\frac{\partial Q_1^*}{\partial P_1} < 0$. *Calculate* $\frac{\partial Q_2^*}{\partial P_1}$. *Use this to determine when* Q_1 *and* Q_2 *will be substitutes or complements.*

Answer: The Lagrangian for the constrained maximization problem is

$$\mathcal{L}(\lambda, Q_1, Q_2) = U_1(Q_1) + U_2(Q_2) + \lambda(Y - P_1Q_1 - P_2Q_2)$$

which yields the first-order conditions

$$\begin{aligned} Y - P_1Q_1^* - P_2Q_2^* &= 0 \\ U_1'(Q_1^*) - \lambda^* P_1 &= 0 \\ U_2'(Q_2^*) - \lambda^* P_2 &= 0. \end{aligned}$$

Taking the total differential we find that

$$\begin{aligned} 0d\lambda^* - P_1dQ_1^* - P_2dQ_2^* + dY - Q_1^*dP_1 - Q_2^*dP_2 &= 0 \\ -P_1d\lambda^* + U_1''dQ_1^* + 0dQ_2^* - \lambda^*dP_1 &= 0 \\ -P_2d\lambda^* + 0dQ_1^* + U_2''dQ_2^* - \lambda^*dP_2 &= 0 \end{aligned}$$

where to simplify the notation we define

$$U_1'' \equiv U_1''(Q_1^*) < 0, U_2'' \equiv U_2''(Q_2^*) < 0.$$

Written in matrix notation this becomes

$$\underbrace{\begin{bmatrix} 0 & -P_1 & -P_2 \\ -P_1 & U_1'' & 0 \\ -P_2 & 0 & U_2'' \end{bmatrix}}_{H} \begin{bmatrix} d\lambda^* \\ dQ_1^* \\ dQ_2^* \end{bmatrix} = \begin{bmatrix} Q_1^*dP_1 + Q_2^*dP_2 - dY \\ \lambda^*dP_1 \\ \lambda^*dP_2 \end{bmatrix}.$$

We have

$$\det[H] = \det \begin{bmatrix} 0 & -P_1 & -P_2 \\ -P_1 & U_1'' & 0 \\ -P_2 & 0 & U_2'' \end{bmatrix} = -(P_1)^2 U_2'' - (P_2)^2 U_1'' > 0$$

since $U_1'' < 0$ and $U_2'' < 0$.

To calculate $\frac{\partial Q_1^*}{\partial Y}$ we 1) set $dP_1 = dP_2 = 0$ and 2) change all the remaining $d's$ to $\partial's$ to obtain

$$\underbrace{\begin{bmatrix} 0 & -P_1 & -P_2 \\ -P_1 & U_1'' & 0 \\ -P_2 & 0 & U_2'' \end{bmatrix}}_{H} \begin{bmatrix} \frac{\partial \lambda^*}{\partial Y} \\ \frac{\partial Q_1^*}{\partial Y} \\ \frac{\partial Q_2^*}{\partial Y} \end{bmatrix} = \begin{bmatrix} -1 \\ 0 \\ 0 \end{bmatrix}$$

so that

$$\frac{\partial Q_1^*}{\partial Y} = \frac{\det \begin{bmatrix} 0 & -1 & -P_2 \\ -P_1 & 0 & 0 \\ -P_2 & 0 & U_2'' \end{bmatrix}}{\det [H]} = -\frac{-P_1 U_2''}{\det [H]} > 0$$

since $\det [H] > 0$ and $U_2'' < 0$.

To calculate $\frac{\partial Q_1^*}{\partial P_1}, \frac{\partial Q_2^*}{\partial P_1}$ we 1) set $dP_2 = dY = 0$ and 2) change all the remaining $d's$ to $\partial's$ to obtain

$$\begin{aligned} 0\partial\lambda^* - P_1\partial Q_1^* - P_2\partial Q_2^* - Q_1^*\partial P_1 &= 0 \\ -P_1\partial\lambda^* + U_1''\partial Q_1^* + 0\partial Q_2^* - \lambda^*\partial P_1 &= 0 \\ -P_2\partial\lambda^* + 0\partial Q_1^* + U_2''\partial Q_2^* &= 0. \end{aligned}$$

Writing this in matrix notation and dividing both sides by ∂P_1 we obtain

$$\underbrace{\begin{bmatrix} 0 & -P_1 & -P_2 \\ -P_1 & U_1'' & 0 \\ -P_2 & 0 & U_2'' \end{bmatrix}}_{H} \begin{bmatrix} \frac{\partial\lambda^*}{\partial P_1} \\ \frac{\partial Q_1^*}{\partial P_1} \\ \frac{\partial Q_2^*}{\partial P_1} \end{bmatrix} = \begin{bmatrix} Q_1^* \\ \lambda^* \\ 0 \end{bmatrix}$$

so that

$$\frac{\partial Q_1^*}{\partial P_1} = \frac{\det \begin{bmatrix} 0 & Q_1^* & -P_2 \\ -P_1 & \lambda^* & 0 \\ -P_2 & 0 & U_2'' \end{bmatrix}}{\det [H]} = \frac{P_1 Q_1^* U_2'' - (P_2)^2 \lambda^*}{\det [H]} < 0$$

since $\det [H] > 0$ and $\lambda^* = \frac{U_1'}{P_1} > 0$.

For $\frac{\partial Q_2^*}{\partial P_1}$ we have

$$\begin{aligned} \frac{\partial Q_2^*}{\partial P_1} &= \frac{\det \begin{bmatrix} 0 & -P_1 & Q_1^* \\ -P_1 & U_1'' & \lambda^* \\ -P_2 & 0 & 0 \end{bmatrix}}{\det [H]} = \frac{-P_2 \det \begin{bmatrix} -P_1 & Q_1^* \\ U_1'' & \lambda^* \end{bmatrix}}{\det [H]} \\ &= \frac{-P_2 \left(-P_1\lambda^* - Q_1^* U_1''\right)}{\det [H]}. \end{aligned}$$

From the second first-order condition we have

$$U_1'(Q_1^*) - \lambda^* P_1 = 0 \Longrightarrow \lambda^* = \frac{U_1'}{P_1}$$

so that

$$\begin{aligned} \frac{\partial Q_2^*}{\partial P_1} &= \frac{-P_2 \left(-P_1\lambda^* - U_1'' Q_1^*\right)}{\det [H]} = \frac{-P_2 \left(-P_1 \frac{U_1'}{P_1} - U_1'' Q_1^*\right)}{\det [H]} \\ &= \frac{-P_2 \left(-U_1' - U_1'' Q_1^*\right)}{\det [H]} = \frac{P_2 U_1' \left(1 + \frac{U_1'' Q_1^*}{U_1'}\right)}{\det [H]} = \frac{P_2 U_1' \left(1 + \eta_{U_1}\right)}{\det [H]} \end{aligned}$$

where $\eta_{U_1} \equiv \frac{U_1'' Q_1^*}{U_1'}$ is the elasticity of $MU_1 \equiv U_1'$, the marginal utility of good 1. Thus $\frac{\partial Q_2^*}{\partial P_1} > 0$ (Q_1 and Q_2 are substitutes) if MU_1 is inelastic ($-1 < \eta_{U_1} < 0$) and $\frac{\partial Q_2^*}{\partial P_1} < 0$ (Q_1 and Q_2 are complements) if MU_1 is elastic ($\eta_{U_1} < -1$).

2.7 General Utility Maximization

The next questions are based on the following information. A household with concave utility function $U(Q_1, Q_2)$ maximizes its utility, subject to the budget constraint $Y = P_1 Q_1 + P_2 Q_2$ at Q_1^*, Q_2^*. Define the marginal utilities MU_1, MU_2 and the second-order partial derivatives U_{11}, U_{12}, U_{22} as

$$\begin{aligned} MU_1 &\equiv \frac{\partial U(Q_1^*, Q_2^*)}{\partial Q_1}, MU_2 \equiv \frac{\partial U(Q_1^*, Q_2^*)}{\partial Q_2} \\ U_{11} &\equiv \frac{\partial^2 U(Q_1^*, Q_2^*)}{\partial Q_1^2}, U_{12} \equiv \frac{\partial^2 U(Q_1^*, Q_2^*)}{\partial Q_1 \partial Q_2}, U_{22} \equiv \frac{\partial^2 U(Q_1^*, Q_2^*)}{\partial Q_2^2}. \end{aligned}$$

Since marginal utilities are positive you may assume $MU_1 > 0, MU_2 > 0$. Since the utility function is concave you may assume $U_{11} < 0, U_{22} < 0$ and that

$$H_U \equiv \begin{bmatrix} U_{11} & U_{12} \\ U_{12} & U_{22} \end{bmatrix}$$

is negative definite so that

$$M_2 = \det[H_U] = U_{11} U_{22} - U_{12}^2 > 0.$$

Problem 2.40 *Construct the Lagrangian for the utility maximization problem and find the first-order conditions. [B***] Show that*

$$\lambda^* = \frac{MU_1}{P_1} = \frac{MU_2}{P_2} > 0.$$

*Let $E_1 \equiv P_1 Q_1, E_2 \equiv P_2 Q_2$ be expenditure (measured in say dollars) on Q_1, Q_2 and let $\tilde{U}(E_1, E_2)$ be utility as a function of the expenditure on each good. [A***] What is the economic interpretation of $\frac{\partial \tilde{U}(E_1,E_2)}{\partial E_1}, \frac{\partial \tilde{U}(E_1,E_2)}{\partial E_2}$? Show that*

$$\frac{\partial \tilde{U}(E_1^*, E_2^*)}{\partial E_1} = \frac{MU_1}{P_1}, \frac{\partial \tilde{U}(E_1^*, E_2^*)}{\partial E_2} = \frac{MU_2}{P_1}.$$

Use the fact that λ^* is the marginal utility of income to interpret

$$\lambda^* = \frac{MU_1}{P_1} = \frac{MU_2}{P_2}.$$

Answer: The Lagrangian for the constrained maximization problem is

$$\mathcal{L}(\lambda, Q_1, Q_2) = U(Q_1, Q_2) + \lambda(Y - P_1 Q_1 - P_2 Q_2).$$

The first-order conditions yield

$$\begin{aligned}
Y - P_1Q_1^* - P_2Q_2^* &= 0 \\
&\Longrightarrow Y = P_1Q_1^* + P_2Q_2^* \\
\frac{\partial U\left(Q_1^*, Q_2^*\right)}{\partial Q_1} - \lambda^* P_1 &= 0 \\
&\Longrightarrow MU_1 = \lambda^* P_1 \\
&\Longrightarrow \lambda^* = \frac{MU_1}{P_1} \\
\frac{\partial U\left(Q_1^*, Q_2^*\right)}{\partial Q_2} - \lambda^* P_2 &= 0 \\
&\Longrightarrow MU_2 = \lambda^* P_2 \\
&\Longrightarrow \lambda^* = \frac{MU_2}{P_2}.
\end{aligned}$$

We have $\lambda^* > 0$ from the last two results as

$$\lambda^* = \frac{MU_1}{P_1} = \frac{MU_2}{P_2} > 0.$$

We have

$$E_1 \equiv P_1Q_1 \Longrightarrow Q_1 = \frac{E_1}{P_1} \text{ and } E_2 \equiv P_2Q_2 \Longrightarrow Q_2 = \frac{E_2}{P_2}$$

so that utility $\tilde{U}\left(E_1, E_2\right)$ written as a function of expenditure E_1, E_2 is

$$\tilde{U}\left(E_1, E_2\right) = U\left(Q_1, Q_2\right) = U\left(\frac{E_1}{P_1}, \frac{E_2}{P_2}\right).$$

Here $\frac{\partial \tilde{U}(E_1,E_2)}{\partial E_1}$ is the marginal utility of spending an extra dollar on Q_1, while $\frac{\partial \tilde{U}(E_1,E_2)}{\partial E_2}$ is the marginal utility of spending an extra dollar on Q_2. Using the chain rule we have

$$\begin{aligned}
\frac{\partial \tilde{U}\left(E_1, E_2\right)}{\partial E_1} &= \frac{\partial}{\partial E_1} U\left(\frac{E_1}{P_1}, \frac{E_2}{P_2}\right) \\
&= \frac{\partial U\left(\frac{E_1}{P_1}, \frac{E_2}{P_2}\right)}{\partial Q_1} \frac{1}{P_1} \\
&= \frac{\partial U\left(Q_1, Q_2\right)}{\partial Q_1} \frac{1}{P_1} \\
\frac{\partial \tilde{U}\left(E_1, E_2\right)}{\partial E_2} &= \frac{\partial}{\partial E_2} U\left(\frac{E_1}{P_1}, \frac{E_2}{P_2}\right) \\
&= \frac{\partial U\left(\frac{E_1}{P_1}, \frac{E_2}{P_2}\right)}{\partial Q_2} \frac{1}{P_2} \\
&= \frac{\partial U\left(Q_1, Q_2\right)}{\partial Q_2} \frac{1}{P_2}.
\end{aligned}$$

If $E_1^* \equiv P_1 Q_1^*, E_2^* \equiv P_2 Q_2^*$ then

$$\begin{aligned}\frac{\partial \tilde{U}(E_1^*, E_2^*)}{\partial E_1} &= \frac{\partial U(Q_1^*, Q_2^*)}{\partial Q_1}\frac{1}{P_1} = \frac{MU_1}{P_1}\\ \frac{\partial \tilde{U}(E_1^*, E_2^*)}{\partial E_2} &= \frac{\partial U(Q_1^*, Q_2^*)}{\partial Q_2}\frac{1}{P_2} = \frac{MU_2}{P_2}\end{aligned}$$

and so $\lambda^* = \frac{MU_1}{P_1} = \frac{MU_2}{P_2}$ states that the marginal utility of the last dollar spent on either good must equal the marginal utility of income λ^*. (In Question 2.42 on page 82 we show λ^* equals the marginal utility of income).

Problem 2.41 *[C***] Write the first-order conditions as three implicit functions and find the total differential. [B***] Use the total differential to find* $\frac{\partial Q_1^*}{\partial Y}$ *and* $\frac{\partial Q_1^*}{\partial P_1}$. *[A ***] Show that* $\frac{\partial Q_1^*}{\partial P_1} + Q_1^* \frac{\partial Q_1^*}{\partial Y} < 0$. *Show that if* $U_{12} > 0$ *then* $\frac{\partial Q_1^*}{\partial Y} > 0$ *(*Q_1 *is a normal good) and that* $\frac{\partial Q_1^*}{\partial P_1} < 0$ *(the demand curve for* Q_1 *is downward sloping). Show more generally that if* $\frac{\partial Q_1^*}{\partial Y} > 0$ *then* $\frac{\partial Q_1^*}{\partial P_1} < 0$.

Answer: The first-order conditions can be written as 3 implicit functions as

$$\begin{aligned}g_1(\lambda^*, Q_1^*, Q_2^*, P_1, P_2, Y) &= Y - P_1 Q_1^* - P_2 Q_2^* = 0\\ g_2(\lambda^*, Q_1^*, Q_2^*, P_1, P_2, Y) &= \frac{\partial U(Q_1^*, Q_2^*)}{\partial Q_1} - \lambda^* P_1 = 0\\ g_3(\lambda^*, Q_1^*, Q_2^*, P_1, P_2, Y) &= \frac{\partial U(Q_1^*, Q_2^*)}{\partial Q_2} - \lambda^* P_2 = 0\end{aligned}$$

which determines the reduced form

$$\lambda^* = \lambda^*(P_1, P_2, Y),\ Q_1^* = Q_1^*(P_1, P_2, Y),\ Q_2^* = Q_2^*(P_1, P_2, Y).$$

Taking the total differential we find that

$$\begin{aligned}0 d\lambda^* - P_1 dQ_1^* - P_2 dQ_2^* + dY - Q_1^* dP_1 - Q_2^* dP_2 &= 0\\ -P_1 d\lambda^* + U_{11} dQ_1^* + U_{12} dQ_2^* - \lambda^* dP_1 &= 0\\ -P_2 d\lambda^* + U_{12} dQ_1^* + U_{22} dQ_2^* - \lambda^* dP_2 &= 0\end{aligned}$$

or in matrix notation

$$\underbrace{\begin{bmatrix} 0 & -P_1 & -P_2 \\ -P_1 & U_{11} & U_{12} \\ -P_2 & U_{12} & U_{22} \end{bmatrix}}_{H} \begin{bmatrix} d\lambda^* \\ dQ_1^* \\ dQ_2^* \end{bmatrix} = \begin{bmatrix} Q_1^* dP_1 + Q_2^* dP_2 - dY \\ \lambda^* dP_1 \\ \lambda^* dP_2 \end{bmatrix}.$$

We will need to show $\det[H] > 0$. We have

$$\begin{aligned}
\det[H] &= \det\begin{bmatrix} 0 & -P_1 & -P_2 \\ -P_1 & U_{11} & U_{12} \\ -P_2 & U_{12} & U_{22} \end{bmatrix} \\
&= -P_1^2U_{22} + 2P_1P_2U_{12} - P_2^2U_{11} \\
&= -\begin{bmatrix} P_2 & -P_1 \end{bmatrix}\begin{bmatrix} U_{11} & U_{12} \\ U_{12} & U_{22} \end{bmatrix}\begin{bmatrix} P_2 \\ -P_1 \end{bmatrix} \\
&= -x^T H_U x
\end{aligned}$$

where

$$H_U = \begin{bmatrix} U_{11} & U_{12} \\ U_{12} & U_{22} \end{bmatrix}, x = \begin{bmatrix} P_2 \\ -P_1 \end{bmatrix} \neq \begin{bmatrix} 0 \\ 0 \end{bmatrix}.$$

Since H_U is negative definite $x^T H_U x < 0$ for any $x \neq 0$ and so $\det[H] = -x^T H_U x > 0$.

To find $\frac{\partial Q_1^*}{\partial Y}$ we 1) set $dP_1 = dP_2 = 0$ and 2) replace the remaining $d's$ with $\partial' s$ as

$$\underbrace{\begin{bmatrix} 0 & -P_1 & -P_2 \\ -P_1 & U_{11} & U_{12} \\ -P_2 & U_{12} & U_{22} \end{bmatrix}}_{H}\begin{bmatrix} \frac{\partial \lambda^*}{\partial Y} \\ \frac{\partial Q_1^*}{\partial Y} \\ \frac{\partial Q_2^*}{\partial Y} \end{bmatrix} = \begin{bmatrix} -1 \\ 0 \\ 0 \end{bmatrix}.$$

Using Cramer's rule

$$\begin{aligned}
\frac{\partial Q_1^*}{\partial Y} &= \frac{\det\begin{bmatrix} 0 & -1 & -P_2 \\ -P_1 & 0 & U_{12} \\ -P_2 & 0 & U_{22} \end{bmatrix}}{\det[H]} \\
&= \frac{-P_1U_{22} + P_2U_{12}}{\det[H]}.
\end{aligned}$$

If $U_{12} > 0$ then since $\det[H] > 0$ and $U_{22} < 0$ it follows that $\frac{\partial Q_1^*}{\partial Y} > 0$.

To find $\frac{\partial Q_1^*}{\partial P_1}$ we 1) set $dP_2 = dY = 0$ and 2) replace the remaining $d's$ with $\partial' s$ as

$$\begin{bmatrix} 0 & -P_1 & -P_2 \\ -P_1 & U_{11} & U_{12} \\ -P_2 & U_{12} & U_{22} \end{bmatrix}\begin{bmatrix} \frac{\partial \lambda^*}{\partial P_1} \\ \frac{\partial Q_1^*}{\partial P_1} \\ \frac{\partial Q_2^*}{\partial P_1} \end{bmatrix} = \begin{bmatrix} Q_1^* \\ \lambda^* \\ 0 \end{bmatrix}.$$

Using Cramer's rule we have

$$\begin{aligned}\frac{\partial Q_1^*}{\partial P_1} &= \frac{\det\begin{bmatrix}0 & Q_1^* & -P_2\\ -P_1 & \lambda^* & U_{12}\\ -P_2 & 0 & U_{22}\end{bmatrix}}{\det[H]}\\ &= -Q_1^*\overbrace{\left(\frac{-P_1U_{22}+P_2U_{12}}{\det[H]}\right)}^{\frac{\partial Q_1^*}{\partial Y}} - \frac{\lambda^* P_2^2}{\det[H]}\\ &= -Q_1^*\frac{\partial Q_1^*}{\partial Y} - \frac{\lambda^* P_2^2}{\det[H]}.\end{aligned}$$

Thus

$$\frac{\partial Q_1^*}{\partial P_1} + Q_1^*\frac{\partial Q_1^*}{\partial Y} = -\frac{\lambda^* P_2^2}{\det[H]} < 0.$$

If $\frac{\partial Q_1^*}{\partial Y} > 0$ then

$$\frac{\partial Q_1^*}{\partial P_1} = -Q_1^*\frac{\partial Q_1^*}{\partial Y} - \frac{\lambda^* P_2^2}{\det[H]} < 0.$$

If $U_{12} > 0$ then we have shown that $\frac{\partial Q_1^*}{\partial Y} > 0$, so that $\frac{\partial Q_1^*}{\partial P_1} < 0$.

Problem 2.42 *[A ***] Consider putting the demand curves $Q_1^*(P_1,P_2,Y), Q_2^*(P_1,P_2,Y)$ back into the utility function $U(Q_1,Q_2)$. This yields the indirect utility function $U^*(P_1,P_2,Y)$, a function of P_1,P_2,Y, as*

$$U^*(P_1,P_2,Y) \equiv U(Q_1^*(P_1,P_2,Y), Q_2^*(P_1,P_2,Y)).$$

Show that the Lagrange multiplier is the marginal utility of income as $\frac{\partial U^*}{\partial Y} = \lambda^*$. Show there is a diminishing marginal utility of income ($\frac{\partial^2 U^*}{\partial Y^2} = \frac{\partial \lambda^*}{\partial Y} < 0$).

Answer: Using the multivariate chain rule the marginal utility of income is

$$\begin{aligned}\frac{\partial U^*(P_1,P_2,Y)}{\partial Y} &= \frac{\partial U(Q_1^*,Q_2^*)}{\partial Q_1}\frac{\partial Q_1^*}{\partial Y} + \frac{\partial U(Q_1^*,Q_2^*)}{\partial Q_2}\frac{\partial Q_2^*}{\partial Y}\\ &= MU_1\frac{\partial Q_1^*}{\partial Y} + MU_2\frac{\partial Q_2^*}{\partial Y}.\end{aligned}$$

We wish to show this reduces to λ^*. From the first-order conditions in Question 2.41 on page 80 we have $MU_1 = \lambda^* P_1, MU_2 = \lambda^* P_2$ so that

$$\begin{aligned}\frac{\partial U^*(P_1,P_2,Y)}{\partial Y} &= MU_1\frac{\partial Q_1^*}{\partial Y} + MU_2\frac{\partial Q_2^*}{\partial Y}\\ &= \lambda^* P_1\frac{\partial Q_1^*}{\partial Y} + \lambda^* P_2\frac{\partial Q_2^*}{\partial Y}\\ &= \lambda^*\left(P_1\frac{\partial Q_1^*}{\partial Y} + P_2\frac{\partial Q_2^*}{\partial Y}\right).\end{aligned}$$

To evaluate the expression in brackets differentiate both sides of

$$Y = P_1 Q_1^* + P_2 Q_2^*$$

with respect to Y to obtain

$$\begin{aligned} 1 &= \frac{\partial}{\partial Y} Y = \frac{\partial}{\partial Y}\left(P_1 Q_1^* + P_2 Q_2^*\right) \\ &= P_1 \frac{\partial Q_1^*}{\partial Y} + P_2 \frac{\partial Q_2^*}{\partial Y} \\ &\Longrightarrow P_1 \frac{\partial Q_1^*}{\partial Y} + P_2 \frac{\partial Q_2^*}{\partial Y} = 1. \end{aligned}$$

Thus

$$\frac{\partial U^*(P_1, P_2, Y)}{\partial Y} = \lambda^* \underbrace{\left(P_1 \frac{\partial Q_1^*}{\partial Y} + P_2 \frac{\partial Q_2^*}{\partial Y}\right)}_{1} = \lambda^*.$$

From Question 2.41 on page 80 we have

$$\underbrace{\begin{bmatrix} 0 & -P_1 & -P_2 \\ -P_1 & U_{11} & U_{12} \\ -P_2 & U_{12} & U_{22} \end{bmatrix}}_{H} \begin{bmatrix} \frac{\partial \lambda^*}{\partial Y} \\ \frac{\partial Q_1^*}{\partial Y} \\ \frac{\partial Q_2^*}{\partial Y} \end{bmatrix} = \begin{bmatrix} -1 \\ 0 \\ 0 \end{bmatrix}$$

so that from Cramer's rule

$$\begin{aligned} \frac{\partial \lambda^*}{\partial Y} &= \frac{\det \begin{bmatrix} -1 & -P_1 & -P_2 \\ 0 & U_{11} & U_{12} \\ 0 & U_{12} & U_{22} \end{bmatrix}}{\det[H]} \\ &= \frac{-\det \begin{bmatrix} U_{11} & U_{12} \\ U_{12} & U_{22} \end{bmatrix}}{\det[H]} \\ &= \frac{-\det[H_U]}{\det[H]} < 0 \end{aligned}$$

since $\det[H] > 0$, and $\det[H_U] > 0$.

Problem 2.43 *[A***] For the indirect utility function $U^*(P_1, P_2, Y)$ defined in the last question, show that $\frac{\partial U^*}{\partial P_1} = -\lambda^* Q_1^*$ and $\frac{\partial U^*}{\partial P_2} = -\lambda^* Q_2^*$. Prove Roy's identity that*

$$\begin{aligned} Q_1^*(P_1, P_2, Y) &= -\frac{\frac{\partial U^*(P_1, P_2, Y)}{\partial P_1}}{\frac{\partial U^*(P_1, P_2, Y)}{\partial Y}} \\ Q_2^*(P_1, P_2, Y) &= -\frac{\frac{\partial U^*(P_1, P_2, Y)}{\partial P_2}}{\frac{\partial U^*(P_1, P_2, Y)}{\partial Y}}. \end{aligned}$$

Answer: To show $\frac{\partial U^*}{\partial P_1} = -\lambda^* Q_1^*$ use the multivariate chain rule as

$$\begin{aligned}\frac{\partial U^*(P_1,P_2,Y)}{\partial P_1} &= \frac{\partial U(Q_1^*,Q_2^*)}{\partial Q_1}\frac{\partial Q_1^*}{\partial P_1} + \frac{\partial U(Q_1^*,Q_2^*)}{\partial Q_2}\frac{\partial Q_2^*}{\partial P_1} \\ &= MU_1\frac{\partial Q_1^*}{\partial P_1} + MU_2\frac{\partial Q_2^*}{\partial P_1}.\end{aligned}$$

From the first-order conditions in Question 2.41 on page 80 we have $MU_1 = \lambda^* P_1, MU_2 = \lambda^* P_2$ so that

$$\begin{aligned}\frac{\partial U^*(P_1,P_2,Y)}{\partial P_1} &= (\lambda^* P_1)\frac{\partial Q_1^*}{\partial P_1} + (\lambda^* P_2)\frac{\partial Q_2^*}{\partial P_1} \\ &= \lambda^*\left(P_1\frac{\partial Q_1^*}{\partial P_1} + P_2\frac{\partial Q_2^*}{\partial P_1}\right).\end{aligned}$$

To evaluate the expression in brackets differentiate both sides of

$$Y = P_1Q_1^* + P_2Q_2^*$$

with respect to P_1 to obtain

$$\begin{aligned}0 &= \frac{\partial}{\partial P_1}Y = \frac{\partial}{\partial P_1}(P_1Q_1^* + P_2Q_2^*) \\ &= Q_1^* + P_1\frac{\partial Q_1^*}{\partial P_1} + P_2\frac{\partial Q_2^*}{\partial P_1} \\ \Longrightarrow & \; P_1\frac{\partial Q_1^*}{\partial P_1} + P_2\frac{\partial Q_2^*}{\partial P_1} = -Q_1^*.\end{aligned}$$

Thus

$$\begin{aligned}\frac{\partial U^*(P_1,P_2,Y)}{\partial P_1} &= \lambda^*\underbrace{\left(P_1\frac{\partial Q_1^*}{\partial P_1} + P_2\frac{\partial Q_2^*}{\partial P_1}\right)}_{-Q_1^*} \\ &= -\lambda^* Q_1^*.\end{aligned}$$

The proof that $\frac{\partial U^*}{\partial P_2} = -\lambda^* Q_2^*$ follows exactly the same lines.

Roy's identity then follows as

$$\begin{aligned}-\frac{\frac{\partial U^*(P_1,P_2,Y)}{\partial P_1}}{\frac{\partial U^*(P_1,P_2,Y)}{\partial Y}} &= -\frac{-\lambda^* Q_1^*}{\lambda^*} = Q_1^* \\ -\frac{\frac{\partial U^*(P_1,P_2,Y)}{\partial P_2}}{\frac{\partial U^*(P_1,P_2,Y)}{\partial Y}} &= -\frac{-\lambda^* Q_2^*}{\lambda^*} = Q_2^*.\end{aligned}$$

Problem 2.44 *[A***] Consider again the indirect utility function $U^*(P_1,P_2,Y)$. Using total differentials and the results from Question 2.42 on page 82, show that*

$$dU^* = \lambda^*(dY - Q_1^* dP_1 - Q_2^* dP_2).$$

Answer: Using the indirect utility function we have

$$U^* = U^*(P_1, P_2, Y) \Longrightarrow g(U^*, P_1, P_2, Y) = U^* - U^*(P_1, P_2, Y) = 0.$$

The total differential for U^* yields

$$dU^* - \left(\frac{\partial U^*}{\partial P_1}dP_1 + \frac{\partial U^*}{\partial P_2}dP_2 + \frac{\partial U^*}{\partial Y}dY\right) = 0.$$

Using the results from Question 2.42 on page 82 we have

$$\frac{\partial U^*}{\partial P_1} = -\lambda^* Q_1^*, \frac{\partial U^*}{\partial P_2} = -\lambda^* Q_2^*, \frac{\partial U^*}{\partial Y} = \lambda^*$$

and so

$$dU^* - (-\lambda^* Q_1^* dP_1 - \lambda^* Q_2^* dP_2 + \lambda^* dY) = 0 \Longrightarrow dU^* = \lambda^*(dY - Q_1^* dP_1 - Q_2^* dP_2).$$

Problem 2.45 *[A ***] Consider the effect of an increase in P_1 on Q_1^*, as measured by $\frac{\partial Q_1^*}{\partial P_1}$. This is itself made up of two effects: 1) The substitution effect and 2) Income effects. Provide an intuitive discussion of the income and substitution effects influence on the slope of the demand curve. Show mathematically how the income effect can be removed from $\frac{\partial Q_1^*}{\partial P_1}$, and that once this is done it follows that $\frac{\partial Q_1^*}{\partial P_1} < 0$.*

Answer: The effect of an increase in P_1 on Q_1^*, as measured by $\frac{\partial Q_1^*}{\partial P_1}$, is itself made up of two effects: 1) The substitution effect: when P_1 increases then Q_1 becomes relatively more expensive, and the household will want to substitute the cheaper Q_2. For an increase in P_1 the substitution effect reduces the demand for Q_1 and so by itself leads to a downward sloping demand curve. 2) The income effect: an increase in P_1 reduces the purchasing power of income Y, and so makes the household poorer. The direction of the income effect is ambiguous: if Q_1 is normal the income effect works in the right direction to insure that $\frac{\partial Q_1^*}{\partial P_1} < 0$ since a poorer household will demand less of a normal good. But if Q_1 is inferior the income effect works against the substitution effect, and in extreme circumstances can lead to an upward sloping demand curve or $\frac{\partial Q_1^*}{\partial P_1} > 0$.

From Question 2.41 on page 80 we have

$$\begin{aligned} 0d\lambda^* - P_1 dQ_1^* - P_2 dQ_2^* + dY - Q_1^* dP_1 - Q_2^* dP_2 &= 0 \\ -P_1 d\lambda^* + U_{11} dQ_1^* + U_{12} dQ_2^* - \lambda^* dP_1 &= 0 \\ -P_2 d\lambda^* + U_{12} dQ_1^* + U_{22} dQ_2^* - \lambda^* dP_2 &= 0. \end{aligned}$$

Normally when we calculate $\frac{\partial Q_1^*}{\partial P_1}$ we set both $dP_2 = 0$ and $dY = 0$. Instead of setting $dY = 0$ let's set $dY = Q_1^* dP_1$ to compensate for the income effect. This holds the household's welfare U^* (or real income) constant as $dU^* = 0$ since from Question 2.44 on page 84 we have

$$dU^* = \lambda^*(dY - Q_1^* dP_1 - Q_2^* dP_2) = \lambda^*(dY - Q_1^* dP_1) = 0.$$

Now replacing the remaining $d's$ with $\partial's$ we have

$$\begin{aligned}
0\partial\lambda^* - P_1\partial Q_1^* - P_2\partial Q_2^* + \overbrace{dY - Q_1^*P_1}^{0} &= 0 \\
-P_1\partial\lambda^* + U_{11}\partial Q_1^* + U_{12}\partial Q_2^* - \lambda^*\partial P_1 &= 0 \\
-P_2\partial\lambda^* + U_{12}\partial Q_1^* + U_{22}\partial Q_2^* &= 0
\end{aligned}$$

so that in matrix notation

$$\underbrace{\begin{bmatrix} 0 & -P_1 & -P_2 \\ -P_1 & U_{11} & U_{12} \\ -P_2 & U_{12} & U_{22} \end{bmatrix}}_{H} \begin{bmatrix} \frac{\partial\lambda^*}{\partial P_1} \\ \frac{\partial Q_1^*}{\partial P_1} \\ \frac{\partial Q_2^*}{\partial P_1} \end{bmatrix} = \begin{bmatrix} 0 \\ \lambda^* \\ 0 \end{bmatrix}.$$

Using Cramer's rule we have

$$\frac{\partial Q_1^*}{\partial P_1} = \frac{\det\begin{bmatrix} 0 & 0 & -P_2 \\ -P_1 & \lambda^* & U_{12} \\ -P_2 & 0 & U_{22} \end{bmatrix}}{\det[H]} = \frac{-\lambda^* P_2^2}{\det[H]} < 0.$$

Problem 2.46 *[A **] Consider a household with the utility function*

$$U(Q_1, Q_2) = 4Q_1^{\frac{1}{4}} - 2Q_2^{-\frac{1}{2}}.$$

Show that the marginal utilities are positive: $\frac{\partial U}{\partial Q_1} > 0$, $\frac{\partial U}{\partial Q_2} > 0$. Show that $U(Q_1, Q_2)$ is globally concave. (This shows that this is a normal utility function.) From the Lagrangian find the first-order conditions and solve. Show that $\frac{\partial Q_1^*}{\partial P_2} < 0$ but that $\frac{\partial Q_2^*}{\partial P_1} > 0$. The fact that the signs of $\frac{\partial Q_1^*}{\partial P_2}$ and $\frac{\partial Q_2^*}{\partial P_1}$ conflict demonstrates that there is a problem with our definition of substitutes and complements since from $\frac{\partial Q_1^*}{\partial P_2} < 0$ would say that Q_1 and Q_2 are complements, while $\frac{\partial Q_2^*}{\partial P_1} > 0$ would say that Q_1 and Q_2 are substitutes. The villain is the income effect. In the next question we show that this conflict is removed if we take out the income effect.

Answer: We have

$$\frac{\partial U(Q_1, Q_2)}{\partial Q_1} = Q_1^{-\frac{3}{4}} > 0, \frac{\partial U(Q_1, Q_2)}{\partial Q_2} = Q_2^{-\frac{3}{2}} > 0.$$

The utility function is concave since the Hessian is a diagonal matrix with negative diagonal elements as

$$\begin{aligned}
H(Q_1, Q_2) &= \begin{bmatrix} \frac{\partial^2 U(Q_1,Q_2)}{\partial Q_1^2} & \frac{\partial^2 U(Q_1,Q_2)}{\partial Q_1 \partial Q_2} \\ \frac{\partial^2 U(Q_1,Q_2)}{\partial Q_1 \partial Q_2} & \frac{\partial^2 U(Q_1,Q_2)}{\partial Q_2^2} \end{bmatrix} \\
&= \begin{bmatrix} -\frac{3}{4}Q_1^{-\frac{7}{4}} & 0 \\ 0 & -\frac{3}{2}Q_2^{-\frac{5}{2}} \end{bmatrix}.
\end{aligned}$$

The Lagrangian is

$$\mathcal{L}(\lambda, Q_1, Q_2) = 4Q_1^{\frac{1}{4}} - 2Q_2^{-\frac{1}{2}} + \lambda(Y - P_1Q_1 - P_2Q_2)$$

so that

$$\begin{aligned}
\frac{\partial \mathcal{L}(\lambda, Q_1, Q_2)}{\partial \lambda} &= Y - P_1Q_1 - P_2Q_2 \\
\frac{\partial \mathcal{L}(\lambda, Q_1, Q_2)}{\partial Q_1} &= Q_1^{-\frac{3}{4}} - \lambda P_1 \\
\frac{\partial \mathcal{L}(\lambda, Q_1, Q_2)}{\partial Q_2} &= Q_2^{-\frac{3}{2}} - \lambda P_2.
\end{aligned}$$

The first-order conditions yield

$$\begin{aligned}
Y - P_1Q_1^* - P_2Q_2^* &= 0 \Longrightarrow Y = P_1Q_1^* + P_2Q_2^* \\
(Q_1^*)^{-\frac{3}{4}} - \lambda^* P_1 &= 0 \\
&\Longrightarrow Q_1^* = (\lambda^* P_1)^{-\frac{4}{3}} = (\lambda^*)^{-\frac{4}{3}} P_1^{-\frac{4}{3}} \\
&\Longrightarrow P_1Q_1^* = (\lambda^*)^{-\frac{4}{3}} P_1^{-\frac{1}{3}} \\
(Q_2^*)^{-\frac{3}{2}} - \lambda^* P_2 &= 0 \\
&\Longrightarrow Q_2^* = (\lambda^* P_2)^{-\frac{2}{3}} = (\lambda^*)^{-\frac{2}{3}} P_2^{-\frac{2}{3}} \\
&\Longrightarrow P_2Q_2^* = (\lambda^*)^{-\frac{2}{3}} P_2^{\frac{1}{3}}
\end{aligned}$$

so that

$$\begin{aligned}
Y &= P_1Q_1^* + P_2Q_2^* = (\lambda^*)^{-\frac{4}{3}} P_1^{-\frac{1}{3}} + (\lambda^*)^{-\frac{2}{3}} P_2^{\frac{1}{3}} \\
&\Longrightarrow (\lambda^*)^{-\frac{4}{3}} P_1^{-\frac{1}{3}} + (\lambda^*)^{-\frac{2}{3}} P_2^{\frac{1}{3}} - Y = 0.
\end{aligned}$$

Setting $x = (\lambda^*)^{-\frac{2}{3}}$ this becomes a quadratic which can be solved for $(\lambda^*)^{-\frac{2}{3}}$ as

$$\begin{aligned}
(\lambda^*)^{-\frac{4}{3}} P_1^{-\frac{1}{3}} + (\lambda^*)^{-\frac{2}{3}} P_2^{\frac{1}{3}} - Y &= 0 \\
&\Longrightarrow P_1^{-\frac{1}{3}} x^2 + P_2^{\frac{1}{3}} x - Y = 0 \\
&\Longrightarrow x = (\lambda^*)^{-\frac{2}{3}} = \frac{-P_2^{\frac{1}{3}} \pm \sqrt{P_2^{\frac{2}{3}} + 4P_1^{-\frac{1}{3}} Y}}{2P_1^{-\frac{1}{3}}}.
\end{aligned}$$

Since $(\lambda^*)^{-\frac{2}{3}} > 0$ we take the positive root

$$(\lambda^*)^{-\frac{2}{3}} = \frac{\sqrt{P_2^{\frac{2}{3}} + 4P_1^{-\frac{1}{3}} Y} - P_2^{\frac{1}{3}}}{2P_1^{-\frac{1}{3}}}.$$

Our analysis will be simpler if we get all prices in the denominator of $(\lambda^*)^{-\frac{2}{3}}$. To this end we have

$$\begin{aligned}
(\lambda^*)^{-\frac{2}{3}} &= \frac{\sqrt{P_2^{\frac{2}{3}} + 4P_1^{-\frac{1}{3}}Y} - P_2^{\frac{1}{3}}}{2P_1^{-\frac{1}{3}}} \\
&= \frac{\sqrt{P_2^{\frac{2}{3}} + 4P_1^{-\frac{1}{3}}Y} - P_2^{\frac{1}{3}}}{2P_1^{-\frac{1}{3}}} \frac{\sqrt{P_2^{\frac{2}{3}} + 4P_1^{-\frac{1}{3}}Y} + P_2^{\frac{1}{3}}}{\sqrt{P_2^{\frac{2}{3}} + 4P_1^{-\frac{1}{3}}Y} + P_2^{\frac{1}{3}}} \\
&= \frac{\left(P_2^{\frac{2}{3}} + 4P_1^{-\frac{1}{3}}Y\right) - P_2^{\frac{2}{3}}}{2P_1^{-\frac{1}{3}}\left(\sqrt{P_2^{\frac{2}{3}} + 4P_1^{-\frac{1}{3}}Y} + P_2^{\frac{1}{3}}\right)} \\
&= \frac{4P_1^{-\frac{1}{3}}Y}{2P_1^{-\frac{1}{3}}\left(\sqrt{P_2^{\frac{2}{3}} + 4P_1^{-\frac{1}{3}}Y} + P_2^{\frac{1}{3}}\right)} \\
&= \frac{2Y}{\sqrt{P_2^{\frac{2}{3}} + 4P_1^{-\frac{1}{3}}Y} + P_2^{\frac{1}{3}}}
\end{aligned}$$

so that

$$\lambda^* = \left(\frac{2Y}{\sqrt{P_2^{\frac{2}{3}} + 4P_1^{-\frac{1}{3}}Y} + P_2^{\frac{1}{3}}}\right)^{-\frac{3}{2}}.$$

Therefore

$$\begin{aligned}
Q_1^* &= (\lambda^*)^{-\frac{4}{3}} P_1^{-\frac{4}{3}} = \left(\frac{2Y}{\sqrt{P_2^{\frac{2}{3}} + 4P_1^{-\frac{1}{3}}Y} + P_2^{\frac{1}{3}}}\right)^2 P_1^{-\frac{4}{3}} \\
Q_2^* &= (\lambda^*)^{-\frac{2}{3}} P_2^{-\frac{2}{3}} = \frac{2Y}{\sqrt{P_2^{\frac{2}{3}} + 4P_1^{-\frac{1}{3}}Y} + P_2^{\frac{1}{3}}} P_2^{-\frac{2}{3}}.
\end{aligned}$$

In the expression for Q_1^* notice how P_2 appears only in the denominator of the squared term with positive exponents as $P_2^{\frac{2}{3}}$ and $P_2^{\frac{1}{3}}$, so that any increase in P_2 makes the denominator larger, and hence Q_1^* smaller. It follows that $\frac{\partial Q_1^*}{\partial P_2} < 0$.

In the expression for Q_2^* notice how P_1 appears only in the denominator as $P_1^{-\frac{1}{3}}$. Since $P_1^{-\frac{1}{3}}$ has a negative exponent, any increase in P_1 makes the denominator smaller, and hence Q_2^* larger. It follows that $\frac{\partial Q_2^*}{\partial P_1} > 0$.

Problem 2.47 *[A **] Show that if the income effect is removed from* $\frac{\partial Q_1^*}{\partial P_2}$ *and* $\frac{\partial Q_2^*}{\partial P_1}$ *then* $\frac{\partial Q_1^*}{\partial P_2} = \frac{\partial Q_2^*}{\partial P_1}$ *so there is no conflict if we define substitutes as goods* *with* $\frac{\partial Q_1^*}{\partial P_2} = \frac{\partial Q_2^*}{\partial P_1} > 0$, *and complements as goods with* $\frac{\partial Q_1^*}{\partial P_2} = \frac{\partial Q_2^*}{\partial P_1} < 0$. *Show that in the case of only two goods, the goods must be substitutes for each other (complements only exist when there are three or more goods).*

Answer: Normally when we calculate $\frac{\partial Q_1^*}{\partial P_2}$ we set $dP_1 = 0$ and $dY = 0$. Following the approach of Question 2.45 on page 85, instead of setting $dY = 0$ let's set $dY = Q_2^* dP_2$ to compensate for income effect so that the household's welfare U^* is held constant as

$$dU^* = \lambda^* \left(dY - Q_1^* dP_1 - Q_2^* dP_2\right) = \lambda^* \left(dY - Q_2^* dP_2\right) = 0.$$

Now using Cramer's rule on the total differential we have

$$\begin{bmatrix} 0 & -P_1 & -P_2 \\ -P_1 & U_{11} & U_{12} \\ -P_2 & U_{12} & U_{22} \end{bmatrix} \begin{bmatrix} \frac{\partial \lambda^*}{\partial P_1} \\ \frac{\partial Q_1^*}{\partial P_1} \\ \frac{\partial Q_2^*}{\partial P_1} \end{bmatrix} = \begin{bmatrix} 0 \\ \lambda^* \\ 0 \end{bmatrix}$$

so that

$$\frac{\partial Q_2^*}{\partial P_1} = \frac{\det \begin{bmatrix} 0 & -P_1 & 0 \\ -P_1 & U_{11} & \lambda^* \\ -P_2 & U_{12} & 0 \end{bmatrix}}{\det [H]} = \frac{\lambda^* P_1 P_2}{\det [H]} > 0.$$

To calculate $\frac{\partial Q_2^*}{\partial P_1}$ set $dP_2 = 0$ but to remove the income effect set $dY = Q_1^* dP_1$ so that

$$dU^* = \lambda^* \left(dY - Q_1^* dP_1 - Q_2^* dP_2\right) = \lambda^* \left(dY - Q_1^* dP_1\right) = 0.$$

Now using Cramer's rule on the total differential we have

$$\begin{bmatrix} 0 & -P_1 & -P_2 \\ -P_1 & U_{11} & U_{12} \\ -P_2 & U_{12} & U_{22} \end{bmatrix} \begin{bmatrix} \frac{\partial \lambda^*}{\partial P_2} \\ \frac{\partial Q_1^*}{\partial P_2} \\ \frac{\partial Q_2^*}{\partial P_2} \end{bmatrix} = \begin{bmatrix} 0 \\ 0 \\ \lambda^* \end{bmatrix}$$

so that

$$\frac{\partial Q_1^*}{\partial P_2} = \frac{\det \begin{bmatrix} 0 & 0 & -P_2 \\ -P_1 & 0 & U_{12} \\ -P_2 & \lambda^* & U_{22} \end{bmatrix}}{\det [H]} = \frac{\lambda^* P_1 P_2}{\det [H]} > 0$$

so that the two goods are substitutes with

$$\frac{\partial Q_2^*}{\partial P_1} = \frac{\partial Q_1^*}{\partial P_2} = \frac{\lambda^* P_1 P_2}{\det [H]} > 0.$$

2.8 Cost Minimization

Problem 2.48 *Consider a firm minimizing costs* $WL + RK$ *given a concave production function* $Q = F(L, K)$ *with a negative definite Hessian. [D***] Write down the Lagrangian for the cost minimization problem and [C***] find the first-order conditions. [B***] Using the total differential show that* $\frac{\partial K^*}{\partial R} < 0$ *and* $\frac{\partial L^*}{\partial R} = \frac{\partial K^*}{\partial W}$*. [A***] Using the homogeneity of* $L^*(Q, W, R)$ *in* W *and* R *and Euler's theorem show that* $\frac{\partial K^*}{\partial W} = \frac{\partial L^*}{\partial R} = -\frac{W}{R}\frac{\partial L^*}{\partial W}$*. If* $C^*(Q, W, R) \equiv WL^* + RK^*$ *is the cost function, use this result to prove Shephard's lemma that* $\frac{\partial C^*}{\partial W} = L^*$*. Show that* $\frac{\partial \lambda^*}{\partial Q} > 0$*.*

Answer: The Lagrangian is

$$\mathcal{L}(\lambda, L, K, Q, W, R) = WL + RK + \lambda(Q - F(L, K))$$

with first-order conditions

$$\begin{aligned} Q - F(L^*, K^*) &= 0 \\ W - \lambda^* \frac{\partial F(L^*, K^*)}{\partial L} &= 0 \\ R - \lambda^* \frac{\partial F(L^*, K^*)}{\partial K} &= 0. \end{aligned}$$

Using the notation

$$\begin{aligned} &F_L \equiv \frac{\partial F(L^*,K^*)}{\partial L} \quad F_k \equiv \frac{\partial F(L^*,K^*)}{\partial K} \\ &F_{LL} \equiv \frac{\partial^2 F(L^*,K^*)}{\partial L^2} \quad F_{LK} \equiv \frac{\partial^2 F(L^*,K^*)}{\partial L \partial K}, F_{KK} \equiv \frac{\partial^2 F(L^*,K^*)}{\partial K^2}. \end{aligned}$$

and taking the total differential we find that

$$\begin{aligned} 0d\lambda^* - F_L dL^* - F_K dK^* + dQ &= 0 \\ -F_L d\lambda^* - \lambda^* F_{LL} dL^* - \lambda^* F_{LK} dK^* + dW &= 0 \\ -F_K d\lambda^* - \lambda^* F_{LK} dL^* - \lambda^* F_{KK} dK^* + dR &= 0 \end{aligned}$$

or in matrix notation

$$\underbrace{\begin{bmatrix} 0 & -F_L & -F_K \\ -F_L & -\lambda^* F_{LL} & -\lambda^* F_{LK} \\ -F_K & -\lambda^* F_{LK} & -\lambda^* F_{KK} \end{bmatrix}}_{H} \begin{bmatrix} d\lambda^* \\ dL^* \\ dK^* \end{bmatrix} = \begin{bmatrix} -dQ \\ -dW \\ -dR \end{bmatrix}.$$

For the 3×3 matrix

$$H = \begin{bmatrix} 0 & -F_L & -F_K \\ -F_L & -\lambda^* F_{LL} & -\lambda^* F_{LK} \\ -F_K & -\lambda^* F_{LK} & -\lambda^* F_{KK} \end{bmatrix}.$$

we have $\det[H] < 0$ as

$$\begin{aligned}
\det[H] &= \det\begin{bmatrix} 0 & -F_L & -F_K \\ -F_L & -\lambda^* F_{LL} & -\lambda^* F_{LK} \\ -F_K & -\lambda^* F_{LK} & -\lambda^* F_{KK} \end{bmatrix} \\
&= F_L \det\begin{bmatrix} -F_L & -\lambda^* F_{LK} \\ -F_K & -\lambda^* F_{KK} \end{bmatrix} - F_K \det\begin{bmatrix} -F_L & -\lambda^* F_{LL} \\ -F_K & -\lambda^* F_{LK} \end{bmatrix} \\
&= \lambda^* F_L^2 F_{KK} - \lambda^* F_L F_K F_{LK} - \lambda^* F_L F_K F_{LK} + \lambda^* F_K^2 F_{LL} \\
&= \lambda^* F_L^2 F_{KK} - 2\lambda^* F_L F_K F_{LK} + \lambda^* F_K^2 F_{LL} \\
&= \lambda^* \begin{bmatrix} F_K & -F_L \end{bmatrix} \begin{bmatrix} F_{LL} & F_{LK} \\ F_{LK} & F_{KK} \end{bmatrix} \begin{bmatrix} F_K \\ -F_L \end{bmatrix} < 0
\end{aligned}$$

since

$$\lambda^* = \frac{W}{F_L} = \frac{R}{F_K} > 0$$

and the Hessian of the production function

$$A \equiv \begin{bmatrix} F_{LL} & F_{LK} \\ F_{LK} & F_{KK} \end{bmatrix}$$

is negative definite so that $x^T A x < 0$ for $x \neq 0$ with

$$x = \begin{bmatrix} F_L \\ -F_K \end{bmatrix} \neq \begin{bmatrix} 0 \\ 0 \end{bmatrix}$$

since the marginal products are positive.

To calculate $\frac{\partial K^*}{\partial W}$ we 1) set $dR = dQ = 0$ and 2) change the remaining $d's$ to $\partial's$ to obtain

$$\begin{aligned}
\begin{bmatrix} 0 & -F_L & -F_K \\ -F_L & -\lambda^* F_{LL} & -\lambda^* F_{LK} \\ -F_K & -\lambda^* F_{LK} & -\lambda^* F_{KK} \end{bmatrix} \begin{bmatrix} \partial\lambda^* \\ \partial L^* \\ \partial K^* \end{bmatrix} &= \begin{bmatrix} 0 \\ \partial W \\ 0 \end{bmatrix} \\
&= \begin{bmatrix} 0 \\ -1 \\ 0 \end{bmatrix} \partial W.
\end{aligned}$$

Dividing both sides by ∂W we obtain

$$\underbrace{\begin{bmatrix} 0 & -F_L & -F_K \\ -F_L & -\lambda^* F_{LL} & -\lambda^* F_{LK} \\ -F_K & -\lambda^* F_{LK} & -\lambda^* F_{KK} \end{bmatrix}}_{H} \begin{bmatrix} \frac{\partial\lambda^*}{\partial W} \\ \frac{\partial L^*}{\partial W} \\ \frac{\partial K^*}{\partial W} \end{bmatrix} = \begin{bmatrix} 0 \\ -1 \\ 0 \end{bmatrix}$$

so that

$$\frac{\partial K^*}{\partial W} = \frac{\det\begin{bmatrix} 0 & -F_L & 0 \\ -F_L & -\lambda^* F_{LL} & -1 \\ -F_K & -\lambda^* F_{LK} & 0 \end{bmatrix}}{\det[H]} = -\frac{F_L F_K}{\det[H]} > 0$$

since $\det[H] < 0, F_L > 0, F_K > 0$.

To calculate $\frac{\partial K^*}{\partial R}, \frac{\partial L^*}{\partial R}$ we 1) set $dW = dQ = 0$ and 2) change the remaining $d's$ to $\partial's$ to obtain

$$\begin{aligned} 0\partial\lambda^* - F_L\partial L^* - F_K\partial K^* &= 0 \\ -F_L\partial\lambda^* - \lambda^* F_{LL}\partial L^* - \lambda^* F_{LK}\partial K^* &= 0 \\ -F_K\partial\lambda^* - \lambda^* F_{LK}\partial L^* - \lambda^* F_{KK}\partial K^* + \partial R &= 0 \end{aligned}$$

or in matrix notation

$$\underbrace{\begin{bmatrix} 0 & -F_L & -F_K \\ -F_L & -\lambda^* F_{LL} & -\lambda^* F_{LK} \\ -F_K & -\lambda^* F_{LK} & -\lambda^* F_{KK} \end{bmatrix}}_{H} \begin{bmatrix} \frac{\partial\lambda^*}{\partial R} \\ \frac{\partial L^*}{\partial R} \\ \frac{\partial K^*}{\partial R} \end{bmatrix} = \begin{bmatrix} 0 \\ 0 \\ -1 \end{bmatrix}.$$

Using Cramer's rule

$$\begin{aligned} \frac{\partial K^*}{\partial R} &= \frac{\det\begin{bmatrix} 0 & -F_L & 0 \\ -F_L & -\lambda^* F_{LL} & 0 \\ -F_K & -\lambda^* F_{LK} & -1 \end{bmatrix}}{\det[H]} \\ &= \frac{(F_L)^2}{\det[H]} < 0 \end{aligned}$$

$$\begin{aligned} \frac{\partial L^*}{\partial R} &= \frac{\det\begin{bmatrix} 0 & 0 & -F_K \\ -F_L & 0 & -\lambda^* F_{LK} \\ -F_K & -1 & -\lambda^* F_{KK} \end{bmatrix}}{\det[H]} \\ &= -\frac{F_L F_K}{\det[H]} > 0 \end{aligned}$$

since $\det[H] < 0, F_L > 0, F_K > 0$. Notice from above that $\frac{\partial L^*}{\partial R} = \frac{\partial K^*}{\partial W}$.

Rationality requires that $L^*(Q, W, R)$ is homogeneous of degree 0 in W and R so that by Euler's theorem

$$\frac{\partial L^*}{\partial W}W + \frac{\partial L^*}{\partial R}R = 0 \Longrightarrow \frac{\partial K^*}{\partial W} = \frac{\partial L^*}{\partial R} = -\frac{W}{R}\frac{\partial L^*}{\partial W}.$$

If

$$C^*(Q, W, R) \equiv WL^*(Q, W, R) + RK^*(Q, W, R)$$

then

$$\begin{aligned} \frac{\partial C^*(Q, W, R)}{\partial W} &= L^*(Q, W, R) + W\frac{\partial L^*(Q, W, R)}{\partial W} + R\frac{\partial K^*(Q, W, R)}{\partial W} \\ &= L^*(Q, W, R) + W\frac{\partial L^*(Q, W, R)}{\partial W} + R\left(-\frac{W}{R}\frac{\partial L^*(Q, W, R)}{\partial W}\right) \\ &= L^*(Q, W, R) + W\frac{\partial L^*(Q, W, R)}{\partial W} - W\frac{\partial L^*(Q, W, R)}{\partial W} = L^*(Q, W, R). \end{aligned}$$

To calculate $\frac{\partial\lambda^*}{\partial Q}$ we 1) set $dW = dR = 0$ and 2) change the remaining $d's$ to $\partial's$ to obtain

$$\begin{bmatrix} 0 & -F_L & -F_K \\ -F_L & -\lambda^* F_{LL} & -\lambda^* F_{LK} \\ -F_K & -\lambda^* F_{LK} & -\lambda^* F_{KK} \end{bmatrix} \begin{bmatrix} \partial\lambda^* \\ \partial L^* \\ \partial K^* \end{bmatrix} = \begin{bmatrix} -\partial Q \\ 0 \\ 0 \end{bmatrix} = \begin{bmatrix} -1 \\ 0 \\ 0 \end{bmatrix} \partial Q.$$

Dividing both sides by ∂Q we obtain

$$\underbrace{\begin{bmatrix} 0 & -F_L & -F_K \\ -F_L & -\lambda^* F_{LL} & -\lambda^* F_{LK} \\ -F_K & -\lambda^* F_{LK} & -\lambda^* F_{KK} \end{bmatrix}}_{=H} \begin{bmatrix} \frac{\partial\lambda^*}{\partial Q} \\ \frac{\partial L^*}{\partial Q} \\ \frac{\partial K^*}{\partial Q} \end{bmatrix} = \begin{bmatrix} -1 \\ 0 \\ 0 \end{bmatrix}$$

so that

$$\begin{aligned} \frac{\partial\lambda^*}{\partial Q} &= \frac{\det \begin{bmatrix} -1 & -F_L & -F_K \\ 0 & -\lambda^* F_{LL} & -\lambda^* F_{LK} \\ 0 & -\lambda^* F_{LK} & -\lambda^* F_{KK} \end{bmatrix}}{\det [H]} \\ &= -\frac{(\lambda^*)^2 \left(F_{LL}F_{KK} - F_{LK}^2\right)}{\det [H]} > 0 \end{aligned}$$

since $\det [H] < 0$ and

$$M_2 = \det \begin{bmatrix} F_{LL} & F_{LK} \\ F_{LK} & F_{KK} \end{bmatrix} = F_{LL}F_{KK} - F_{LK}^2 > 0$$

by the concavity of $F(L, K)$.

Problem 2.49 *Consider a firm minimizing costs $WL+RK$ given a production function $Q = F(L, K)$. [D***] Find the first-order conditions for cost minimizing L^*, K^*. Now consider the cost function*

$$C^*(Q, W, R) \equiv WL^* + RK^*.$$

*[A***] Prove that the Lagrange multiplier is marginal cost; that is $\frac{\partial C^*(Q,W,R)}{\partial Q} = \lambda^*$. Prove Shephard's lemma: $\frac{\partial C^*(Q,W,R)}{\partial W} = L^*$ and $\frac{\partial C^*(Q,W,R)}{\partial R} = K^*$.*

Answer: The Lagrangian is

$$\mathcal{L}(\lambda, L, K, Q, W, R) = WL + RK + \lambda (Q - F(L, K)).$$

The first-order conditions yield

$$\begin{aligned}
Q-F(L^*,K^*) &= 0 \\
&\Longrightarrow Q=F(L^*,K^*) \\
W-\lambda^*\frac{\partial F(L^*,K^*)}{\partial L} &= 0 \\
&\Longrightarrow W=\lambda^* F_L \Longrightarrow F_L=\frac{W}{\lambda^*} \\
R-\lambda^*\frac{\partial F(L^*,K^*)}{\partial K} &= 0 \\
&\Longrightarrow R=\lambda^* F_K \Longrightarrow F_K=\frac{R}{\lambda^*}
\end{aligned}$$

where $F_L \equiv \frac{\partial F(L^*,K^*)}{\partial L}$, $F_K \equiv \frac{\partial F(L^*,K^*)}{\partial K}$.

The cost function is

$$C^*(Q,W,R) \equiv WL^*(Q,W,R)+RK^*(Q,W,R)$$

and so

$$\frac{\partial C^*(Q,W,R)}{\partial Q}=W\frac{\partial L^*}{\partial Q}+R\frac{\partial K^*}{\partial Q}.$$

From $Q=F(L^*,K^*)$ and differentiating both sides with respect to Q we obtain

$$1=\frac{\partial}{\partial Q}Q=\frac{\partial}{\partial Q}F(L^*,K^*)=F_L\frac{\partial L^*}{\partial Q}+F_K\frac{\partial K^*}{\partial Q}.$$

But from the first-order conditions we saw that $F_L=\frac{W}{\lambda^*}$, $F_K=\frac{R}{\lambda^*}$ and so

$$\begin{aligned}
1 &= F_L\frac{\partial L^*}{\partial Q}+F_K\frac{\partial K^*}{\partial Q} \\
&= \left(\frac{W}{\lambda^*}\right)\frac{\partial L^*}{\partial Q}+\left(\frac{R}{\lambda^*}\right)\frac{\partial K^*}{\partial Q} \\
&= \frac{1}{\lambda^*}\left(W\frac{\partial L^*}{\partial Q}+R\frac{\partial K^*}{\partial Q}\right) \\
&\Longrightarrow W\frac{\partial L^*}{\partial Q}+R\frac{\partial K^*}{\partial Q}=\lambda^*.
\end{aligned}$$

Thus λ^* is marginal cost as

$$\frac{\partial C^*(Q,W,R)}{\partial Q}=W\frac{\partial L^*}{\partial Q}+R\frac{\partial K^*}{\partial Q}=\lambda^*.$$

Let's now evaluate $\frac{\partial C^*(Q,W,R)}{\partial W}$. We have

$$\frac{\partial C^*(Q,W,R)}{\partial W}=W\frac{\partial L^*(Q,W,R)}{\partial W}+R\frac{\partial K^*(Q,W,R)}{\partial W}+L^*.$$

Differentiating both sides of $Q = F(L^*, K^*)$ and using results from the first-order conditions $F_L = \frac{W}{\lambda^*}, F_K = \frac{R}{\lambda^*}$ we have

$$\begin{aligned} 0 &= \frac{\partial}{\partial W} Q = \frac{\partial}{\partial W} F(L^*, K^*) = F_L \frac{\partial L^*}{\partial W} + F_K \frac{\partial K^*}{\partial W} \\ &= \left(\frac{W}{\lambda^*}\right) \frac{\partial L^*}{\partial W} + \left(\frac{R}{\lambda^*}\right) \frac{\partial K^*}{\partial W} = \frac{1}{\lambda^*} \left(W \frac{\partial L^*}{\partial W} + R \frac{\partial K^*}{\partial W} \right) \\ &\Longrightarrow W \frac{\partial L^*}{\partial W} + R \frac{\partial K^*}{\partial W} = 0 \end{aligned}$$

and so

$$\frac{\partial C^*(Q, W, R)}{\partial W} = W \frac{\partial L^*(Q, W, R)}{\partial W} + R \frac{\partial K^*(Q, W, R)}{\partial W} + L^* = L^*.$$

The same argument applies to $\frac{\partial C^*(Q,W,R)}{\partial R}$. We have

$$\frac{\partial C^*(Q, W, R)}{\partial R} = W \frac{\partial L^*(Q, W, R)}{\partial R} + R \frac{\partial K^*(Q, W, R)}{\partial R} + K^*.$$

Differentiating both sides of $Q = F(L^*, K^*)$ and using results from the first-order conditions $F_L = \frac{W}{\lambda^*}, F_K = \frac{R}{\lambda^*}$ we have

$$\begin{aligned} 0 &= \frac{\partial}{\partial R} Q = \frac{\partial}{\partial R} F(L^*, K^*) = F_L \frac{\partial L^*}{\partial R} + F_K \frac{\partial K^*}{\partial R} \\ &= \left(\frac{W}{\lambda^*}\right) \frac{\partial L^*}{\partial R} + \left(\frac{R}{\lambda^*}\right) \frac{\partial K^*}{\partial R} \\ &= \frac{1}{\lambda^*} \left(W \frac{\partial L^*}{\partial R} + R \frac{\partial K^*}{\partial R} \right) \\ &\Longrightarrow W \frac{\partial L^*}{\partial W} + R \frac{\partial K^*}{\partial W} = 0 \end{aligned}$$

and so

$$\frac{\partial C^*(Q, W, R)}{\partial R} = W \frac{\partial L^*(Q, W, R)}{\partial R} + R \frac{\partial K^*(Q, W, R)}{\partial R} + K^* = K^*.$$

Chapter 3

The Envelope Theorem

3.1 Profit Maximization

Problem 3.1 *[D***] Suppose (naive) real profits are given by*

$$\pi_R(L, w) = f(L) - wL$$

*with Cobb-Douglas production function $Q = f(L) = 3L^{\frac{1}{3}}$. Find the clever profit function $\pi_R^*(w)$. Show that $\frac{d\pi_R^*(w)}{dw} = -L^*(w)$. [C***] Show that $\pi_R^*(w)$ is convex, and that its convexity implies that labour demand slopes downwards.*

Answer: The first-order condition yields

$$\frac{\partial \pi_R(L^*, w)}{\partial L} = 0 \Longrightarrow (L^*)^{-\frac{2}{3}} - w = 0 \Longrightarrow L^* = w^{-\frac{3}{2}}.$$

We then have $\pi_R^*(w) = 2w^{-\frac{1}{2}}$ as

$$\pi_R^*(w) = 3(L^*)^{\frac{1}{3}} - wL^* = 3\left(w^{-\frac{3}{2}}\right)^{\frac{1}{3}} - w \times w^{-\frac{3}{2}} = 3w^{-\frac{1}{2}} - w^{-\frac{1}{2}} = 2w^{-\frac{1}{2}}.$$

It follows then that

$$\frac{d\pi_R^*(w)}{dw} = \frac{d}{dw}\left(2w^{-\frac{1}{2}}\right) = -w^{-\frac{3}{2}} = -L^*(w).$$

Furthermore $\pi_R^*(w) = 2w^{-\frac{1}{2}}$ is convex as

$$\frac{d^2\pi_R^*(w)}{dw^2} = \frac{d}{dw}\left(-w^{-\frac{3}{2}}\right) = \frac{3}{2}w^{-\frac{5}{2}} > 0.$$

Since $\frac{d\pi_R^*(w)}{dw} = -L^*(w)$ and $\frac{d^2\pi_R^*(w)}{dw^2} > 0$ we have

$$\frac{d^2\pi_R^*(w)}{dw^2} > 0 \Longrightarrow -\frac{dL^*(w)}{dw} > 0 \Longrightarrow \frac{dL^*(w)}{dw} < 0$$

and so labour demand slopes downwards.

Problem 3.2 *[C***] Suppose (naive) real profits are given by*

$$\pi_R(L,w) = f(L) - wL$$

with Cobb-Douglas production function $Q = \frac{1}{\alpha}L^{\alpha}$ *with* $0 < \alpha < 1$. *Find the clever profit function* $\pi_R^*(w)$ *and show that* $\pi_R^*(w) > 0$. *Show that* $\frac{d\pi_R^*(w)}{dw} = -L^*(w)$. *Show that* $\pi_R^*(w)$ *is convex, and that its convexity implies that labour demand slopes downwards.*

Answer: The first-order condition yields

$$\frac{\partial \pi_R(L^*,w)}{\partial L} = 0 \Longrightarrow (L^*)^{\alpha-1} - w = 0 \Longrightarrow L^* = w^{-\frac{1}{1-\alpha}}.$$

We then have $\pi_R^*(w) = \left(\frac{1-\alpha}{\alpha}\right) w^{-\frac{\alpha}{1-\alpha}}$ as

$$\begin{aligned}
\pi_R^*(w) &= \frac{1}{\alpha}(L^*)^{\alpha} - wL^* = \frac{1}{\alpha}\left(w^{-\frac{1}{1-\alpha}}\right)^{\alpha} - w \times w^{-\frac{1}{1-\alpha}} \\
&= \frac{1}{\alpha}w^{-\frac{\alpha}{1-\alpha}} - w^{-\frac{1}{1-\alpha}+1} = \frac{1}{\alpha}w^{-\frac{\alpha}{1-\alpha}} - w^{-\frac{1}{1-\alpha}+\frac{1-\alpha}{1-\alpha}} \\
&= \frac{1}{\alpha}w^{-\frac{\alpha}{1-\alpha}} - w^{-\frac{\alpha}{1-\alpha}} = \left(\frac{1}{\alpha} - 1\right) w^{-\frac{\alpha}{1-\alpha}} \\
&= \left(\frac{1-\alpha}{\alpha}\right) w^{-\frac{\alpha}{1-\alpha}}.
\end{aligned}$$

We have positive profits as

$$\begin{aligned}
0 &< \alpha < 1 \\
&\Longrightarrow \frac{1-\alpha}{\alpha} > 0 \\
&\Longrightarrow \pi_R^*(w) = \left(\frac{1-\alpha}{\alpha}\right) w^{-\frac{\alpha}{1-\alpha}} > 0.
\end{aligned}$$

Now

$$\frac{d\pi_R^*(w)}{dw} = \frac{d}{dw}\left(\left(\frac{1-\alpha}{\alpha}\right) w^{-\frac{\alpha}{1-\alpha}}\right) = -w^{-\frac{\alpha}{1-\alpha}-1} = -w^{-\frac{1}{1-\alpha}} = -L^*(w).$$

Furthermore $\pi_R^*(w) = \left(\frac{1-\alpha}{\alpha}\right) w^{-\frac{\alpha}{1-\alpha}}$ is convex as

$$\frac{d^2\pi_R^*(w)}{dw^2} = \frac{d}{dw}\left(-w^{-\frac{1}{1-\alpha}}\right) = \frac{1}{1-\alpha}w^{-\frac{1}{1-\alpha}-1} > 0$$

since $0 < \alpha < 1$. Since $\frac{d\pi_R^*(w)}{dw} = -L^*(w)$ and $\frac{d^2\pi_R^*(w)}{dw^2} > 0$ we have

$$\frac{d^2\pi_R^*(w)}{dw^2} > 0 \Longrightarrow -\frac{dL^*(w)}{dw} > 0 \Longrightarrow \frac{dL^*(w)}{dw} < 0$$

and so labour demand slopes downwards.

Problem 3.3 *Suppose real profits $\pi_R \equiv Q - wL - rK$ are given by*

$$\pi_R(L, K, w, r) = 3L^{\frac{1}{3}}K^{\frac{1}{3}} - wL - rK$$

with Cobb-Douglas production function $Q = 3L^{\frac{1}{3}}K^{\frac{1}{3}}$. Find the first-order conditions for maximizing real profits and the profit maximizing levels of labour and capital L^*, K^*. Solve for L^*, K^*. Verify that $\frac{\partial L^*}{\partial w} < 0$ and $\frac{\partial K^*}{\partial r} < 0$. Verify that $\frac{\partial L^*}{\partial r} = \frac{\partial K^*}{\partial w}$. Now show the same results using total differentials.

Answer: The first-order conditions yield

$$\begin{aligned}
(L^*)^{-\frac{2}{3}}(K^*)^{\frac{1}{3}} - w &= 0 \\
&\Longrightarrow (L^*)^{-\frac{2}{3}}(K^*)^{\frac{1}{3}} = w \\
(L^*)^{\frac{1}{3}}(K^*)^{-\frac{2}{3}} - r &= 0 \\
&\Longrightarrow (L^*)^{\frac{1}{3}}(K^*)^{-\frac{2}{3}} = r.
\end{aligned}$$

Dividing we have

$$\begin{aligned}
\frac{(L^*)^{-\frac{2}{3}}(K^*)^{\frac{1}{3}}}{(L^*)^{\frac{1}{3}}(K^*)^{-\frac{2}{3}}} &= \frac{w}{r} \\
&\Longrightarrow \frac{K^*}{L^*} = wr^{-1} \\
&\Longrightarrow K^* = wr^{-1}L^*
\end{aligned}$$

so that

$$\begin{aligned}
(L^*)^{-\frac{2}{3}}(K^*)^{\frac{1}{3}} &= w \\
&\Longrightarrow (L^*)^{-\frac{2}{3}}\left(wr^{-1}L^*\right)^{\frac{1}{3}} = w \\
&\Longrightarrow (L^*)^{-\frac{1}{3}}w^{\frac{1}{3}}r^{-\frac{1}{3}} \\
&\Longrightarrow L^* = w^{-2}r^{-1}
\end{aligned}$$

and

$$\begin{aligned}
K^* &= wr^{-1}L^* \\
&\Longrightarrow K^* = wr^{-1} \times \left(w^{-2}r^{-1}\right) = w^{-1}r^{-2}.
\end{aligned}$$

Thus

$$\frac{\partial L^*(w,r)}{\partial w} = -2w^{-3}r^{-1} < 0 \text{ and } \frac{\partial K^*(w,r)}{\partial r} = -2w^{-1}r^{-3} < 0.$$

So we have $\frac{\partial L^*}{\partial r} = \frac{\partial K^*}{\partial w}$ as

$$\frac{\partial L^*(w,r)}{\partial r} = -w^{-2}r^{-2} = \frac{\partial K^*(w,r)}{\partial w}.$$

Now instead let's use total differentials. The first-order conditions yield two implicit functions

$$\begin{aligned} g_1(L^*, K^*, w, r) &= (L^*)^{-\frac{2}{3}} (K^*)^{\frac{1}{3}} - w = 0 \\ g_2(L^*, K^*, w, r) &= (L^*)^{\frac{1}{3}} (K^*)^{-\frac{2}{3}} - r = 0 \end{aligned}$$

with total differential

$$\begin{aligned} -\frac{2}{3}(L^*)^{-\frac{5}{3}} (K^*)^{\frac{1}{3}} dL^* + \frac{1}{3}(L^*)^{-\frac{2}{3}} (K^*)^{-\frac{2}{3}} dK^* - dw &= 0 \\ \frac{1}{3}(L^*)^{-\frac{2}{3}} (K^*)^{-\frac{2}{3}} dL^* - \frac{2}{3}(L^*)^{\frac{1}{3}} (K^*)^{-\frac{5}{3}} dK^* - dr &= 0. \end{aligned}$$

To find $\frac{\partial L^*}{\partial w}, \frac{\partial K^*}{\partial w}$ set $dr = 0$ and the remaining $d's$ to $\partial's$ as

$$\begin{aligned} -\frac{2}{3}(L^*)^{-\frac{5}{3}} (K^*)^{\frac{1}{3}} \partial L^* + \frac{1}{3}(L^*)^{-\frac{2}{3}} (K^*)^{-\frac{2}{3}} \partial K^* - \partial w &= 0 \\ \frac{1}{3}(L^*)^{-\frac{2}{3}} (K^*)^{-\frac{2}{3}} \partial L^* - \frac{2}{3}(L^*)^{\frac{1}{3}} (K^*)^{-\frac{5}{3}} \partial K^* &= 0 \end{aligned}$$

and write this in matrix notation as

$$\begin{bmatrix} -\frac{2}{3}(L^*)^{-\frac{5}{3}} (K^*)^{\frac{1}{3}} & \frac{1}{3}(L^*)^{-\frac{2}{3}} (K^*)^{-\frac{2}{3}} \\ \frac{1}{3}(L^*)^{-\frac{2}{3}} (K^*)^{-\frac{2}{3}} & -\frac{2}{3}(L^*)^{\frac{1}{3}} (K^*)^{-\frac{5}{3}} \end{bmatrix} \begin{bmatrix} \frac{\partial L^*}{\partial w} \\ \frac{\partial K^*}{\partial w} \end{bmatrix} = \begin{bmatrix} 1 \\ 0 \end{bmatrix}.$$

Here

$$\begin{aligned} \det[H] &= \det \begin{bmatrix} -\frac{2}{3}(L^*)^{-\frac{5}{3}} (K^*)^{\frac{1}{3}} & \frac{1}{3}(L^*)^{-\frac{2}{3}} (K^*)^{-\frac{2}{3}} \\ \frac{1}{3}(L^*)^{-\frac{2}{3}} (K^*)^{-\frac{2}{3}} & -\frac{2}{3}(L^*)^{\frac{1}{3}} (K^*)^{-\frac{5}{3}} \end{bmatrix} \\ &= \frac{4}{9}(L^*)^{-\frac{4}{3}} (K^*)^{-\frac{4}{3}} - \frac{1}{9}(L^*)^{-\frac{4}{3}} (K^*)^{-\frac{4}{3}} = \frac{3}{9}(L^*)^{-\frac{4}{3}} (K^*)^{-\frac{4}{3}} > 0. \end{aligned}$$

Thus using Cramer's rule

$$\begin{aligned} \frac{\partial L^*(w,r)}{\partial w} &= \frac{\det \begin{bmatrix} 1 & \frac{1}{3}(L^*)^{-\frac{2}{3}} (K^*)^{-\frac{2}{3}} \\ 0 & -\frac{2}{3}(L^*)^{\frac{1}{3}} (K^*)^{-\frac{5}{3}} \end{bmatrix}}{\det[H]} \\ &= \frac{-\frac{2}{3}(L^*)^{\frac{1}{3}} (K^*)^{-\frac{5}{3}}}{\det[H]} < 0 \\ \frac{\partial K^*(w,r)}{\partial w} &= \frac{\det \begin{bmatrix} -\frac{2}{3}(L^*)^{-\frac{5}{3}} (K^*)^{\frac{1}{3}} & 1 \\ \frac{1}{3}(L^*)^{-\frac{2}{3}} (K^*)^{-\frac{2}{3}} & 0 \end{bmatrix}}{\det[H]} \\ &= \frac{-\frac{1}{3}(L^*)^{-\frac{2}{3}} (K^*)^{-\frac{2}{3}}}{\det[H]} < 0. \end{aligned}$$

To find $\frac{\partial L^*}{\partial r}, \frac{\partial K^*}{\partial r}$ we 1) set $dw = 0$ and 2) change the remaining $d's$ to $\partial's$ as

$$\begin{aligned} -\frac{2}{3}(L^*)^{-\frac{5}{3}}(K^*)^{\frac{1}{3}}\partial L^* + \frac{1}{3}(L^*)^{-\frac{2}{3}}(K^*)^{-\frac{2}{3}}\partial K^* &= 0 \\ \frac{1}{3}(L^*)^{-\frac{2}{3}}(K^*)^{-\frac{2}{3}}\partial L^* - \frac{2}{3}(L^*)^{\frac{1}{3}}(K^*)^{-\frac{5}{3}}\partial K^* - \partial r &= 0. \end{aligned}$$

Writing this in matrix notation we obtain

$$\begin{bmatrix} -\frac{2}{3}(L^*)^{-\frac{5}{3}}(K^*)^{\frac{1}{3}} & \frac{1}{3}(L^*)^{-\frac{2}{3}}(K^*)^{-\frac{2}{3}} \\ \frac{1}{3}(L^*)^{-\frac{2}{3}}(K^*)^{-\frac{2}{3}} & -\frac{2}{3}(L^*)^{\frac{1}{3}}(K^*)^{-\frac{5}{3}} \end{bmatrix} \begin{bmatrix} \frac{\partial L^*}{\partial w} \\ \frac{\partial K^*}{\partial w} \end{bmatrix} = \begin{bmatrix} 0 \\ 1 \end{bmatrix}.$$

Thus by Cramer's rule

$$\begin{aligned} \frac{\partial L^*(w,r)}{\partial r} &= \frac{\det\begin{bmatrix} 0 & \frac{1}{3}(L^*)^{-\frac{2}{3}}(K^*)^{-\frac{2}{3}} \\ 1 & -\frac{2}{3}(L^*)^{\frac{1}{3}}(K^*)^{-\frac{5}{3}} \end{bmatrix}}{\det[H]} \\ &= \frac{-\frac{1}{3}(L^*)^{-\frac{2}{3}}(K^*)^{-\frac{2}{3}}}{\det[H]} < 0 \\ \frac{\partial K^*(w,r)}{\partial r} &= \frac{\det\begin{bmatrix} -\frac{2}{3}(L^*)^{-\frac{5}{3}}(K^*)^{\frac{1}{3}} & 0 \\ \frac{1}{3}(L^*)^{-\frac{2}{3}}(K^*)^{-\frac{2}{3}} & 1 \end{bmatrix}}{\det[H]} \\ &= \frac{-\frac{2}{3}(L^*)^{-\frac{5}{3}}(K^*)^{\frac{1}{3}}}{\det[H]} < 0. \end{aligned}$$

So we have $\frac{\partial L^*}{\partial r} = \frac{\partial K^*}{\partial w}$ as

$$\frac{\partial L^*(w,r)}{\partial r} = \frac{-\frac{1}{3}(L^*)^{-\frac{2}{3}}(K^*)^{-\frac{2}{3}}}{\det[H]} = \frac{\partial K^*(w,r)}{\partial w}.$$

Problem 3.4 *[D***] Suppose (naive) real profits are given by*

$$\pi_R(L,K,w,r) = F(L,K) - wL - rK$$

with Cobb-Douglas production function

$$Q = F(L,K) = 3L^{\frac{1}{3}}K^{\frac{1}{3}}.$$

Find the clever profit function

$$\pi_R^*(w,r) = \pi_R(L^*(w,r), K^*(w,r), w, r).$$

Show that

$$\frac{\partial \pi_R^*(w,r)}{\partial w} = -L^*(w,r) \text{ and } \frac{\partial \pi_R^*(w,r)}{\partial r} = -K^*(w,r).$$

Show that $\pi_R^*(w,r)$ is convex.

Answer: From Problem 2.31 on page 63 it was shown that

$$L^*(w,r) = w^{-2}r^{-1} \text{ and } K^*(w,r) = w^{-1}r^{-2}.$$

We then have $\pi_R^*(w,r) = w^{-1}r^{-1}$ as

$$\begin{aligned}
\pi_R^*(w,r) &= 3\left(L^*\right)^{\frac{1}{3}}\left(K^*\right)^{\frac{1}{3}} - wL^* - rK^* \\
&= 3\left(w^{-2}r^{-1}\right)^{\frac{1}{3}}\left(w^{-1}r^{-2}\right)^{\frac{1}{3}} - ww^{-2}r^{-1} - rw^{-1}r^{-2} \\
&= 3w^{-1}r^{-1} - w^{-1}r^{-1} - w^{-1}r^{-1} = w^{-1}r^{-1}.
\end{aligned}$$

It follows then that

$$\begin{aligned}
\frac{\partial \pi_R^*(w,r)}{\partial w} &= -w^{-2}r^{-1} = -L^*(w,r) \\
\frac{\partial \pi_R^*(w,r)}{\partial r} &= -w^{-1}r^{-2} = -K^*(w,r).
\end{aligned}$$

The Hessian of $\pi_R^*(w,r) = w^{-1}r^{-1}$ is

$$H(w,r) = \begin{bmatrix} \frac{\partial^2 \pi_R^*(w,r)}{\partial w^2} & \frac{\partial^2 \pi_R^*(w,r)}{\partial r \partial w} \\ \frac{\partial^2 \pi_R^*(w,r)}{\partial r \partial w} & \frac{\partial^2 \pi_R^*(w,r)}{\partial r^2} \end{bmatrix} = \begin{bmatrix} 2w^{-3}r^{-1} & w^{-2}r^{-2} \\ w^{-2}r^{-2} & 2w^{-1}r^{-3} \end{bmatrix}$$

which is positive definite for all w, r since

$$\begin{aligned}
M_1 &= 2w^{-3}r^{-1} > 0 \\
M_2 &= \det\begin{bmatrix} 2w^{-3}r^{-1} & w^{-2}r^{-2} \\ w^{-2}r^{-2} & 2w^{-1}r^{-3} \end{bmatrix} = 4w^{-4}r^{-4} - w^{-4}r^{-4} = 3w^{-4}r^{-4} > 0.
\end{aligned}$$

Problem 3.5 *Suppose naive real profits are given by*

$$\pi_R(L,K,w,r) = F(L,K) - wL - rK$$

with Cobb-Douglas production function

$$Q = F(L,K) = \frac{1}{\alpha}L^{\alpha}K^{\alpha}$$

where $\alpha < \frac{1}{2}$. Find the clever profit function

$$\pi_R^*(w,r) = \pi_R(L^*(w,r), K^*(w,r), w, r).$$

Show that

$$\frac{\partial \pi_R^*(w,r)}{\partial w} = -L^*(w,r) \text{ and } \frac{\partial \pi_R^*(w,r)}{\partial r} = -K^*(w,r).$$

Show that $\pi_R^*(w,r)$ is convex.

Answer: From Problem 2.32 on page 64 it was shown that

$$L^*(w,r) = w^{-\frac{1-\alpha}{1-2\alpha}} r^{-\frac{\alpha}{1-2\alpha}} \text{ and } K^*(w,r) = w^{-\frac{\alpha}{1-2\alpha}} r^{-\frac{1-\alpha}{1-2\alpha}}.$$

The clever profit function then is

$$\begin{aligned}
\pi_R^*(w,r) &= \frac{1}{\alpha}(L^*)^\alpha (K^*)^\alpha - wL^* - rK^* \\
&= \frac{1}{\alpha}\left(w^{-\frac{1-\alpha}{1-2\alpha}} r^{-\frac{\alpha}{1-2\alpha}}\right)^\alpha \left(w^{-\frac{\alpha}{1-2\alpha}} r^{-\frac{1-\alpha}{1-2\alpha}}\right)^\alpha \\
&\quad - ww^{-\frac{1-\alpha}{1-2\alpha}} r^{-\frac{\alpha}{1-2\alpha}} - rw^{-\frac{\alpha}{1-2\alpha}} r^{-\frac{1-\alpha}{1-2\alpha}} \\
&= \frac{1}{\alpha} w^{-\frac{(1-\alpha)\alpha}{1-2\alpha}} r^{-\frac{\alpha^2}{1-2\alpha}} w^{-\frac{\alpha^2}{1-2\alpha}} r^{-\frac{(1-\alpha)\alpha}{1-2\alpha}} \\
&\quad - w^{-\frac{1-\alpha}{1-2\alpha}+1} r^{-\frac{\alpha}{1-2\alpha}} - w^{-\frac{\alpha}{1-2\alpha}} r^{-\frac{1-\alpha}{1-2\alpha}+1} \\
&= \frac{1}{\alpha} w^{-\frac{(1-\alpha)\alpha+\alpha^2}{1-2\alpha}} r^{\frac{(1-\alpha)\alpha+\alpha^2}{1-2\alpha}} - w^{-\frac{1-\alpha}{1-2\alpha}+\frac{1-2\alpha}{1-2\alpha}} r^{-\frac{\alpha}{1-2\alpha}} - w^{-\frac{\alpha}{1-2\alpha}} r^{-\frac{1-\alpha}{1-2\alpha}+\frac{1-2\alpha}{1-2\alpha}} \\
&= \frac{1}{\alpha} w^{-\frac{\alpha}{1-2\alpha}} r^{-\frac{\alpha}{1-2\alpha}} - w^{-\frac{\alpha}{1-2\alpha}} r^{-\frac{\alpha}{1-2\alpha}} - w^{-\frac{\alpha}{1-2\alpha}} r^{-\frac{\alpha}{1-2\alpha}} \\
&= \left(\frac{1}{\alpha} - 2\right) w^{-\frac{\alpha}{1-2\alpha}} r^{-\frac{\alpha}{1-2\alpha}} = \frac{1-2\alpha}{\alpha} w^{-\frac{\alpha}{1-2\alpha}} r^{-\frac{\alpha}{1-2\alpha}}.
\end{aligned}$$

Now

$$\begin{aligned}
\frac{\partial \pi_R^*(w,r)}{\partial w} &= -w^{-\frac{\alpha}{1-2\alpha}-1} r^{-\frac{\alpha}{1-2\alpha}} = -w^{-\frac{1-\alpha}{1-2\alpha}} r^{-\frac{\alpha}{1-2\alpha}} = -L^*(w,r) \\
\frac{\partial \pi_R^*(w,r)}{\partial r} &= -w^{-\frac{\alpha}{1-2\alpha}} r^{-\frac{\alpha}{1-2\alpha}-1} = -w^{-\frac{\alpha}{1-2\alpha}} r^{-\frac{1-\alpha}{1-2\alpha}} = -K^*(w,r).
\end{aligned}$$

The Hessian of $\pi_R^*(w,r)$ is

$$\begin{aligned}
\tilde{H}(w,r) &= \begin{bmatrix} \frac{\partial^2 \pi_R^*(w,r)}{\partial w^2} & \frac{\partial^2 \pi_R^*(w,r)}{\partial r \partial w} \\ \frac{\partial^2 \pi_R^*(w,r)}{\partial r \partial w} & \frac{\partial^2 \pi_R^*(w,r)}{\partial r^2} \end{bmatrix} \\
&= \begin{bmatrix} \frac{1-\alpha}{1-2\alpha} w^{-\frac{1-\alpha}{1-2\alpha}-1} r^{-\frac{\alpha}{1-2\alpha}} & \frac{\alpha}{1-2\alpha} w^{-\frac{1-\alpha}{1-2\alpha}} r^{-\frac{1-\alpha}{1-2\alpha}} \\ \frac{\alpha}{1-2\alpha} w^{-\frac{1-\alpha}{1-2\alpha}} r^{-\frac{1-\alpha}{1-2\alpha}} & \frac{1-\alpha}{1-2\alpha} w^{-\frac{\alpha}{1-2\alpha}} r^{-\frac{1-\alpha}{1-2\alpha}-1} \end{bmatrix}
\end{aligned}$$

which is positive definite for all w, r since

$$\begin{aligned}
M_1 &= \frac{1-\alpha}{1-2\alpha} w^{-\frac{1-\alpha}{1-2\alpha}-1} r^{-\frac{\alpha}{1-2\alpha}} > 0 \\
M_2 &= \det \begin{bmatrix} \frac{1-\alpha}{1-2\alpha} w^{-\frac{1-\alpha}{1-2\alpha}-1} r^{-\frac{\alpha}{1-2\alpha}} & \frac{\alpha}{1-2\alpha} w^{-\frac{1-\alpha}{1-2\alpha}} r^{-\frac{1-\alpha}{1-2\alpha}} \\ \frac{\alpha}{1-2\alpha} w^{-\frac{1-\alpha}{1-2\alpha}} r^{-\frac{1-\alpha}{1-2\alpha}} & \frac{1-\alpha}{1-2\alpha} w^{-\frac{\alpha}{1-2\alpha}} r^{-\frac{1-\alpha}{1-2\alpha}-1} \end{bmatrix} \\
&= \left(\frac{1-\alpha}{1-2\alpha}\right)^2 w^{-\frac{1-\alpha}{1-2\alpha}-1-\frac{\alpha}{1-2\alpha}} r^{-\frac{1-\alpha}{1-2\alpha}-1-\frac{\alpha}{1-2\alpha}} \\
&\quad -\left(\frac{\alpha}{1-2\alpha}\right)^2 \left(w^{-\frac{1-\alpha}{1-2\alpha}} r^{-\frac{1-\alpha}{1-2\alpha}}\right)^2 \\
&= \left(\frac{1-\alpha}{1-2\alpha}\right)^2 w^{-\frac{1-\alpha-(1-2\alpha)-\alpha}{1-2\alpha}} r^{-\frac{1-\alpha-(1-2\alpha)-\alpha}{1-2\alpha}} \\
&\quad -\left(\frac{\alpha}{1-2\alpha}\right)^2 \left(w^{-\frac{1-\alpha}{1-2\alpha}} r^{-\frac{1-\alpha}{1-2\alpha}}\right)^2 \\
&= \left(\frac{1-\alpha}{1-2\alpha}\right)^2 w^{-2\left(\frac{1-\alpha}{1-2\alpha}\right)} r^{-2\left(\frac{1-\alpha}{1-2\alpha}\right)} \\
&\quad -\left(\frac{\alpha}{1-2\alpha}\right)^2 \left(w^{-\frac{1-\alpha}{1-2\alpha}} r^{-\frac{1-\alpha}{1-2\alpha}}\right)^2 \\
&= \left(\frac{1-\alpha}{1-2\alpha}\right)^2 \left(w^{-\frac{1-\alpha}{1-2\alpha}} r^{-\frac{1-\alpha}{1-2\alpha}}\right)^2 - \left(\frac{\alpha}{1-2\alpha}\right)^2 \left(w^{-\frac{1-\alpha}{1-2\alpha}} r^{-\frac{1-\alpha}{1-2\alpha}}\right)^2 \\
&= \left(\left(\frac{1-\alpha}{1-2\alpha}\right)^2 - \left(\frac{\alpha}{1-2\alpha}\right)^2\right) \left(w^{-\frac{1-\alpha}{1-2\alpha}} r^{-\frac{1-\alpha}{1-2\alpha}}\right)^2 \\
&= \frac{1}{1-2\alpha} \left(w^{-\frac{1-\alpha}{1-2\alpha}} r^{-\frac{1-\alpha}{1-2\alpha}}\right)^2 > 0
\end{aligned}$$

where both inequalities use $0 < \alpha < \frac{1}{2}$.

Problem 3.6 *Suppose (naive) real profits are given by*

$$\pi_R (L, K, w, r) = F (L, K) - wL - rK$$

*with concave production function $Q = F(L, K)$. [D***] Using the envelope theorem, show for the clever real profit function*

$$\pi_R^*(w, r) \equiv \pi_R(L^*(w, r), K^*(w, r), w, r)$$

that

$$\frac{\partial \pi_R^*(w, r)}{\partial w} = -L^*(w, r) \text{ and } \frac{\partial \pi_R^*(w, r)}{\partial r} = -K^*(w, r).$$

*[A***] Using results from total differentials, show that $\pi_R^*(w, r)$ is convex. Show that*

$$Q^* (w, r) \equiv F (L^* (w, r), K^* (w, r))$$

can be found from $\pi_R^*(w,r)$ *as*

$$\begin{aligned} Q^*(w,r) &= \pi_R^*(w,r) - \frac{\partial \pi_R^*(w,r)}{\partial w} w - \frac{\partial \pi_R^*(w,r)}{\partial r} r \\ &= \pi_R^*(w,r) \times (1 - \eta_w - \eta_r) \end{aligned}$$

where η_w, η_r *are the corresponding elasticities of* $\pi_R^*(w,r)$ *with respect to* w *and* r. *Show the the firm's supply curve* $Q^*(P,W,R)$ *expressed in terms of nominal prices* P,W,R *can be derived from* $\pi_R^*(w,r)$. *Use the convexity of* $\pi_R^*(w,r)$ *to prove that the supply curve slopes upwards.*

Answer: Using the envelope theorem we have

$$\begin{aligned} \frac{\partial \pi_R^*(w,r)}{\partial w} &= \left(\frac{\partial \pi_R(L,K,w,r)}{\partial w}\right)\Big|_{\substack{L=L^* \\ K=K^*}} \\ &= \frac{\partial}{\partial w}\left(F(L,K) - wL - rK\right)\Big|_{\substack{L=L^* \\ K=K^*}} \\ &= \frac{\partial}{\partial w}(-L)\Big|_{\substack{L=L^* \\ K=K^*}} = -L^* \\ \frac{\partial \pi_R^*(w,r)}{\partial r} &= \left(\frac{\partial \pi_R(L,K,w,r)}{\partial r}\right)\Big|_{\substack{L=L^* \\ K=K^*}} \\ &= \frac{\partial}{\partial r}\left(F(L,K) - wL - rK\right)\Big|_{\substack{L=L^* \\ K=K^*}} \\ &= \frac{\partial}{\partial r}(-K)\Big|_{\substack{L=L^* \\ K=K^*}} = -K^*. \end{aligned}$$

The Hessian of $\pi_R^*(w,r)$ is

$$\begin{aligned} \tilde{H}(w,r) &= \begin{bmatrix} \frac{\partial^2 \pi_R^*(w,r)}{\partial w^2} & \frac{\partial^2 \pi_R^*(w,r)}{\partial r \partial w} \\ \frac{\partial^2 \pi_R^*(w,r)}{\partial r \partial w} & \frac{\partial^2 \pi_R^*(w,r)}{\partial r^2} \end{bmatrix} \\ &= \begin{bmatrix} -\frac{\partial L^*(w,r)}{\partial w} & -\frac{\partial L^*(w,r)}{\partial r} \\ -\frac{\partial K^*(w,r)}{\partial w} & -\frac{\partial K^*(w,r)}{\partial r} \end{bmatrix}. \end{aligned}$$

In Problem 2.5 on page 66 it was shown that

$$\frac{\partial L^*(w,r)}{\partial w} = \frac{F_{KK}}{\det[H]}, \frac{\partial K^*(w,r)}{\partial r} = \frac{F_{LL}}{\det[H]}, \frac{\partial L^*(w,r)}{\partial r} = \frac{\partial K^*(w,r)}{\partial w} = -\frac{F_{LK}}{\det[H]}$$

where

$$H = \begin{bmatrix} F_{LL} & F_{LK} \\ F_{LK} & F_{KK} \end{bmatrix}$$

is negative definite since the production function is concave. Thus

$$\begin{aligned} \tilde{H}(w,r) &= \begin{bmatrix} -\frac{F_{KK}}{\det[H]} & \frac{F_{LK}}{\det[H]} \\ \frac{F_{LK}}{\det[H]} & -\frac{F_{LL}}{\det[H]} \end{bmatrix} = -\frac{1}{\det[H]}\begin{bmatrix} F_{KK} & -F_{LK} \\ -F_{LK} & F_{LL} \end{bmatrix} \\ &= -\begin{bmatrix} F_{LL} & F_{LK} \\ F_{LK} & F_{KK} \end{bmatrix}^{-1} = -H^{-1}. \end{aligned}$$

Since H is negative definite and the inverse of a negative definite matrix is also negative definite, it follows that $\tilde{H}(w,r) = -H^{-1}$ is positive definite, and so $\pi_R^*(w,r)$ is convex.

To find $Q^*(w,r)$ from $\pi_R^*(w,r)$ use

$$\begin{aligned}
\pi_R^*(w,r) &= F(L^*(w,r), K^*(w,r)) - wL^*(w,r) - rK^*(w,r) \\
&= Q^*(w,r) - w\left(-\frac{\partial \pi_R^*(w,r)}{\partial w}\right) - r\left(-\frac{\partial \pi_R^*(w,r)}{\partial r}\right) \\
&= Q^*(w,r) + \frac{\partial \pi_R^*(w,r)}{\partial w} w + \frac{\partial \pi_R^*(w,r)}{\partial r} r
\end{aligned}$$

so that solving for $Q^*(w,r)$ yields

$$\begin{aligned}
Q^*(w,r) &= \pi_R^*(w,r) - \frac{\partial \pi_R^*(w,r)}{\partial w} w - \frac{\partial \pi_R^*(w,r)}{\partial r} r \\
&= \pi_R^*(w,r)\left(1 - \frac{\partial \pi_R^*(w,r)}{\partial w}\frac{w}{\pi_R^*(w,r)} - \frac{\partial \pi_R^*(w,r)}{\partial r}\frac{r}{\pi_R^*(w,r)}\right) \\
&= \pi_R^*(w,r)(1 - \eta_w - \eta_r).
\end{aligned}$$

Then

$$\begin{aligned}
\frac{\partial Q^*(w,r)}{\partial w} &= \frac{\partial \pi_R^*(w,r)}{\partial w} - \frac{\partial \pi_R^*(w,r)}{\partial w} - \frac{\partial^2 \pi_R^*(w,r)}{\partial w^2} w - \frac{\partial \pi_R^*(w,r)}{\partial r \partial w} r \\
&= -\frac{\partial^2 \pi_R^*(w,r)}{\partial w^2} w - \frac{\partial \pi_R^*(w,r)}{\partial r \partial w} r
\end{aligned}$$

and

$$\begin{aligned}
\frac{\partial Q^*(w,r)}{\partial r} &= \frac{\partial \pi_R^*(w,r)}{\partial r} - \frac{\partial^2 \pi_R^*(w,r)}{\partial w \partial r} w - \frac{\partial \pi_R^*(w,r)}{\partial r} - \frac{\partial \pi_R^*(w,r)}{\partial r^2} r \\
&= -\frac{\partial^2 \pi_R^*(w,r)}{\partial w \partial r} w - \frac{\partial \pi_R^*(w,r)}{\partial r^2} r.
\end{aligned}$$

The supply curve $Q^*(P,W,R)$ is found as

$$Q^*(P,W,R) = Q^*(w,r) = Q^*\left(\frac{W}{P}, \frac{R}{P}\right)$$

so that

$$\begin{aligned}
\frac{\partial Q^*}{\partial P} &= \frac{\partial Q^*}{\partial w}\left(-\frac{W}{P^2}\right) + \frac{\partial Q^*}{\partial r}\left(-\frac{R}{P^2}\right) = -\frac{1}{P}\left(\frac{\partial Q^*}{\partial w} w + \frac{\partial Q^*}{\partial r} r\right) \\
&= -\frac{1}{P}\left(\begin{array}{l} \left(-\frac{\partial^2 \pi_R^*(w,r)}{\partial w^2} w - \frac{\partial \pi_R^*(w,r)}{\partial r \partial w} r\right) w \\ + \left(-\frac{\partial^2 \pi_R^*(w,r)}{\partial w \partial r} w - \frac{\partial \pi_R^*(w,r)}{\partial r^2} r\right) r \end{array}\right) \\
&= \frac{1}{P}\begin{bmatrix} w & r \end{bmatrix} \begin{bmatrix} \frac{\partial^2 \pi_R^*(w,r)}{\partial w^2} & \frac{\partial \pi_R^*(w,r)}{\partial r \partial w} \\ \frac{\partial \pi_R^*(w,r)}{\partial r \partial w} & \frac{\partial \pi_R^*(w,r)}{\partial r^2} \end{bmatrix} \begin{bmatrix} w \\ r \end{bmatrix} > 0
\end{aligned}$$

since the by convexity of $\pi_R^*(w,r)$ the 2×2 matrix is positive definite.

Problem 3.7 *[D***] Suppose the (nominal) clever profit function is given by*

$$\pi^*(P,W,R) = 30P^{\frac{8}{3}}W^{-\frac{4}{3}}R^{-\frac{1}{3}}.$$

Using this information calculate the firm's supply curve $Q^*(P,W,R)$ as well as the factor demands $L^*(P,W,R)$ and $K^*(P,W,R)$.

Answer: The desired functions can be obtained from differentiating $\pi^*(P,W,R)$ using Hotelling's lemma as

$$\begin{aligned}
Q^*(P,W,R) &= \frac{\partial}{\partial P}\left(30P^{\frac{8}{3}}W^{-\frac{4}{3}}R^{-\frac{1}{3}}\right) = 80P^{\frac{5}{3}}W^{-\frac{4}{3}}R^{-\frac{1}{3}} \\
L^*(P,W,R) &= -\frac{\partial}{\partial W}\left(30P^{\frac{8}{3}}W^{-\frac{4}{3}}R^{-\frac{1}{3}}\right) = 40P^{\frac{8}{3}}W^{-\frac{7}{3}}R^{-\frac{1}{3}} \\
K^*(P,W,R) &= -\frac{\partial}{\partial R}\left(30P^{\frac{8}{3}}W^{-\frac{4}{3}}R^{-\frac{1}{3}}\right) = 10P^{\frac{8}{3}}W^{-\frac{4}{3}}R^{-\frac{4}{3}}.
\end{aligned}$$

Problem 3.8 *[D***] Suppose the clever profit function is given by*

$$\pi^*(P,W,R) = 10P^{\frac{5}{2}}W^{-\frac{3}{4}}R^{-\frac{3}{4}}.$$

Calculate the firm's supply curve $Q^*(P,W,R)$, demand curve for labour $L^*(P,W,R)$ and demand curve for capital $K^*(P,W,R)$.

Answer: We have

$$\begin{aligned}
Q^*(P,W,R) &= \frac{\partial\pi^*(P,W,R)}{\partial P} = \frac{\partial}{\partial P}\left(10P^{\frac{5}{2}}W^{-\frac{3}{4}}R^{-\frac{3}{4}}\right) = 25P^{\frac{3}{2}}W^{-\frac{3}{4}}R^{-\frac{3}{4}} \\
L^*(P,W,R) &= -\frac{\partial\pi^*(P,W,R)}{\partial W} = -\frac{\partial}{\partial W}\left(10P^{\frac{5}{2}}W^{-\frac{3}{4}}R^{-\frac{3}{4}}\right) = \frac{15}{2}P^{\frac{5}{2}}W^{-\frac{7}{4}}R^{-\frac{3}{4}} \\
K^*(P,W,R) &= -\frac{\partial\pi^*(P,W,R)}{\partial R} = -\frac{\partial}{\partial R}\left(10P^{\frac{5}{2}}W^{-\frac{3}{4}}R^{-\frac{3}{4}}\right) = \frac{15}{2}P^{\frac{5}{2}}W^{-\frac{3}{4}}R^{-\frac{7}{4}}.
\end{aligned}$$

Problem 3.9 *[C***] Suppose that the profit function is given by*

$$\pi^*(P,W,R) = P^{1+\alpha}\left(a_1W^\rho + a_2R^\rho\right)^{-\frac{\alpha}{\rho}}$$

where $\alpha > 0$, $a_1 > 0$ and $a_2 > 0$. Show that $\pi^*(P,W,R)$ is homogeneous of degree 1. Find the firm's supply curve and factor demand curves.

Answer: We have

$$\begin{aligned}
\pi^*(\tau P,\tau W,\tau R) &= (\tau P)^{1+\alpha}\left(a_1(\tau W)^\rho + a_2(\tau R)^\rho\right)^{-\frac{\alpha}{\rho}} = \tau^{1+\alpha}P^{1+\alpha}\left(a_1\tau^\rho W^\rho + a_2\tau^\rho R^\rho\right)^{-\frac{\alpha}{\rho}} \\
&= \tau^{1+\alpha}P^{1+\alpha}(\tau^\rho)^{-\frac{\alpha}{\rho}}\left(a_1W^\rho + a_2R^\rho\right)^{-\frac{\alpha}{\rho}} = \tau^{1+\alpha}\tau^{-\alpha}P^{1+\alpha}\left(a_1W^\rho + a_2R^\rho\right)^{-\frac{\alpha}{\rho}} \\
&= \tau^1P^{1+\alpha}\left(a_1W^\rho + a_2R^\rho\right)^{-\frac{\alpha}{\rho}} = \tau^1\pi^*(P,W,R)
\end{aligned}$$

so that $\pi^*(P,W,R)$ is homogeneous of degree 1.

From Hotelling's lemma we have

$$\begin{aligned}
Q^*(P,W,R) &= \frac{\partial \pi^*(P,W,R)}{\partial P} = \frac{\partial}{\partial P}\left(P^{1+\alpha}(a_1W^\rho + a_2R^\rho)^{-\frac{\alpha}{\rho}}\right) \\
&= (1+\alpha)P^\alpha (a_1W^\rho + a_2R^\rho)^{-\frac{\alpha}{\rho}} \\
L^*(P,W,R) &= -\frac{\partial \pi^*(P,W,R)}{\partial W} = -\frac{\partial}{\partial W}\left(P^{1+\alpha}(a_1W^\rho + a_2R^\rho)^{-\frac{\alpha}{\rho}}\right) \\
&= -\left(-\frac{\alpha}{\rho}P^{1+\alpha}(a_1W^\rho + a_2R^\rho)^{-\frac{\alpha}{\rho}-1}\rho a_1 W^{\rho-1}\right) \\
&= \alpha P^{1+\alpha}(a_1W^\rho + a_2R^\rho)^{-\frac{\alpha}{\rho}-1} a_1 W^{\rho-1} \\
K^*(P,W,R) &= -\frac{\partial \pi^*(P,W,R)}{\partial R} = -\frac{\partial}{\partial R}\left(P^{1+\alpha}(a_1W^\rho + a_2R^\rho)^{-\frac{\alpha}{\rho}}\right) \\
&= -\left(-\frac{\alpha}{\rho}P^{1+\alpha}(a_1W^\rho + a_2R^\rho)^{-\frac{\alpha}{\rho}-1}\rho a_2 R^{\rho-1}\right) \\
&= \alpha P^{1+\alpha}(a_1W^\rho + a_2R^\rho)^{-\frac{\alpha}{\rho}-1} a_2 R^{\rho-1}.
\end{aligned}$$

Problem 3.10 *[D***] For the profit function*

$$\pi^*(P,W,R) = P^3W^{-1}R^{-1}$$

find the firm's supply curve and demand for labour and capital.

Answer: We have

$$\begin{aligned}
Q^*(P,W,R) &= \frac{\partial}{\partial P}\left(P^3W^{-1}R^{-1}\right) = 3P^2W^{-1}R^{-1} \\
L^*(P,W,R) &= -\frac{\partial}{\partial W}\left(P^3W^{-1}R^{-1}\right) = P^3W^{-2}R^{-1} \\
K^*(P,W,R) &= -\frac{\partial}{\partial R}\left(P^3W^{-1}R^{-1}\right) = P^3W^{-1}R^{-2}.
\end{aligned}$$

Problem 3.11 *Suppose a firm has a production function $Q = F(L,K)$ that is homogeneous of degree k where $0 < k < 1$. Use Euler's theorem and Hotelling's lemma to show that the clever profit function $\pi^*(P,W,R)$ must take the form*

$$\pi^*(P,W,R) = P^{\frac{1}{1-k}}A(W,R)$$

where $A(W,R)$ is homogeneous of degree $-\frac{k}{1-k}$. What then is the elasticity of supply?

Answer: Since $Q = F(L,K)$ is homogeneous of degree k it follows from Euler's theorem that

$$kQ^* = \frac{\partial F(L^*,K^*)}{\partial L}L^* + \frac{\partial F(L^*,K^*)}{\partial K}K^*.$$

By the first-order conditions for profit maximization

$$\frac{\partial F(L^*,K^*)}{\partial L} = \frac{W}{P}, \frac{\partial F(L^*,K^*)}{\partial R} = \frac{R}{P}$$

so that

$$\begin{aligned} kQ^* &= \frac{W}{P}L^* + \frac{R}{P}K^* \Longrightarrow kPQ^* = WL^* + RK^* \\ &\Longrightarrow \pi^*(P,W,R) = PQ^* - (WL^* + RK^*) = PQ^* - kPQ^* \\ &\Longrightarrow \pi^*(P,W,R) = (1-k)PQ^*. \end{aligned}$$

By Hotelling's lemma $Q^*(P,W,R) = \frac{\partial \pi^*(P,W,R)}{\partial P}$ so that

$$\begin{aligned} \pi^*(P,W,R) &= (1-k)PQ^*(P,W,R) = (1-k)P\frac{\partial \pi^*(P,W,R)}{\partial P} \\ &\Longrightarrow \frac{\partial \pi^*(P,W,R)}{\partial P}\frac{P}{\pi^*(P,W,R)} = \frac{1}{1-k}. \end{aligned}$$

Since $\frac{\partial \pi^*}{\partial P}\frac{P}{\pi^*}$ is the elasticity with respect to P and is a constant $\frac{1}{1-k}$ it follows that

$$\pi^*(P,W,R) = A(W,R)P^{\frac{1}{1-k}}.$$

Since $\pi^*(P,W,R)$ is homogeneous of degree 1 we have

$$\pi^*(\tau P, \tau W, \tau R) = \tau\pi^*(P,W,R) = \tau A(W,R)P^{\frac{1}{1-k}}$$

and

$$\pi^*(\tau P, \tau W, \tau R) = A(\tau W, \tau R)(\tau P)^{\frac{1}{1-k}} = \tau^{\frac{1}{1-k}}A(\tau W, \tau R)P^{\frac{1}{1-k}}$$

so that

$$\tau^{\frac{1}{1-k}}A(\tau W,\tau R)P^{\frac{1}{1-k}} = \tau A(W,R)P^{\frac{1}{1-k}} \Longrightarrow A(\tau W,\tau R) = \tau^{1-\frac{1}{1-k}}A(W,R) = \tau^{-\frac{k}{1-k}}A(W,R)$$

so that $A(W,R)$ is homogeneous of degree $-\frac{k}{1-k}$.

The supply curve is given by

$$Q^* = \frac{\partial \pi^*(P,W,R)}{\partial P} = \frac{\partial}{\partial P}\left(A(W,R)P^{\frac{1}{1-k}}\right) = \frac{1}{1-k}A(W,R)P^{\frac{1}{1-k}-1} = \frac{1}{1-k}A(W,R)P^{\frac{k}{1-k}}$$

so that the elasticity of supply is the exponent on P or $\frac{k}{1-k}$.

Problem 3.12 *[B***] If we were to add another exogenous variable temperature T to the production function so that $Q = F(L,K,T)$ with $\frac{\partial F(L,K,T)}{\partial T} > 0$ (higher temperatures increase output), show that*

$$\frac{\partial \pi^*(P,W,R,T)}{\partial T} = P\frac{\partial F(L^*,K^*,T)}{\partial T} > 0.$$

Answer: We have the naive profit function

$$\pi(L,K,P,W,R,T) = PF(L,K,T) - WL - RK.$$

From the envelope theorem

$$\begin{aligned} \frac{\partial \pi^*(P,W,R,T)}{\partial T} &= \frac{\partial \pi(L,K,P,W,R,T)}{\partial T}|_{L=L^*,K=K^*} = P\frac{\partial F(L,K,T)}{\partial T}|_{L=L^*,K=K^*} \\ &= P\frac{\partial F(L^*,K^*,T)}{\partial T} > 0. \end{aligned}$$

Problem 3.13 *[B***] Using total differentials prove that*

$$\pi^*(w) = f(L^*(w)) - wL^*(w)$$

is convex given that $Q = f(L)$ *with* $f'(L) > 0$ *and* $f''(L) < 0$.

Answer: From the first-order conditions for profit maximization we have

$$\begin{aligned} f'(L^*) - w &= 0 \\ &\Longrightarrow f''(L^*)\,dL^* - dw = 0 \\ &\Longrightarrow \frac{dL^*}{dw} = \frac{1}{f''(L^*)} < 0. \end{aligned}$$

Now using the chain rule we have

$$\begin{aligned} \frac{d\pi^*(w)}{dw} &= f'(L^*(w))\frac{dL^*(w)}{dw} - w\frac{dL^*(w)}{dw} - L^*(w) \\ &= \overbrace{(f'(L^*(w)) - w)}^{0}\frac{dL^*(w)}{dw} - L^*(w) = -L^*(w) \end{aligned}$$

so that

$$\frac{d^2\pi^*(w)}{dw^2} = -\frac{dL^*(w)}{dw} > 0$$

so that $\pi^*(w)$ is convex.

Problem 3.14 *[D***] Consider a firm with a production function*

$$Q = F(L,K) = L^{\frac{1}{2}}K^{\frac{1}{3}}.$$

Derive the firm's demand for labour $L^*(P,W,R)$, *the firm's demand for capital* $K^*(P,W,R)$ *and the firm's supply curve* $Q^*(P,W,R)$ *from the first-order conditions. Find the clever profit function* $\pi^*(P,W,R)$ *and show that it is homogeneous of degree* 1. *Now use Hotelling's lemma to find* L^*, K^*, Q^* *from* $\pi^*(P,W,R)$ *and verify that you get the same answers as in the above problem.*

Answer: We have

$$\begin{aligned} \frac{\partial\pi(L,K)}{\partial L} &= \frac{1}{2}PL^{-\frac{1}{2}}K^{\frac{1}{3}} - W \\ \frac{\partial\pi(L,K)}{\partial K} &= \frac{1}{3}PL^{\frac{1}{2}}K^{-\frac{2}{3}} - R \end{aligned}$$

so that the first-order conditions yield

$$\begin{aligned} \frac{1}{2}P(L^*)^{-\frac{1}{2}}(K^*)^{\frac{1}{3}} - W &= 0 \\ &\Longrightarrow (L^*)^{-\frac{1}{2}}(K^*)^{\frac{1}{3}} = 2\frac{W}{P} \\ \frac{1}{3}P(L^*)^{\frac{1}{2}}(K^*)^{-\frac{2}{3}} - R &= 0 \\ &\Longrightarrow (L^*)^{\frac{1}{2}}(K^*)^{-\frac{2}{3}} = 3\frac{R}{P}. \end{aligned}$$

Combining these two results yields

$$\frac{(L^*)^{-\frac{1}{2}}(K^*)^{\frac{1}{3}}}{(L^*)^{\frac{1}{2}}(K^*)^{-\frac{2}{3}}} = \frac{2\frac{W}{P}}{3\frac{R}{P}} \Longrightarrow \frac{K^*}{L^*} = 2\times 3^{-1}WR^{-1} \Longrightarrow K^* = 2\times 3^{-1}WR^{-1}L^*$$

so that

$$\begin{aligned}
(L^*)^{-\frac{1}{2}}(K^*)^{\frac{1}{3}} &= 2\frac{W}{P} \Longrightarrow (L^*)^{-\frac{1}{2}}\left(2\times 3^{-1}WR^{-1}L^*\right)^{1/3} = 2P^{-1}W \\
&\Longrightarrow (L^*)^{-1/2}2^{\frac{1}{3}}3^{-\frac{1}{3}}W^{\frac{1}{3}}R^{-\frac{1}{3}}(L^*)^{\frac{1}{3}} = 2P^{-1}W \\
&\Longrightarrow (L^*)^{-\frac{1}{6}} = 2^{\frac{2}{3}}3^{\frac{1}{3}}P^{-1}W^{\frac{2}{3}}R^{\frac{1}{3}} \\
&\Longrightarrow L^* = \left(2^{\frac{2}{3}}3^{\frac{1}{3}}\right)^{-6}P^6W^{-4}R^{-2} = \frac{1}{144}P^6W^{-4}R^{-2}.
\end{aligned}$$

Now solving for K^* we have

$$K^* = 2\times 3^{-1}WR^{-1}L^* = 2\times 3^{-1}WR^{-1}\frac{1}{144}P^6W^{-4}R^{-2} = \frac{1}{216}P^6W^{-3}R^{-3}$$

so that

$$L^*(P,W,R) = \frac{1}{144}P^6W^{-4}R^{-2} \text{ and } K^*(P,W,R) = \frac{1}{216}P^6W^{-3}R^{-3}.$$

Therefore

$$Q^* = (L^*)^{\frac{1}{2}}(K^*)^{\frac{1}{3}} = \left(\frac{1}{144}P^6W^{-4}R^{-2}\right)^{\frac{1}{2}}\left(\frac{1}{216}P^6W^{-3}R^{-3}\right)^{\frac{1}{3}} = \frac{1}{72}P^5W^{-3}R^{-2}.$$

(Alternatively use

$$\begin{aligned}
(L^*)^{-1/2}(K^*)^{1/3} &= 2\frac{W}{P} \Longrightarrow -\frac{1}{2}\ln(L^*) + \frac{1}{3}\ln(K^*) = \ln\left(2\frac{W}{P}\right) \\
(L^*)^{1/2}(K^*)^{-2/3} &= 3\frac{R}{P} \Longrightarrow \frac{1}{2}\ln(L^*) - \frac{2}{3}\ln(K^*) = \ln\left(3\frac{R}{P}\right),
\end{aligned}$$

solve for $\ln(L^*)$ and $\ln(K^*)$ using Cramer's rule, and use $L^* = e^{\ln(L^*)}$ and $K^* = e^{\ln(K^*)}$.)

We have

$$\begin{aligned}
\pi^*(P,W,R) &= PQ^*(P,W,R) - WL^*(P,W,R) - RK^*(P,W,R) \\
&= P\frac{1}{72}P^5W^{-3}R^{-2} - W\frac{1}{144}P^6W^{-4}R^{-2} - R\frac{1}{216}P^6W^{-3}R^{-3} \\
&= \left(\frac{1}{72} - \frac{1}{144} - \frac{1}{216}\right)P^6W^{-3}R^{-2} = \frac{1}{432}P^6W^{-3}R^{-2}.
\end{aligned}$$

Adding the exponents we see that $\pi^*(P,W,R)$ is homogeneous of degree

$$k = 6 + -3 + -2 = 1.$$

Now we verify Hotelling's lemma as

$$\begin{aligned}
\frac{\partial \pi^*(P,W,R)}{\partial P} &= \frac{\partial}{\partial P}\left(\frac{1}{432}P^6W^{-3}R^{-2}\right) \\
&= \frac{6}{432}P^5W^{-3}R^{-2} = \frac{1}{72}P^5W^{-3}R^{-2} \\
&= Q^*(P,W,R) \\
\frac{\partial \pi^*(P,W,R)}{\partial W} &= \frac{\partial}{\partial W}\left(\frac{1}{432}P^6W^{-3}R^{-2}\right) \\
&= -\frac{3}{432}P^6W^{-4}R^{-2} = -\frac{1}{144}P^6W^{-4}R^{-2} \\
&= -L^*(P,W,R) \\
\frac{\partial \pi^*(P,W,R)}{\partial R} &= \frac{\partial}{\partial R}\left(\frac{1}{432}P^6W^{-3}R^{-2}\right) \\
&= -\frac{2}{432}P^6W^{-3}R^{-3} = -\frac{1}{216}P^6W^{-3}R^{-3} \\
&= -K^*(P,W,R)
\end{aligned}$$

Problem 3.15 *[D***] Consider a firm with a clever profit function*

$$\pi^*(P,W,R) = P^2\left(W^{\frac{1}{3}} + R^{\frac{1}{3}}\right)^{-3}.$$

Prove that $\pi^*(P,W,R)$ is homogeneous of degree 1. What relationship is therefore satisfied according to Euler's theorem? Using Hotelling's lemma, find the firm's supply curve Q^*, its demand for labour L^* and its demand for capital K^* and prove that these are all homogeneous of degree 0. Show that $\frac{\partial K^*}{\partial W} = \frac{\partial L^*}{\partial R}$.

Answer: We have

$$\begin{aligned}
\pi^*(\tau P,\tau W,\tau R) &= (\tau P)^2\left((\tau W)^{\frac{1}{3}} + (\tau R)^{\frac{1}{3}}\right)^{-3} = \tau^2P^2\left(\tau^{\frac{1}{3}}\left(W^{\frac{1}{3}} + R^{\frac{1}{3}}\right)\right)^{-3} \\
&= \tau^2P^2\left(\tau^{\frac{1}{3}}\right)^{-3}\left(W^{\frac{1}{3}} + R^{\frac{1}{3}}\right)^{-3} = \tau^2\tau^{-1}P^2\left(W^{\frac{1}{3}} + R^{\frac{1}{3}}\right)^{-3} \\
&= \tau^1\pi^*(P,W,R)
\end{aligned}$$

so that $\pi^*(P,W,R)$ is homogeneous of degree 1. Euler's theorem then implies that

$$1 \times \pi^*(P,W,R) = \frac{\partial \pi^*(P,W,R)}{\partial P}P + \frac{\partial \pi^*(P,W,R)}{\partial W}W + \frac{\partial \pi^*(P,W,R)}{\partial R}R.$$

Using Hotelling's lemma we have

$$\begin{aligned}
Q^*(P,W,R) &= \frac{\partial \pi^*(P,W,R)}{\partial P} = 2P\left(W^{\frac{1}{3}} + R^{\frac{1}{3}}\right)^{-3} \\
L^*(P,W,R) &= -\frac{\partial \pi^*(P,W,R)}{\partial W} = P^2\left(W^{\frac{1}{3}} + R^{\frac{1}{3}}\right)^{-4}W^{-\frac{2}{3}} \\
K^*(P,W,R) &= -\frac{\partial \pi^*(P,W,R)}{\partial R} = P^2\left(W^{\frac{1}{3}} + R^{\frac{1}{3}}\right)^{-4}R^{-\frac{2}{3}}.
\end{aligned}$$

A partial derivative of a function which is homogeneous of degree k is homogeneous of degree $k-1$. Since Q^*, L^*, K^* are derivatives of π^*, they are homogeneous of degree 0. Finally we have via a direct calculation of the two derivatives $\frac{\partial K^*}{\partial W}$ and $\frac{\partial L^*}{\partial R}$ that

$$\frac{\partial K^*}{\partial W} = \frac{\partial L^*}{\partial R} = -\frac{4}{3}P^2\left(W^{\frac{1}{3}} + R^{\frac{1}{3}}\right)^{-5} W^{-\frac{2}{3}}R^{-\frac{2}{3}}.$$

A quicker derivation is to note that

$$\begin{aligned}\frac{\partial K^*}{\partial W} &= -\frac{\partial}{\partial W}\frac{\partial \pi^*(P,W,R)}{\partial R} = -\frac{\partial^2 \pi^*(P,W,R)}{\partial W \partial R} \\ \frac{\partial L^*}{\partial R} &= -\frac{\partial}{\partial R}\frac{\partial \pi^*(P,W,R)}{\partial W} = -\frac{\partial^2 \pi^*(P,W,R)}{\partial R \partial W}\end{aligned}$$

so that $\frac{\partial K^*}{\partial W} = \frac{\partial L^*}{\partial R}$ follows by Young's theorem.

Problem 3.16 *Suppose the production function is given by*

$$Q = F(L,K) = 2\left(L^{\frac{1}{2}} + K^{\frac{1}{2}}\right).$$

*[D***] Find the first-order conditions for the profit maximizing L^* and K^*. Show that the clever profit function will be given by*

$$\pi^*(P,W,R) = P^2\left(W^{-1} + R^{-1}\right).$$

Show that $\pi^*(P,W,R)$ is homogeneous and determine the degree of the homogeneity. Write down Euler's theorem for $\pi^*(P,W,R)$. Use the envelope theorem to find the supply curve $Q^*(P,W,R)$.

Answer: If

$$\pi(L,K) = P \times 2\left(L^{\frac{1}{2}} + K^{\frac{1}{2}}\right) - WL - RK$$

then the first-order conditions are

$$\begin{aligned}\frac{\partial \pi(L^*,K^*)}{\partial L} &= P(L^*)^{-\frac{1}{2}} - W = 0 \Longrightarrow L^*(P,W,R) = \left(\frac{W}{P}\right)^{-2} \\ \frac{\partial \pi(L^*,K^*)}{\partial K} &= P(K^*)^{-\frac{1}{2}} - R = 0 \Longrightarrow K^*(P,W,R) = \left(\frac{R}{P}\right)^{-2}\end{aligned}$$

so that

$$\begin{aligned}\pi^*(P,W,R) &= \pi(L^*,K^*) \\ &= P \times 2\left(\left(\left(\frac{W}{P}\right)^{-2}\right)^{\frac{1}{2}} + \left(\left(\frac{R}{P}\right)^{-2}\right)^{\frac{1}{2}}\right) - W\left(\frac{W}{P}\right)^{-2} - R\left(\frac{R}{P}\right)^{-2} \\ &= P \times 2\left(\left(\frac{W}{P}\right)^{-1} + \left(\frac{R}{P}\right)^{-1}\right) - P^2W^{-1} - P^2R^{-1} \\ &= 2P^2\left(W^{-1} + R^{-1}\right) - P^2\left(W^{-1} + R^{-1}\right) = P^2\left(W^{-1} + R^{-1}\right).\end{aligned}$$

The clever profit function $\pi^* (P, W, R)$ is homogeneous of degree 1 since

$$\begin{aligned}\pi^* (\lambda P, \lambda W, \lambda R) &= (\lambda P)^2 \left((\lambda W)^{-1} + (\lambda R)^{-1} \right) \\ &= \lambda^2 \lambda^{-1} P^2 \left(W^{-1} + R^{-1} \right) = \lambda^1 P^2 \left(W^{-1} + R^{-1} \right) = \lambda^1 \pi^* (P, W, R) .\end{aligned}$$

Thus by Euler's theorem

$$\pi^* = \frac{\partial \pi^*}{\partial P} P + \frac{\partial \pi^*}{\partial W} W + \frac{\partial \pi^*}{\partial R} R$$

or

$$\begin{aligned}P^2 \left(W^{-1} + R^{-1} \right) &= 2P \left(W^{-1} + R^{-1} \right) P + P^2 \left(-W^{-2} \right) W + P^2 \left(-R^{-2} \right) R \\ &= P^2 \left(W^{-1} + R^{-1} \right) .\end{aligned}$$

Using the envelope theorem we have

$$\begin{aligned}\frac{\partial \pi^* (P, W, R)}{\partial P} &= \frac{\partial}{\partial P} \left(P \times 2 \left(L^{\frac{1}{2}} + K^{\frac{1}{2}} \right) - WL - RK \right) |_{\substack{L=L^* \\ K=K^*}} \\ &= \left(2 \left(L^{\frac{1}{2}} + K^{\frac{1}{2}} \right) \right) |_{\substack{L=L^* \\ K=K^*}} = 2 \left((L^*)^{\frac{1}{2}} + (K^*)^{\frac{1}{2}} \right) \\ &= Q^* (P, W, R)\end{aligned}$$

and so

$$Q^* (P, W, R) = \frac{\partial \pi^* (P, W, R)}{\partial P} = \frac{\partial}{\partial P} \left(P^2 \left(W^{-1} + R^{-1} \right) \right) = 2P \left(W^{-1} + R^{-1} \right) .$$

Problem 3.17 *[C***] Suppose the firm's production function* $Q = F(L, K)$ *is Cobb-Douglas or* $Q = L^{\alpha} K^{\beta}$ *where* $\alpha + \beta < 1$ *(the production function is homogeneous of degree* $\alpha + \beta$*.) Shown that the clever profit function* $\pi^* (P, W, R)$ *also has the Cobb-Douglas form*

$$\pi^* (P, W, R) = B P^c W^d R^e$$

that $c > 0,\ d < 0, e < 0$ *and that* $\pi^* (P, W, R)$ *is homogeneous of degree 1 or* $c + d + e = 1$.

Answer: Solving the first-order conditions for L^* and K^* from

$$\pi (L, K, P, W, R) = P L^{\alpha} K^{\beta} - WL - RK$$

we find that

$$\begin{aligned}L^* &= \alpha^{\frac{1-\beta}{1-\alpha-\beta}} \beta^{\frac{\beta}{1-\alpha-\beta}} P^{\frac{1}{1-\alpha-\beta}} W^{\frac{\beta-1}{1-\alpha-\beta}} R^{\frac{-\beta}{1-\alpha-\beta}} \\ K^* &= \alpha^{\frac{\alpha}{1-\alpha-\beta}} \beta^{\frac{\alpha-1}{1-\alpha-\beta}} P^{\frac{1}{1-\alpha-\beta}} W^{\frac{-\alpha}{1-\alpha-\beta}} R^{\frac{\alpha-1}{1-\alpha-\beta}}\end{aligned}$$

so that after some work

$$\begin{aligned}
\pi^*(P,W,R) &= \pi(L^*,K^*,P,W,R) \\
&= P\left(\alpha^{\frac{1-\beta}{1-\alpha-\beta}}\beta^{\frac{\beta}{1-\alpha-\beta}}P^{\frac{1}{1-\alpha-\beta}}W^{\frac{\beta-1}{1-\alpha-\beta}}R^{\frac{-\beta}{1-\alpha-\beta}}\right)^{\alpha} \\
&\quad\times\left(\alpha^{\frac{\alpha}{1-\alpha-\beta}}\beta^{\frac{\alpha-1}{1-\alpha-\beta}}P^{\frac{1}{1-\alpha-\beta}}W^{\frac{-\alpha}{1-\alpha-\beta}}R^{\frac{\alpha-1}{1-\alpha-\beta}}\right)^{\beta} \\
&\quad -W\alpha^{\frac{1-\beta}{1-\alpha-\beta}}\beta^{\frac{\beta}{1-\alpha-\beta}}P^{\frac{1}{1-\alpha-\beta}}W^{\frac{\beta-1}{1-\alpha-\beta}}R^{\frac{-\beta}{1-\alpha-\beta}} \\
&\quad -R\alpha^{\frac{\alpha}{1-\alpha-\beta}}\beta^{\frac{\alpha-1}{1-\alpha-\beta}}P^{\frac{1}{1-\alpha-\beta}}W^{\frac{-\alpha}{1-\alpha-\beta}}R^{\frac{\alpha-1}{1-\alpha-\beta}} \\
&= (1-\alpha-\beta)\,\alpha^{\frac{\alpha}{1-\alpha-\beta}}\beta^{\frac{\beta}{1-\alpha-\beta}}P^{\frac{1}{1-\alpha-\beta}}W^{\frac{-\alpha}{1-\alpha-\beta}}R^{\frac{-\beta}{1-\alpha-\beta}}
\end{aligned}$$

so that

$$\pi^*(P,W,R) = (1-\alpha-\beta)\,\alpha^{\frac{\alpha}{1-\alpha-\beta}}\beta^{\frac{\beta}{1-\alpha-\beta}}P^{\frac{1}{1-\alpha-\beta}}W^{\frac{-\alpha}{1-\alpha-\beta}}R^{\frac{-\beta}{1-\alpha-\beta}}.$$

Thus

$$B = (1-\alpha-\beta)\,\alpha^{\frac{\alpha}{1-\alpha-\beta}}\beta^{\frac{\beta}{1-\alpha-\beta}},\ c = \frac{1}{1-\alpha-\beta} > 0,\ d = \frac{-\alpha}{1-\alpha-\beta} < 0,\ e = \frac{-\beta}{1-\alpha-\beta} < 0$$

where the inequalities follow from $0 < \alpha+\beta < 1$. Also $\pi^*(P,W,R)$ is homogeneous of degree 1 since

$$c+d+e = \frac{1}{1-\alpha-\beta} + \frac{-\alpha}{1-\alpha-\beta} + \frac{-\beta}{1-\alpha-\beta} = 1.$$

Note that we can rewrite this as

$$\pi^*(P,W,R) = BP^cW^dR^e$$

where

$$B = (1-\alpha-\beta)\,\alpha^{\frac{\alpha}{1-\alpha-\beta}}\beta^{\frac{\beta}{1-\alpha-\beta}},\ c = \frac{1}{1-\alpha-\beta},\ d = \frac{-\alpha}{1-\alpha-\beta},\ e = \frac{-\beta}{1-\alpha-\beta}.$$

which also has the Cobb-Douglas functional form.

Note that the profit function is homogeneous of degree 1 since

$$c+d+e = \frac{1}{1-\alpha-\beta} + \frac{-\alpha}{1-\alpha-\beta} + \frac{-\beta}{1-\alpha-\beta} = 1$$

with $c \geq 0, d \leq 0, e \leq 0$.

Sometimes we start with the production function, derive $L^*(P,W,R)$, $K^*(P,W,R)$ and $Q^*(P,W,R)$ and then $\pi^*(P,W,R)$. However, we can also start with $\pi^*(P,W,R)$ and from there derive $L^*(P,W,R)$, $K^*(P,W,R)$ and $Q^*(P,W,R)$.

Problem 3.18 *[A***] A function $f(x)$ with a Taylor series*

$$\bar{f}(x) = f(x^0) + \nabla f(x^0)^T (x - x^0)$$

is convex if and only if for all x^0

$$f(x) \geq \bar{f}(x) = f(x^0) + \nabla f(x^0)^T (x - x^0).$$

Use this result and Hotelling's lemma to prove that the profit function $\pi^*(P, W, R)$ *is convex.*

Answer: Think of P_0, W_0, R_0 as Period 0 prices and P, W, R as Period 1 prices. Define profits π^*, Q^*, L^*, K^* for both periods as

$$\begin{aligned}
\pi_0^* &\equiv \pi^*(P_0, W_0, R_0), \pi_1^* \equiv \pi^*(P, W, R) \\
Q_0^* &\equiv Q^*(P_0, W_0, R_0), Q_1^* \equiv Q^*(P, W, R) \\
L_0^* &\equiv L^*(P_0, W_0, R_0), L_1^* \equiv L^*(P, W, R) \\
K_0^* &\equiv K^*(P_0, W_0, R_0), K_1^* \equiv K^*(P, W, R).
\end{aligned}$$

The Taylor series for $\pi^*(P, W, R) \equiv \pi_1^*$ at P_0, W_0, R_0 then is

$$\begin{aligned}
\bar{\pi}^*(P, W, R) &= \pi^*(P_0, W_0, R_0) + \frac{\partial \pi^*(P_0, W_0, R_0)}{\partial P}(P - P_0) \\
&\quad + \frac{\partial \pi^*(P_0, W_0, R_0)}{\partial W}(W - W_0) + \frac{\partial \pi^*(P_0, W_0, R_0)}{\partial R}(R - R_0) \\
&= \pi_0^* + \frac{\partial \pi^*(P_0, W_0, R_0)}{\partial P}(P - P_0) \\
&\quad + \frac{\partial \pi^*(P_0, W_0, R_0)}{\partial W}(W - W_0) + \frac{\partial \pi^*(P_0, W_0, R_0)}{\partial R}(R - R_0).
\end{aligned}$$

By Hotelling's lemma

$$\begin{aligned}
\frac{\partial \pi^*(P_0, W_0, R_0)}{\partial P} &= Q^*(P_0, W_0, R_0) \equiv Q_0^* \\
\frac{\partial \pi^*(P_0, W_0, R_0)}{\partial W} &= -L^*(P_0, W_0, R_0) \equiv -L_0^* \\
\frac{\partial \pi^*(P_0, W_0, R_0)}{\partial R} &= -K^*(P_0, W_0, R_0) \equiv -K_0^*
\end{aligned}$$

so that

$$\begin{aligned}
\bar{\pi}^*(P, W, R) &= \pi_0^* + Q_0^* \times (P - P_0) - L_0^* \times (W - W_0) - K_0^* \times (R - R_0) \\
&= \pi_0^* - (P_0 \times Q_0^* - W_0 \times L_0^* - R_0 \times K_0^*) + P \times Q_0^* - W \times L_0^* - R \times K_0^* \\
&= P \times Q_0^* - W \times L_0^* - R \times K_0^*
\end{aligned}$$

since

$$\pi_0^* \equiv P_0 \times Q_0^* - W_0 \times L_0^* - R_0 \times K_0^*.$$

Now

$$\bar{\pi}^*(P, W, R) = P \times Q_0^* - W \times L_0^* - R \times K_0^*$$

is profits when there are Period 1 prices P, W, R but Q_0^*, L_0^*, K_0^* are optimal for Period 0 prices P_0, W_0, R_0. Picking Q, L, K optimal for the wrong period will generally lower profits, and certainly cannot increase profits and so

$$\pi^*(P,W,R) = P\times Q_1^* - W\times L_1^* - R\times K_1^* \geq \bar{\pi}^*(P,W,R) = P\times Q_0^* - W\times L_0^* - R\times K_0^*$$

and so the profit function is convex.

Problem 3.19 *[A ***] For many economics problems the exogenous variables are prices that appear linearly in the objective function. Suppose then that the naive function $f(y,x)$ takes the form $f(y,x) = x^T\phi(y)$ where x and $\phi(y)$ are $m\times 1$ vectors and y is an $n\times 1$ vector. Prove that the clever function $f^*(x)$ given by $f^*(x) = x^T\phi(y^*(x))$ is homogeneous of degree 1 and prove that $y^*(x)$ homogeneous of degree 0. Prove that $f^*(x)$ is convex when $f(y,x)$ is being maximized. Use these results to prove that the profit function $\pi^*(P,W,R)$ is homogeneous of degree 1 and convex.*

Answer: The naive function is homogeneous of degree 1 in x since

$$f(y,\tau x) = (\tau x)^T\phi(y) = \tau\left(x^T\phi(y)\right) = \tau f(y,x).$$

Therefore since $\tau > 0$ it follows that $y^*(x)$ also maximizes or minimizes $f(y,\tau x)$ and so

$$y^*(\tau x) = y^*(x).$$

Thus $y^*(x)$ is homogeneous of degree 0. Now

$$\begin{aligned} f^*(\tau x) &= (\tau x)^T\phi(y^*(\tau x)) = (\tau x)^T\phi(y^*(x)) \\ &= \tau x^T\phi(y^*(x)) = \tau f^*(x) \end{aligned}$$

This proves that $f^*(x)$ is homogeneous of degree 1.

Now suppose the naive function $f(y,x)$ is being maximized. Let x_1 and x_2 (which are $m\times 1$ vectors) be two values of x and let $x_3 = \lambda x_1 + (1-\lambda)x_2$ (where $0\leq\lambda\leq 1$) be a convex combination of x_1 and x_2. In order to prove that $f^*(x)$ is convex we need to show that

$$f^*(x_3) \leq \lambda f^*(x_1) + (1-\lambda) f^*(x_2).$$

Now

$$f^*(x_3) = (\lambda x_1 + (1-\lambda)x_2)^T\phi(y^*(x_3)) = \lambda x_1^T\phi(y^*(x_3)) + (1-\lambda)x_2^T\phi(y^*(x_3)).$$

Since $y^*(x_1)$ maximizes $f(y,x_1)$ it follows that $f(y^*(x_3),x_1) \leq f(y^*(x_1),x_1)$ so that

$$x_1^T\phi(y^*(x_3)) = f(y^*(x_3),x_1) \leq f(y^*(x_1),x_1) = x_1^T\phi(y^*(x_1)) = f^*(x_1)$$

Similarly since $y^*(x_2)$ maximizes $f(y,x_2)$ it follows that $f(y^*(x_3),x_2) \leq f(y^*(x_2),x_2)$ so that

$$x_2^T\phi(y^*(x_3)) = f(y^*(x_3),x_2) \leq f(y^*(x_2),x_2) = x_2^T\phi(y^*(x_2)) = f^*(x_2)$$

Combining these last two results we conclude

$$f^*(x_3) = \lambda x_1^T \phi(y^*(x_3)) + (1-\lambda) x_2^T \phi(y^*(x_3)) \leq \lambda f^*(x_1) + (1-\lambda) f^*(x_2)$$

so that $f^*(x)$ is by definition convex.

The naive profit function $\pi(L, K, P, W, R)$ can be written as $f(y, x) = x^T \phi(y)$ as

$$\pi(L, K, P, W, R) = PF(K, L) - WL - RK = x^T \phi(y)$$

where

$$x = \begin{bmatrix} P \\ W \\ R \end{bmatrix}, \; y = \begin{bmatrix} L \\ K \end{bmatrix}, \; \phi(y) = \phi(L, K) = \begin{bmatrix} F(K, L) \\ -L \\ -K \end{bmatrix}.$$

It follows then that

$$f^*(x) = x^T \phi(y^*(x)) = \pi^*(P, W, R)$$

is homogeneous of degree 1 and convex.

Problem 3.20 *[A **] Suppose the clever profit function is given by*

$$\pi^*(P, W) = \pi^*(P, W_1, W_2, \ldots, W_n) = P^\alpha W_1^{-\beta_1} W_2^{-\beta_2} \times \cdots \times W_2^{-\beta_n}$$

where $\alpha > 0, \beta_1 > 0, \beta_2 > 0, \ldots, \beta_n > 0$. *The clever profit function is homogeneous of degree* 1 *so that*

$$\alpha - \beta_1 - \beta_2 - \cdots - \beta_n = 1.$$

Show that the clever profit function is convex by showing that its Hessian is positive semi-definite.

Answer: (This is a sketch of the proof.) To calculate the Hessian we need the second-order partial derivatives

$$\begin{aligned}
\frac{\partial^2 \pi^*}{\partial P^2} &= \pi^*(P, W) \frac{\alpha(\alpha - 1)}{P^2} = \pi^*(P, W)\left(\frac{\alpha^2}{P^2} - \frac{\alpha}{P^2}\right) \\
\frac{\partial^2 \pi^*}{\partial P \partial W_i} &= -\pi^*(P, W) \times \frac{\alpha \beta_i}{P \times W_i} = -\pi^*(P, W) \times \left(\frac{\alpha^2}{P^2} - \frac{\alpha}{P^2}\right) \\
\frac{\partial^2 \pi^*}{\partial W_i^2} &= \pi^*(P, W) \times \frac{\beta_i(\beta_i + 1)}{W_i^2} = \pi^*(P, W) \times \left(\frac{\beta_i^2}{W_i^2} + \frac{\beta_i}{W_i^2}\right) \\
\frac{\partial^2 \pi^*}{\partial W_i \partial W_j} &= \pi^*(P, W) \times \frac{\beta_i \beta_j}{W_i \times W_j} \text{ for } i \neq j.
\end{aligned}$$

The Hessian $H(P, W)$ therefore can be written as

$$H(P, W) = \pi^*(P, W) \times \left(A + aa^T\right)$$

where A is a diagonal matrix and a is a column vector as

$$A = \begin{bmatrix} -\frac{\alpha}{P^2} & 0 & \cdots & 0 \\ 0 & \frac{\beta_1}{W_1^2} & \ddots & 0 \\ \vdots & \ddots & \ddots & 0 \\ 0 & \cdots & 0 & \frac{\beta_n}{W_n^2} \end{bmatrix}, \quad a = \begin{bmatrix} -\frac{\alpha}{P} \\ \frac{\beta_1}{W_1} \\ \vdots \\ \frac{\beta_n}{W_n} \end{bmatrix}.$$

Since $\pi^* (P, W) > 0$, to show that $H (P, W)$ is positive semi-definite we need to show that $A + aa^T$ is positive semi-definite. Let λ and be an eigenvalue of $A + aa^T$ and let x be an eigenvector so that

$$\left(A + aa^T\right) x = \lambda x.$$

Normalize x so that $a^T x = 1$ so that

$$\begin{aligned} \left(A + aa^T\right) x &= \lambda x \Longrightarrow Ax + aa^T x = \lambda x \Longrightarrow Ax + a = \lambda x \\ &\Longrightarrow (\lambda I - A) x = a \Longrightarrow x = (\lambda I - A)^{-1} a. \end{aligned}$$

Since $a^T x = 1$ this means that

$$\begin{aligned} a^T x &= a^T (\lambda I - A)^{-1} a = 1 \\ &\Longrightarrow f(\lambda) = a^T (\lambda I - A)^{-1} a - 1 = 0 \end{aligned}$$

or

$$f(\lambda) = a^T (\lambda I - A)^{-1} a - 1 = 0$$

where

$$\begin{aligned} f(\lambda) &= a^T (\lambda I - A)^{-1} a - 1 \\ &= \frac{\left(\frac{\alpha}{P}\right)^2}{\lambda - \left(-\frac{\alpha}{P^2}\right)} + \frac{\left(\frac{\beta_1}{W_1}\right)^2}{\lambda - \left(\frac{\beta_1}{W_1^2}\right)} + \cdots + \frac{\left(\frac{\beta_n}{W_n}\right)^2}{\lambda - \left(\frac{\beta_n}{W_n^2}\right)} - 1. \end{aligned}$$

Without loss of generality assume that

$$-\frac{\alpha}{P^2} < \frac{\beta_1}{W_1^2} \leq \frac{\beta_2}{W_2^2} \leq \cdots \leq \frac{\beta_n}{W_n^2}.$$

Then $f(\lambda) < 0$ foi $\lambda < -\frac{\alpha}{P^2}$, and $f'(\lambda) < 0$ in the interval $-\frac{\alpha}{P} < \lambda < \frac{\beta_1}{W_1}$ with $f(\lambda)$ going from ∞ to $-\infty$. Similarly $f'(\lambda) < 0$ in the interval $\frac{\beta_1}{W_1} < \lambda < \frac{\beta_2}{W_2}$ with $f(\lambda)$ going from ∞ to $-\infty$, $f'(\lambda) < 0$ in the interval $\frac{\beta_2}{W_2} < \lambda < \frac{\beta_3}{W_3}$ with $f(\lambda)$ going from ∞ to $-\infty$, and so on. Since $f(\lambda)$ is globally decreasing, going from ∞ to $-\infty$ in each interval, it follows that each interval contains a unique root $f(\lambda) = 0$. So there is a unique eigenvalue λ_1 satisfying $f(\lambda_1) = 0$ in the interval $-\frac{\alpha}{P} < \lambda < \frac{\beta_1}{W_1}$, a unique eigenvalue λ_2 satisfying $f(\lambda_2) = 0$

in the interval $\frac{\beta_1}{W_1} < \lambda < \frac{\beta_2}{W_2}$, a unique eigenvalue λ_3 satisfying $f(\lambda_3) = 0$ in the interval $\frac{\beta_2}{W_2} < \lambda < \frac{\beta_3}{W_3}$ and so on. Since $\frac{\beta_i}{Wi} > 0$ it follows that all eigenvalues of $A + aa^T$ are positive except possibly for the smallest λ_1 in the interval $-\frac{\alpha}{P} < \lambda_1 < \frac{\beta_1}{W_1}$. But $\lambda_1 = 0$ is zero since

$$\begin{aligned} f(0) &= \frac{\left(\frac{\alpha}{P}\right)^2}{0 - \left(-\frac{\alpha}{P^2}\right)} + \frac{\left(\frac{\beta_1}{W_1}\right)^2}{0 - \left(\frac{\beta_1}{W_1^2}\right)} + \cdots + \frac{\left(\frac{\beta_n}{W_n}\right)^2}{0 - \left(\frac{\beta_n}{W_n^2}\right)} - 1 \\ &= \alpha - W_1 - W_2 - \cdots - W_n - 1 = 1 - 1 = 0. \end{aligned}$$

Thus $\lambda_1 = 0, \lambda_2 > 0, \ldots, \lambda_{n+1} > 0$, so $A + aa^T$ is positive semi-definite, and hence $\pi^*(P, W)$ is convex.

3.2 Utility Maximization

Problem 3.21 *[D***] Suppose a household has utility function*

$$U(Q_1, Q_2) = 2Q_1^{\frac{1}{2}} + 2Q_2^{\frac{1}{2}}$$

with the usual budget constraint $Y = P_1Q_1 + P_2Q_2$. *Find* λ^*, Q_1^* *and* Q_2^* *from the first-order conditions. Calculate the clever or indirect utility function* $U^*(P_1, P_2, Y)$ *and verify that* $\lambda^* = \frac{\partial U^*}{\partial Y}$, $Q_1^* = -\frac{\partial U^*}{\partial P_1} / \frac{\partial U^*}{\partial Y}$ *and* $Q_2^* = -\frac{\partial U^*}{\partial P_2} / \frac{\partial U^*}{\partial Y}$.

Answer: We have

$$\mathcal{L} = 2Q_1^{\frac{1}{2}} + 2Q_2^{\frac{1}{2}} + \lambda(Y - P_1Q_1 - P_2Q_2)$$

so that the first-order conditions yield

$$\begin{aligned} Y - P_1Q_1^* - P_2Q_2^* &= 0 \Longrightarrow Y = P_1Q_1^* + P_2Q_2^* \\ (Q_1^*)^{-\frac{1}{2}} - \lambda^* P_1 &= 0 \Longrightarrow Q_1^* = (\lambda^*)^{-2} P_1^{-2} \\ (Q_2^*)^{-\frac{1}{2}} - \lambda^* P_2 &= 0 \Longrightarrow Q_2^* = (\lambda^*)^{-2} P_2^{-2}. \end{aligned}$$

Putting the second and third results into the budget constraint then yields

$$\begin{aligned} Y &= P_1Q_1^* + P_2Q_2^* = P_1\left((\lambda^*)^{-2} P_1^{-2}\right) + P_2\left((\lambda^*)^{-2} P_2^{-2}\right) \\ &= (\lambda^*)^{-2}\left(P_1^{-1} + P_2^{-1}\right) \end{aligned}$$

so that

$$(\lambda^*)^{-2} = \left(P_1^{-1} + P_2^{-1}\right)^{-1} Y \Longrightarrow \lambda^*(P_1, P_2, Y) = \left(P_1^{-1} + P_2^{-1}\right)^{\frac{1}{2}} Y^{-\frac{1}{2}}.$$

Thus

$$\begin{aligned} Q_1^*(P_1, P_2, Y) &= (\lambda^*)^{-2} P_1^{-2} = \left(P_1^{-1} + P_2^{-1}\right)^{-1} Y P_1^{-2} \\ Q_2^*(P_1, P_2, Y) &= (\lambda^*)^{-2} P_2^{-2} = \left(P_1^{-1} + P_2^{-1}\right)^{-1} Y P_2^{-2}. \end{aligned}$$

Now to find $U^*(P_1, P_2, Y)$ substitute Q_1^* and Q_2^* into $U(Q_1, Q_2) = 2Q_1^{\frac{1}{2}} + 2Q_2^{\frac{1}{2}}$ so that

$$\begin{aligned} U^*(P_1, P_2, Y) &= 2Q_1^{*\frac{1}{2}} + 2Q_2^{*\frac{1}{2}} \\ &= 2\left(\left(P_1^{-1} + P_2^{-1}\right)^{-1} Y P_1^{-2}\right)^{\frac{1}{2}} + 2\left(\left(P_1^{-1} + P_2^{-1}\right)^{-1} Y P_2^{-2}\right)^{\frac{1}{2}} \\ &= 2\left(P_1^{-1} + P_2^{-1}\right)^{\frac{1}{2}} Y^{\frac{1}{2}} \end{aligned}$$

where the last result requires a little bit of algebra.

Now $\lambda^* = \frac{\partial U^*}{\partial Y}$ since

$$\begin{aligned} \frac{\partial U^*}{\partial Y} &= \frac{\partial}{\partial Y}\left(2\left(P_1^{-1} + P_2^{-1}\right)^{\frac{1}{2}} Y^{\frac{1}{2}}\right) \\ &= \left(P_1^{-1} + P_2^{-1}\right)^{\frac{1}{2}} Y^{-\frac{1}{2}} = \lambda^* \end{aligned}$$

and $Q_1^* = -\frac{\partial U^*}{\partial P_1} / \frac{\partial U^*}{\partial Y}$ since

$$\begin{aligned} \frac{\partial U^*}{\partial P_1} &= \frac{\partial}{\partial P_1}\left(2\left(P_1^{-1} + P_2^{-1}\right)^{\frac{1}{2}} Y^{\frac{1}{2}}\right) \\ &= -\left(P_1^{-1} + P_2^{-1}\right)^{-\frac{1}{2}} Y^{\frac{1}{2}} P_1^{-2} \end{aligned}$$

and so

$$\begin{aligned} -\frac{\partial U^*}{\partial P_1} / \frac{\partial U^*}{\partial Y} &= -\frac{-\left(P_1^{-1} + P_2^{-1}\right)^{-\frac{1}{2}} Y^{\frac{1}{2}} P_1^{-2}}{\left(P_1^{-1} + P_2^{-1}\right)^{\frac{1}{2}} Y^{-\frac{1}{2}}} \\ &= \left(P_1^{-1} + P_2^{-1}\right)^{-1} Y P_1^{-2} = Q_1^*. \end{aligned}$$

Furthermore $Q_2^* = -\frac{\partial U^*}{\partial P_2} / \frac{\partial U^*}{\partial Y}$ since

$$\begin{aligned} \frac{\partial U^*}{\partial P_2} &= \frac{\partial}{\partial P_2}\left(2\left(P_1^{-1} + P_2^{-1}\right)^{\frac{1}{2}} Y^{\frac{1}{2}}\right) \\ &= -\left(P_1^{-1} + P_2^{-1}\right)^{-\frac{1}{2}} Y^{\frac{1}{2}} P_2^{-2} \end{aligned}$$

and so

$$\begin{aligned} -\frac{\partial U^*}{\partial P_2} / \frac{\partial U^*}{\partial Y} &= -\frac{-\left(P_1^{-1} + P_2^{-1}\right)^{-\frac{1}{2}} Y^{\frac{1}{2}} P_2^{-2}}{\left(P_1^{-1} + P_2^{-1}\right)^{\frac{1}{2}} Y^{-\frac{1}{2}}} \\ &= \left(P_1^{-1} + P_2^{-1}\right)^{-1} Y P_2^{-2} = Q_2^*. \end{aligned}$$

Problem 3.22 *[B***] Show that together the envelope theorem and Euler's theorem applied to $U^*(P_1, P_2, Y)$ imply the budget constraint.*

Answer: Rationality requires that $U^*(P_1, P_2, Y)$ be homogeneous of degree 0. Euler's theorem and the envelope theorem therefore imply

$$\begin{aligned} 0 \times U^*(P_1, P_2, Y) &= \frac{\partial U^*}{\partial P_1} P_1 + \frac{\partial U^*}{\partial P_2} P_2 + \frac{\partial U^*}{\partial Y} Y \\ &= -\lambda^* Q_1^* P_1 + -\lambda^* Q_2^* P_2 + \lambda^* Y \\ &= \lambda^* (Y - P_1 Q_1^* - P_2 Q_2^*) \Longrightarrow Y - P_1 Q_1^* - P_2 Q_2^* = 0 \end{aligned}$$

which is the budget constraint.

Problem 3.23 *[D***] Suppose the indirect (or clever) utility function is*

$$U^*(P_1, P_2, Y) = 3Y^{\frac{1}{3}} \left(P_1^{-\frac{1}{2}} + P_2^{-\frac{1}{2}}\right)^{\frac{2}{3}}.$$

Show that U^* is homogeneous and determine the degree of the homogeneity. Find the Lagrange multiplier λ^*, Q_1^* and $\frac{\partial U(Q_1^*, Q_2^*)}{\partial Q_1}/P_1$. What is the income elasticity for Q_1?

Answer: We have

$$\begin{aligned} U^*(\tau P_1, \tau P_2, \tau Y) &= 3(\tau Y)^{\frac{1}{3}} \left((\tau P_1)^{-\frac{1}{2}} + (\tau P_2)^{-\frac{1}{2}}\right)^{\frac{2}{3}} \\ &= 3\tau^{\frac{1}{3}} Y^{\frac{1}{3}} \left(\tau^{-\frac{1}{2}} P_1^{-\frac{1}{2}} + \tau^{-\frac{1}{2}} P_2^{-\frac{1}{2}}\right)^{\frac{2}{3}} \\ &= 3\tau^{\frac{1}{3}} Y^{\frac{1}{3}} \left(\tau^{-\frac{1}{2}} \left(P_1^{-\frac{1}{2}} + P_2^{-\frac{1}{2}}\right)\right)^{\frac{2}{3}} \\ &= 3\tau^{\frac{1}{3}} Y^{\frac{1}{3}} \tau^{-\frac{1}{3}} \left(P_1^{-\frac{1}{2}} + P_2^{-\frac{1}{2}}\right)^{\frac{2}{3}} \\ &= \tau^0 \times 3Y^{\frac{1}{3}} \left(P_1^{-\frac{1}{2}} + P_2^{-\frac{1}{2}}\right)^{\frac{2}{3}} \\ &= \tau^0 U^*(P_1, P_2, Y) \end{aligned}$$

so that $U^*(P_1, P_2, Y)$ is homogeneous of degree 0.

We have

$$\begin{aligned} \lambda^*(P_1, P_2, Y) &= \frac{\partial U^*(P_1, P_2, Y)}{\partial Y} = Y^{-\frac{2}{3}} \left(P_1^{-\frac{1}{2}} + P_2^{-\frac{1}{2}}\right)^{\frac{2}{3}} \\ \frac{\partial U^*(P_1, P_2, Y)}{\partial P_1} &= -\frac{1}{2} \times \frac{2}{3} \times 3Y^{\frac{1}{3}} \left(P_1^{-\frac{1}{2}} + P_2^{-\frac{1}{2}}\right)^{-\frac{1}{3}} P_1^{-\frac{3}{2}} \\ &= -Y^{\frac{1}{3}} \left(P_1^{-\frac{1}{2}} + P_2^{-\frac{1}{2}}\right)^{-\frac{1}{3}} P_1^{-\frac{3}{2}} \end{aligned}$$

so that

$$\begin{aligned} Q_1^*(P_1,P_2,Y) &= -\frac{\frac{\partial U^*(P_1,P_2,Y)}{\partial P_1}}{\frac{\partial U^*(P_1,P_2,Y)}{\partial Y}} \\ &= -\frac{-Y^{\frac{1}{3}}\left(P_1^{-\frac{1}{2}}+P_2^{-\frac{1}{2}}\right)^{-\frac{1}{3}}P_1^{-\frac{3}{2}}}{Y^{-\frac{2}{3}}\left(P_1^{-\frac{1}{2}}+P_2^{-\frac{1}{2}}\right)^{\frac{2}{3}}} \\ &= \frac{YP_1^{-\frac{3}{2}}}{P_1^{-\frac{1}{2}}+P_2^{-\frac{1}{2}}}. \end{aligned}$$

The income elasticity is the exponent on Y or 1.

From the first-order conditions for utility maximization

$$\begin{aligned} \frac{\partial U(Q_1^*,Q_2^*)}{\partial Q_1} - \lambda^* P_1 &= 0 \\ &\implies \lambda^* = \frac{\partial U(Q_1^*,Q_2^*)}{\partial Q_1}/P_1 = Y^{-\frac{2}{3}}\left(P_1^{-\frac{1}{2}}+P_2^{-\frac{1}{2}}\right)^{\frac{2}{3}}. \end{aligned}$$

Problem 3.24 *[A **] Suppose you are running a prison. Inmates in the prison consume only two goods Q_1 and Q_2 (say bread and water) which have prices P_1 and P_2 respectively. You as the warden, having no control over prices, wish to choose Q_1 and Q_2 so as to minimize the expenditure E on each prisoner given by*

$$E = P_1Q_1 + P_2Q_2.$$

Prison regulations stipulate that each prisoner's utility be a certain minimum amount, say U_0, so that your constraint as a warden is that Q_1 and Q_2 must satisfy

$$U_0 = U(Q_1,Q_2) = \alpha\ln(Q_1) + (1-\alpha)\ln(Q_2).$$

From the Lagrangian find the first -order conditions for the prison allocation problem and solve for the optimal values. Find

$$E^*(P_1,P_2,U_0) = P_1Q_1^*(P_1,P_2,U_0) + P_2Q_2^*(P_1,P_2,U_0).$$

Verify that

$$\begin{aligned} Q_1^*(P_1,P_2,U_0) &= \frac{\partial E^*(P_1,P_2,U_0)}{\partial P_1},\ Q_2^*(P_1,P_2,U_0) = \frac{\partial E^*(P_1,P_2,U_0)}{\partial P_2} \\ \lambda^*(P_1,P_2,U_0) &= \frac{\partial E^*(P_1,P_2,U_0)}{\partial U_0} \end{aligned}$$

and show that $\frac{\partial Q_1^*}{\partial P_2} = \frac{\partial Q_2^*}{\partial P1}$.

Answer: We have

$$\mathcal{L}(\lambda,Q_1,Q_2,P_1,P_2,U_0) = P_1Q_1 + P_2Q_2 + \lambda(U_0 - \alpha\ln(Q_1) - (1-\alpha)\ln(Q_2))$$

so that the first-order conditions yield

$$\begin{aligned}
U_0 - \alpha \ln (Q_1^*) - (1-\alpha) \ln (Q_2^*) &= 0 \\
&\Longrightarrow U_0 = \alpha \ln (Q_1^*) + (1-\alpha) \ln (Q_2^*) \\
P_1 - \lambda^* \frac{\alpha}{Q_1^*} &= 0 \\
&\Longrightarrow Q_1^* = \frac{\lambda^* \alpha}{P_1} \\
P_2 - \lambda^* \frac{(1-\alpha)}{Q_2^*} &= 0 \\
&\Longrightarrow Q_2^* = \frac{\lambda^* (1-\alpha)}{P_2}.
\end{aligned}$$

Placing the second and third equation in the first yields

$$U_0 = \alpha \ln \left(\frac{\lambda^* \alpha}{P_1} \right) + (1-\alpha) \ln \left(\frac{\lambda^* (1-\alpha)}{P_2} \right)$$

so that solving for λ^* we have

$$\lambda^* = \frac{e^{U_0} P_1^{\alpha} P_2^{1-\alpha}}{\alpha^{\alpha} (1-\alpha)^{1-\alpha}} = \alpha^{-\alpha} (1-\alpha)^{\alpha-1} e^{U_0} P_1^{\alpha} P_2^{1-\alpha}.$$

Thus

$$\begin{aligned}
Q_1^* &= \frac{\lambda^* \alpha}{P_1} = \alpha^{-\alpha} (1-\alpha)^{\alpha-1} e^{U_0} P_1^{\alpha} P_2^{1-\alpha} \frac{\alpha}{P_1} \\
&= \alpha^{1-\alpha} (1-\alpha)^{\alpha-1} e^{U_0} P_1^{\alpha-1} P_2^{1-\alpha} \\
Q_2^* &= \frac{\lambda^* (1-\alpha)}{P_2} = \alpha^{-\alpha} (1-\alpha)^{\alpha-1} e^{U_0} P_1^{\alpha} P_2^{1-\alpha} \frac{1-\alpha}{P_2} \\
&= \alpha^{-\alpha} (1-\alpha)^{\alpha} e^{U_0} P_1^{\alpha} P_2^{-\alpha}.
\end{aligned}$$

Now

$$\begin{aligned}
E^* (P_1, P_2, U_0) &\equiv P_1 Q_1^* + P_2 Q_2^* = P_1 \frac{\lambda^* \alpha}{P_1} + P_2 \frac{\lambda^* (1-\alpha)}{P_2} = \lambda^* (\alpha + (1-\alpha)) \\
&= \lambda^* = \alpha^{-\alpha} (1-\alpha)^{\alpha-1} e^{U_0} P_1^{\alpha} P_2^{1-\alpha}.
\end{aligned}$$

We have

$$\begin{aligned}
\frac{\partial E^*}{\partial P_1} &= \frac{\partial}{\partial P_1} \left(\alpha^{-\alpha} (1-\alpha)^{\alpha-1} e^{U_0} P_1^{\alpha} P_2^{1-\alpha} \right) = \alpha \alpha^{-\alpha} (1-\alpha)^{\alpha-1} e^{U_0} P_1^{\alpha-1} P_2^{1-\alpha} \\
&= \alpha^{1-\alpha} (1-\alpha)^{\alpha-1} e^{U_0} P_1^{\alpha-1} P_2^{1-\alpha} = Q_1^* \\
\frac{\partial E^*}{\partial P_2} &= \frac{\partial}{\partial P_2} \left(\alpha^{-\alpha} (1-\alpha)^{\alpha-1} e^{U_0} P_1^{\alpha} P_2^{1-\alpha} \right) = (1-\alpha) \alpha^{-\alpha} (1-\alpha)^{\alpha-1} e^{U_0} P_1^{\alpha} P_2^{1-\alpha-1} \\
&= \alpha^{-\alpha} (1-\alpha)^{\alpha} e^{U_0} P_1^{\alpha} P_2^{-\alpha} = Q_2^* \\
\frac{\partial E^*}{\partial U_0} &= \frac{\partial}{\partial U_o} \left(\alpha^{-\alpha} (1-\alpha)^{\alpha-1} e^{U_0} P_1^{\alpha} P_2^{1-\alpha} \right) = \alpha^{-\alpha} (1-\alpha)^{\alpha-1} e^{U_0} P_1^{\alpha} P_2^{1-\alpha} = \lambda^*.
\end{aligned}$$

Furthermore

$$\frac{\partial Q_1^*}{\partial P_2} = \frac{\partial}{\partial P_2}\left(\alpha^{1-\alpha}(1-\alpha)^{\alpha-1}e^{U_0}P_1^{\alpha-1}P_2^{1-\alpha}\right) = \alpha^{1-\alpha}(1-\alpha)^{\alpha}e^{U_0}P_1^{\alpha-1}P_2^{-\alpha}$$
$$\frac{\partial Q_2^*}{\partial P_1} = \frac{\partial}{\partial P_1}\left(\alpha^{-\alpha}(1-\alpha)^{\alpha}e^{U_0}P_1^{\alpha}P_2^{-\alpha}\right) = \alpha^{1-\alpha}(1-\alpha)^{\alpha}e^{U_0}P_1^{\alpha-1}P_2^{-\alpha}$$

and so $\frac{\partial Q_1^*}{\partial P_2} = \frac{\partial Q_2^*}{\partial P_1}$.

Problem 3.25 *[B***] Suppose that the household has to pay income tax on Y of t_y, and a tax on Q_1 of t_1 so that the budget constraint becomes*

$$Y \times (1 - t_y) = P_1 \times (1 + t_1) Q_1 + P_2 Q_2.$$

Let $U^*(P_1, P_2, Y, t_y, t_1)$ be the resulting clever utility function. Find the appropriate modification of Roy's identity that would allow you to calculate Q_1^* from $U^*(P_1, P_2, Y, t_y, t_1)$. Show that $\frac{\partial U^*}{\partial t_1} / \frac{\partial U^*}{\partial t_y}$ is the budget share of good 1 or equivalently that $Q_1^* = \frac{\frac{\partial U^*}{\partial t_1}}{\frac{\partial U^*}{\partial t_y}} \frac{Y}{P_1}$.

Answer: The Lagrangian for this problem is

$$\mathcal{L}(\lambda, Q_1, Q_2, P_1, P_2, Y, t_y, t_1) = U(Q_1, Q_2) + \lambda(Y \times (1 - t_y) - P_1 \times (1 + t_1) Q_1 - P_2 Q_2).$$

Thus

$$\begin{aligned}\frac{\partial U^*(P_1, P_2, Y, t_y, t_1)}{\partial Y} &= \frac{\partial}{\partial Y}\mathcal{L}(\lambda, Q_1, Q_2, P_1, P_2, Y, t_y, t_1)\,|_{\lambda=\lambda^*, Q_1=Q_1^*, Q_2=Q_2^*} \\ &= \lambda(1 - t_y)\,|_{\lambda=\lambda^*, Q_1=Q_1^*, Q_2=Q_2^*} = \lambda^*(1 - t_y)\end{aligned}$$

and

$$\begin{aligned}\frac{\partial U^*(P_1, P_2, Y, t_y, t_1)}{\partial P_1} &= \frac{\partial}{\partial P_1}\mathcal{L}(\lambda, Q_1, Q_2, P_1, P_2, Y, t_y, t_1)\,|_{\lambda=\lambda^*, Q_1=Q_1^*, Q_2=Q_2^*} \\ &= -\lambda(1 + t_1) Q_1|_{\lambda=\lambda^*, Q_1=Q_1^*, Q_2=Q_2^*} = -\lambda^*(1 + t_1) Q_1^*.\end{aligned}$$

Thus

$$\begin{aligned}\frac{\frac{\partial U^*(P_1, P_2, Y, t_y, t_1)}{\partial P_1}}{\frac{\partial U^*(P_1, P_2, Y, t_y, t_1)}{\partial Y}} &= \frac{-\lambda^*(1 + t_1) Q_1^*}{\lambda^*(1 - t_y)} = \frac{-(1 + t_1) Q_1^*}{(1 - t_y)} \\ &\implies Q_1^* = -\frac{(1 - t_y)}{(1 + t_1)} \frac{\frac{\partial U^*(P_1, P_2, Y, t_y, t_1)}{\partial P_1}}{\frac{\partial U^*(P_1, P_2, Y, t_y, t_1)}{\partial Y}}.\end{aligned}$$

We have

$$\begin{aligned}\frac{\partial U^*(P_1, P_2, Y, t_y, t_1)}{\partial t_y} &= \frac{\partial}{\partial t_y}\mathcal{L}(\lambda, Q_1, Q_2, P_1, P_2, Y, t_y, t_1)\,|_{\lambda=\lambda^*, Q_1=Q_1^*, Q_2=Q_2^*} \\ &= -\lambda Y|_{\lambda=\lambda^*, Q_1=Q_1^*, Q_2=Q_2^*} = -\lambda^* Y\end{aligned}$$

and

$$\begin{aligned}\frac{\partial U^*(P_1,P_2,Y,t_y,t_1)}{\partial t_1} &= \frac{\partial}{\partial t_1}\mathcal{L}(\lambda,Q_1,Q_2,P_1,P_2,Y,t_y,t_1)\,|_{\lambda=\lambda^*,Q_1=Q_1^*,Q_2=Q_2^*}\\ &= -\lambda P_1 Q_1|_{\lambda=\lambda^*,Q_1=Q_1^*,Q_2=Q_2^*} = -\lambda^* P_1 Q_1^*.\end{aligned}$$

Thus

$$\frac{\frac{\partial U^*(P_1,P_2,Y,t_y,t_1)}{\partial t_1}}{\frac{\partial U^*(P_1,P_2,Y,t_y,t_1)}{\partial t_y}} = \frac{-\lambda^* P_1 Q_1^*}{-\lambda^* Y} = \frac{P_1 Q_1^*}{Y} = \text{budget share of } Q_1$$

or

$$Q_1^* = \frac{\frac{\partial U^*}{\partial t_1}}{\frac{\partial U^*}{\partial t_y}}\frac{Y}{P_1}.$$

Problem 3.26 *[B***] Consider the household's problem of maximizing utility as a function of three goods $U(Q_1,Q_2,Q_3)$ subject to the budget constraint*

$$Y = P_1Q_1 + P_2Q_2 + P_3Q_3.$$

Set up the Lagrangian for this problem and find the first-order conditions. Show from the first-order conditions that

$$\frac{MU_1}{P_1} = \frac{MU_2}{P_2} = \frac{MU_3}{P_3} = \lambda^*$$

Using the envelope theorem, prove that λ^* is the marginal utility of income. Prove Roy's identity; that is how to calculate $Q_1^*(P_1,P_2,P_3,Y)$, from the clever utility function $U^*(P_1,P_2,P_3,Y)$.

Answer: We have

$$\mathcal{L} = U(Q_1,Q_2,Q_3) + \lambda(Y - P_1Q_1 - P_2Q_2 - P_3Q_3)$$

so that the first-order conditions are

$$\begin{aligned} Y - P_1Q_1^* - P_2Q_2^* - P_3Q_3^* &= 0\\ \frac{\partial U(Q_1^*,Q_2^*,Q_3^*)}{\partial Q_1} - \lambda^* P_1 &= 0\\ \frac{\partial U(Q_1^*,Q_2^*,Q_3^*)}{\partial Q_2} - \lambda^* P_2 &= 0\\ \frac{\partial U(Q_1^*,Q_2^*,Q_3^*)}{\partial Q_3} - \lambda^* P_3 &= 0.\end{aligned}$$

Since $MU_i = \frac{\partial U(Q_1^*,Q_2^*,Q_3^*)}{\partial Q_i}$ we have the desired result.

We have

$$\begin{aligned}\frac{\partial U^*(P_1,P_2,P_3,Y)}{\partial Y} &= \frac{\partial \mathcal{L}(\lambda,Q_1,Q_2,Q_3,P_1,P_2,P_3,Y)}{\partial Y}|_{Q_i=Q_i^*,\lambda=\lambda^*}\\ &= (\lambda)\,|_{Q_i=Q_i^*,\lambda=\lambda^*} = \lambda^*\end{aligned}$$

and

$$\begin{aligned}\frac{\partial U^*(P_1,P_2,P_3,Y)}{\partial P_1} &= \frac{\partial \mathcal{L}(\lambda,Q_1,Q_2,Q_3,P_1,P_2,P_3,Y)}{\partial P_1}|_{Q_i=Q_i^*,\lambda=\lambda^*} \\ &= (-\lambda Q_1)|_{Q_i=Q_i^*,\lambda=\lambda^*} = -\lambda^* Q_1^*\end{aligned}$$

so that

$$-\frac{\frac{\partial U^*(P_1,P_2,P_3,Y)}{\partial P_1}}{\frac{\partial U^*(P_1,P_2,P_3,Y)}{\partial Y}} = -\frac{-\lambda^* Q_1^*}{\lambda^*} = Q_1^*.$$

Problem 3.27 *[B**] Consider making this problem an unconstrained maximization problem by using the budget constraint to solve out Q_3 as*

$$Q_3 = \frac{Y - P_1Q_1 - P_2Q_2}{P_3}.$$

Thus maximizing $U(Q_1,Q_2,Q_3)$ subject to the income constraint is equivalent to the unconstrained problem of maximizing

$$V(Q_1,Q_2,P_1,P_2,P_3,Y)$$

over Q_1 and Q_2 where

$$V(Q_1,Q_2,P_1,P_2,P_3,Y) = U\left(Q_1,Q_2,\frac{Y-P_1Q_1-P_2Q_2}{P_3}\right).$$

Let

$$V^*(P_1,P_2,P_3,Y) = V(Q_1^*,Q_2^*,P_1,P_2,P_3,Y)$$

be the clever version of V. Use the envelope theorem to show how to calculate $Q_1^*(P_1,P_2,P_3,Y)$ from $V^*(P_1,P_2,P_3,Y)$.

Answer: We have

$$\begin{aligned}\frac{\partial V^*(P_1,P_2,P_3,Y)}{\partial Y} &= \frac{\partial U\left(Q_1,Q_2,\frac{Y-P_1Q_1-P_2Q_2}{P_3}\right)}{\partial Y}|_{Q_1=Q_1^*,Q_2=Q_2^*} \\ &= \frac{1}{P_3}\frac{\partial U\left(Q_1,Q_2,\frac{Y-P_1Q_1-P_2Q_2}{P_3}\right)}{\partial Q_3}|_{Q_1=Q_1^*,Q_2=Q_2^*} \\ &= \frac{1}{P_3}\frac{\partial U\left(Q_1^*,Q_2^*,\frac{Y-P_1Q_1^*-P_2Q_2^*}{P_3}\right)}{\partial Q_3} = \frac{1}{P_3}\frac{\partial U(Q_1^*,Q_2^*,Q_3^*)}{\partial Q_3} = \frac{MU_3}{P_3}\end{aligned}$$

and

$$\begin{aligned}
\frac{\partial V^*(P_1,P_2,P_3,Y)}{\partial P_1} &= \frac{\partial U\left(Q_1,Q_2,\frac{Y-P_1Q_1-P_2Q_2}{P_3}\right)}{\partial P_1}|_{Q_1=Q_1^*,Q_2=Q_2^*} \\
&= -\frac{Q_1}{P_3}\frac{\partial U\left(Q_1,Q_2,\frac{Y-P_1Q_1-P_2Q_2}{P_3}\right)}{\partial Q_3}|_{Q_1=Q_1^*,Q_2=Q_2^*} \\
&= -\frac{Q_1^*}{P_3}\frac{\partial U\left(Q_1^*,Q_2^*,\frac{Y-P_1Q_1^*-P_2Q_2^*}{P_3}\right)}{\partial Q_3} \\
&= -\frac{Q_1^*}{P_3}\frac{\partial U\left(Q_1^*,Q_2^*,Q_3^*\right)}{\partial Q_3} = -\frac{Q_1^*}{P_3}MU_3.
\end{aligned}$$

Therefore

$$-\frac{\frac{\partial V^*(P_1,P_2,P_3,Y)}{\partial P_1}}{\frac{\partial V^*(P_1,P_2,P_3,Y)}{\partial Y}} = -\frac{-\frac{Q_1^*}{P_3}MU_3}{\frac{MU_3}{P_3}} = Q_1^*.$$

Problem 3.28 *[A ***] Consider a Cobb-Douglas utility function with n goods*

$$U = \alpha_1 \ln(Q_1) + \alpha_2 \ln(Q_2) + \cdots + \alpha_n \ln(Q_n)$$

where $\alpha_i \geq 0$ and

$$\alpha_1 + \alpha_2 + \cdots + \alpha_n = 1.$$

The budget constraint is

$$Y = P_1Q_1 + P_2Q_2 + \cdots + P_nQ_n.$$

Find the utility maximizing values of Q_i^* and use this to construct the indirect utility function $U^*(P_1,P_2,\ldots,P_n,Y)$. Verify Roy's identify

$$Q_i^* = -\frac{\frac{\partial U^*(P_1,P_2,\ldots,P_n,Y)}{\partial P_i}}{\frac{\partial U^*(P_1,P_2,\ldots,P_n,Y)}{\partial Y}}.$$

Verify that $U^*(P_1,P_2,\ldots,P_n,Y)$ is homogeneous of degree 0.

Answer: The Lagrangian is

$$\begin{aligned}
\mathcal{L}(\lambda,Q_1,Q_2,\ldots,Q_n) &= \alpha_1\ln(Q_1)+\alpha_2\ln(Q_2)+\cdots+\alpha_n\ln(Q_n) \\
&\quad +\lambda(Y-P_1Q_1-P_2Q_2-\cdots-P_nQ_n)
\end{aligned}$$

so that the first-order conditions yield

$$\begin{aligned}
& Y - P_1 Q_1^* - P_2 Q_2^* - \cdots - P_n Q_n^* \\
\implies & Y = P_1 Q_1^* + P_2 Q_2^* + \cdots + P_n Q_n^* \\
& \frac{\alpha_1}{Q_1^*} - \lambda^* P_1 \\
\implies & \alpha_1 = \lambda^* P_1 Q_1^* \\
& \frac{\alpha_2}{Q_2^*} - \lambda^* P_2 \\
\implies & \alpha_2 = \lambda^* P_2 Q_2^* \\
& \vdots \\
& \frac{\alpha_n}{Q_n^*} - \lambda^* P_n \\
\implies & \alpha_n = \lambda^* P_n Q_n^*.
\end{aligned}$$

Adding up the last n first-order conditions we have

$$\begin{aligned}
\alpha_1 + \alpha_2 + \cdots + \alpha_n &= \lambda^* P_1 Q_1^* + \lambda^* P_2 Q_2^* + \cdots + \lambda^* P_n Q_n^* \\
&\implies \underbrace{\alpha_1 + \alpha_2 + \cdots + \alpha_n}_{1} = \lambda^* \underbrace{(P_1 Q_1^* + P_2 Q_2^* + \cdots + P_n Q_n^*)}_{Y} \\
&\implies \lambda^* = \frac{1}{Y}.
\end{aligned}$$

It then follows that

$$\begin{aligned}
\alpha_i &= \lambda^* P_i Q_i^* = \frac{P_i Q_i^*}{Y} \\
&\implies Q_i^* = \frac{\alpha_i Y}{P_i}.
\end{aligned}$$

We therefore have

$$\begin{aligned}
U^*(P_1, P_2, \ldots, P_n, Y) &= \alpha_1 \ln(Q_1^*) + \alpha_2 \ln(Q_2^*) + \cdots + \alpha_n \ln(Q_n^*) \\
&= \alpha_1 \ln\left(\frac{\alpha_1 Y}{P_1}\right) + \alpha_2 \ln\left(\frac{\alpha_2 Y}{P_2}\right) + \cdots + \alpha_n \ln\left(\frac{\alpha_n Y}{P_n}\right) \\
&= \alpha_1 (\ln(\alpha_1) + \ln(Y) - \ln(P_1)) + \cdots + \alpha_n (\ln(\alpha_n) + \ln(Y) - \ln(P_n)) \\
&= (\alpha_1 + \alpha_2 + \cdots + \alpha_n) \ln(Y) - \alpha_1 \ln(P_1) - \cdots - \alpha_n \ln(P_n) + c \\
&= \ln(Y) - \alpha_1 \ln(P_1) - \cdots - \alpha_n \ln(P_n) + c
\end{aligned}$$

where the coefficient on $\ln(Y)$ is one since the α_i $'s$ sum to 1 and the constant c is

$$c = \alpha_1 \ln(\alpha_1) + \alpha_2 \ln(\alpha_2) + \cdots + \alpha_n \ln(\alpha_n).$$

We therefore have

$$\begin{aligned}
\frac{\partial U^*(P_1, P_2, \ldots, P_n, Y)}{\partial Y} &= \frac{1}{Y} = \lambda^* \\
\frac{\partial U^*(P_1, P_2, \ldots, P_n, Y)}{\partial P_i} &= -\frac{\alpha_i}{P_i}
\end{aligned}$$

so that

$$-\frac{\frac{\partial U^*(P_1,P_2,\ldots,P_n,Y)}{\partial P_i}}{\frac{\partial U^*(P_1,P_2,\ldots,P_n,Y)}{\partial Y}} = -\frac{-\frac{\alpha_i}{P_i}}{\frac{1}{Y}} = \frac{\alpha_i Y}{P_i} = Q_i^*.$$

The clever utility function is homogeneous of degree 0 since

$$\begin{aligned}
U^*(\lambda P_1, \lambda P_2, \ldots, \lambda P_n, \lambda Y) &= \ln(\lambda Y) - \alpha_1 \ln(\lambda P_1) + \alpha_2 \ln(\lambda P_2) + \cdots + \alpha_n \ln(\lambda P_n) - c \\
&= \ln(\lambda)\left(1 - \underbrace{(\alpha_1 + \alpha_2 + \cdots + \alpha_n)}_{1}\right) + \ln(Y) \\
&\quad -\alpha_1 \ln(P_1) - \alpha_2 \ln(P_2) \\
&\quad - \cdots - \alpha_n \ln(P_n) + c \\
&= \underbrace{\lambda^0}_{1}\left(\begin{array}{c} \ln(Y) - \alpha_1 \ln(P_1) - \alpha_2 \ln(P_2) \\ - \cdots - \alpha_n \ln(P_n) + c \end{array}\right) \\
&= \lambda^0 U^*(P_1, P_2, \ldots, P_n, Y).
\end{aligned}$$

Problem 3.29 *[A ***] The CES utility function (Constant Elasticity of Substitution) is given by*

$$U = \sum_{i=1}^{n} \alpha_i \left(\frac{Q_i^{1+\rho} - 1}{1+\rho}\right)$$

where $\rho < 0$. This includes the Cobb-Douglas as a special case where $\rho = -1$ since from L'Hôpital's rule it follows that

$$\lim_{\rho \to -1} \frac{Q_i^{1+\rho} - 1}{1+\rho} = \ln(Q_i).$$

Calculate the demand functions Q_i^* for the CES utility function and the indirect utility function $U^*(P_1, P_2, , \ldots, P_n, Y)$. Verify Roy's identity. Verify that $U^*(P_1, P_2, \ldots, P_n, Y)$ is homogeneous of degree 0. (Note: in this problem we use the summation notation, covered in the next chapter, to make things more compact. If you are uncomfortable with using this notation write everything as a conventional sum, for example

$$\begin{aligned}
U &= \sum_{i=1}^{n} \alpha_i \left(\frac{Q_i^{1+\rho} - 1}{1+\rho}\right) \\
&= \alpha_1 \left(\frac{Q_1^{1+\rho} - 1}{1+\rho}\right) + \alpha_2 \left(\frac{Q_2^{1+\rho} - 1}{1+\rho}\right) + \cdots + \alpha_n \left(\frac{Q_n^{1+\rho} - 1}{1+\rho}\right).
\end{aligned}$$

Answer: The Lagrangian is

$$\mathcal{L}(\lambda, Q_1, Q_2, \ldots, Q_n) = \sum_{i=1}^{n} \alpha_i \left(\frac{Q_i^{1+\rho} - 1}{1+\rho}\right) + \lambda\left(Y - \sum_{i=1}^{n} P_i Q_i\right)$$

so that the first-order conditions are

$$\begin{aligned}
\frac{\partial \mathcal{L}\left(\lambda^*, Q_1^*, Q_2^*, \ldots, Q_n^*\right)}{\partial \lambda} &= 0 = Y - \sum_{i=1}^{n} P_i Q_i^* = 0 \Longrightarrow Y = \sum_{i=1}^{n} P_i Q_i^* \\
\frac{\partial \mathcal{L}\left(\lambda^*, Q_1^*, Q_2^*, \ldots, Q_n^*\right)}{\partial Q_i} &= 0 = \alpha_i \left(Q_i^*\right)^{\rho} - \lambda^* P_i \Longrightarrow P_i Q_i^* = \left(\lambda^*\right)^{\frac{1}{\rho}} \alpha_i^{-\frac{1}{\rho}} P_i^{\frac{1}{\rho}+1} \\
\text{for } i &= 1, 2, \ldots, n.
\end{aligned}$$

Adding up the implications of the last n first-order conditions we have

$$\begin{aligned}
Y &= \sum_{i=1}^{n} P_i Q_i^* = \sum_{i=1}^{n} \left(\lambda^*\right)^{\frac{1}{\rho}} \alpha_i^{-\frac{1}{\rho}} P_i^{\frac{1}{\rho}+1} \\
&= \left(\lambda^*\right)^{\frac{1}{\rho}} \left(\sum_{i=1}^{n} \alpha_i^{-\frac{1}{\rho}} P_i^{\frac{1}{\rho}+1}\right) \\
&\Longrightarrow \left(\lambda^*\right)^{\frac{1}{\rho}} = \frac{Y}{\sum_{i=1}^{n} \alpha_i^{-\frac{1}{\rho}} P_i^{\frac{1}{\rho}+1}}.
\end{aligned}$$

Once we have the Lagrange multiplier we can calculate the demand curves using

$$\begin{aligned}
\alpha_i \left(Q_i^*\right)^{\rho} - \lambda^* P_i &= 0 \\
&\Longrightarrow Q_i^* = \left(\lambda^*\right)^{\frac{1}{\rho}} \alpha_i^{-\frac{1}{\rho}} P_i^{\frac{1}{\rho}} = \frac{\alpha_i^{-\frac{1}{\rho}} P_i^{\frac{1}{\rho}} Y}{\sum_{i=1}^{n} \alpha_i^{-\frac{1}{\rho}} P_i^{\frac{1}{\rho}+1}}.
\end{aligned}$$

Now

$$\begin{aligned}
U^*\left(P_1, P_2, \ldots, P_n, Y\right) &= \sum_{i=1}^{n} \alpha_i \left(\frac{Q_i^{*1+\rho} - 1}{1+\rho}\right) \\
&= \frac{1}{1+\rho} \sum_{i=1}^{n} \alpha_i Q_i^{*1+\rho} - \frac{\sum_{i=1}^{n} \alpha_i}{1+\rho} \\
&= \frac{1}{1+\rho} \sum_{i=1}^{n} \alpha_i \left(\frac{\alpha_i^{-\frac{1}{\rho}} P_i^{\frac{1}{\rho}} Y}{\sum_{i=1}^{n} \alpha_i^{-\frac{1}{\rho}} P_i^{\frac{1}{\rho}+1}}\right)^{1+\rho} - \frac{\sum_{i=1}^{n} \alpha_i}{1+\rho} \\
&= \frac{Y^{1+\rho}}{1+\rho} \left(\sum_{i=1}^{n} \alpha_i^{-\frac{1}{\rho}} P_i^{\frac{1}{\rho}+1}\right)^{-\rho} - \frac{\sum_{i=1}^{n} \alpha_i}{1+\rho}.
\end{aligned}$$

Thus

$$\frac{\partial U^*(P_1,P_2,\ldots,P_n,Y)}{\partial Y} = Y^\rho\left(\sum_{i=1}^n \alpha_i^{-\frac{1}{\rho}}P_i^{\frac{1}{\rho}+1}\right)^{-\rho}$$

$$\begin{aligned}\frac{\partial U^*(P_1,P_2,\ldots,P_n,Y)}{\partial P_i} &= \frac{-\rho Y^{1+\rho}}{1+\rho}\left(\sum_{i=1}^n \alpha_i^{-\frac{1}{\rho}}P_i^{\frac{1}{\rho}+1}\right)^{-\rho-1}\alpha_i^{-\frac{1}{\rho}}\left(\frac{1}{\rho}+1\right)P_i^{\frac{1}{\rho}} \\ &= -Y^{1+\rho}\left(\sum_{i=1}^n \alpha_i^{-\frac{1}{\rho}}P_i^{\frac{1}{\rho}+1}\right)^{-(1+\rho)}\alpha_i^{-\frac{1}{\rho}}P_i^{\frac{1}{\rho}}\end{aligned}$$

so that

$$\begin{aligned}-\frac{\frac{\partial U^*(P_1,P_2,\ldots,P_n,Y)}{\partial P_i}}{\frac{\partial U^*(P_1,P_2,\ldots,P_n,Y)}{\partial Y}} &= -\frac{-Y^{1+\rho}\left(\sum_{i=1}^n \alpha_i^{-\frac{1}{\rho}}P_i^{\frac{1}{\rho}+1}\right)^{-(1+\rho)}\alpha_i^{-\frac{1}{\rho}}P_i^{\frac{1}{\rho}}}{Y^\rho\left(\sum_{i=1}^n \alpha_i^{-\frac{1}{\rho}}P_i^{\frac{1}{\rho}+1}\right)^{-\rho}} \\ &= \frac{\alpha_i^{-\frac{1}{\rho}}P_i^{\frac{1}{\rho}}Y}{\sum_{i=1}^n \alpha_i^{-\frac{1}{\rho}}P_i^{\frac{1}{\rho}+1}} = Q_i^*.\end{aligned}$$

Problem 3.30 *[A ***] Consider a household with concave utility function*

$$U(Q_1,Q_2) = U_1(Q_1) + U_2(Q_2)$$

with clever (or indirect) utility function

$$U^*(P_1,P_2,Y) = U_1(Q_1^*(P_1,P_2,Y)) + U_2(Q_2^*(P_1,P_2,Y)).$$

Use the multivariate chain rule and the first-order conditions to prove that

$$\frac{\partial U^*(P_1,P_2,Y)}{\partial P_1} = \lambda^*\left(P_1\frac{\partial Q_1^*}{\partial P_1} + P_2\frac{\partial Q_2^*}{\partial P_1}\right).$$

Now use the envelope theorem to find $\frac{\partial U^*(P_1,P_2,Y)}{\partial P_1}$ *and use this to prove that*

$$Q_1^* = -\left(P_1\frac{\partial Q_1^*}{\partial P_1} + P_2\frac{\partial Q_2^*}{\partial P_1}\right).$$

Answer: From multivariate chain rule and the first-order conditions

$$U_1'(Q_1^*(P_1,P_2,Y)) = \lambda^* P_1 \text{ and } U_2'(Q_2^*(P_1,P_2,Y)) = \lambda^* P_2$$

we have

$$\begin{aligned}\frac{\partial U^*(P_1,P_2,Y)}{\partial P_1} &= U_1'(Q_1^*(P_1,P_2,Y))\frac{\partial Q_1^*(P_1,P_2,Y)}{\partial P_1} \\ &\quad + U_2'(Q_2^*(P_1,P_2,Y))\frac{\partial Q_2^*(P_1,P_2,Y)}{\partial P_1} \\ &= \lambda^* P_1\frac{\partial Q_1^*}{\partial P_1} + \lambda^* P_2\frac{\partial Q_2^*}{\partial P_1} = \lambda^*\left(P_1\frac{\partial Q_1^*}{\partial P_1} + P_2\frac{\partial Q_2^*}{\partial P_1}\right).\end{aligned}$$

From the envelope theorem we have

$$\begin{aligned}
\frac{\partial U^*(P_1,P_2,Y)}{\partial P_1} &= \frac{\partial \mathcal{L}(\lambda,Q_1,Q_2,P_1,P_2,Y)}{\partial P_1}\Big|_{\substack{\lambda=\lambda^*\\ Q_1=Q_1^*\\ Q_2=Q_2^*}} \\
&= \frac{\partial}{\partial P_1}\left(U_1(Q_1)+U_2(Q_2)+\lambda(Y-P_1Q_1-P_2Q_2)\right)\Big|_{\substack{\lambda=\lambda^*\\ Q_1=Q_1^*\\ Q_2=Q_2^*}} \\
&= -\lambda Q_1\Big|_{\substack{\lambda=\lambda^*\\ Q_1=Q_1^*\\ Q_2=Q_2^*}} = -\lambda^* Q_1^*.
\end{aligned}$$

Therefore

$$\begin{aligned}
-\lambda^* Q_1^* &= \frac{\partial U^*(P_1,P_2,Y)}{\partial P_1} = \lambda^*\left(P_1\frac{\partial Q_1^*}{\partial P_1}+P_2\frac{\partial Q_2^*}{\partial P_1}\right) \\
&\implies Q_1^* = -\left(P_1\frac{\partial Q_1^*}{\partial P_1}+P_2\frac{\partial Q_2^*}{\partial P_1}\right).
\end{aligned}$$

Problem 3.31 *[B***] Suppose*

$$U(Q_1,Q_2)=U_1(Q_1)+U_2(Q_2)$$

and a mathematical bureaucrat in the government imposes a tax on Q_1 at a rate of $1+e^{-T}$ so that the new budget constraint becomes

$$Y = P_1\times\left(1+e^{-T}\right)Q_1+P_2Q_2.$$

Let $U^*(P_1,P_2,Y,T)$ be the new clever utility function. Find $\frac{\partial U^*(P_1,P_2,Y,T)}{\partial T}$ and show that increasing T makes the household better off.

Answer: With the tax the Lagrangian becomes

$$\mathcal{L}(\lambda,Q_1,Q_2,P_1,P_2,Y,T) = U_1(Q_1)+U_2(Q_2)+\lambda\left(Y-P_1\times\left(1+e^{-T}\right)Q_1-P_2Q_2\right)$$

so that using the envelope theorem we have

$$\begin{aligned}
\frac{\partial U^*(P_1,P_2,Y,T)}{\partial T} &= \frac{\partial \mathcal{L}(\lambda,Q_1,Q_2,P_1,P_2,Y,T)}{\partial T}\Big|_{\substack{\lambda=\lambda^*\\ Q_1=Q_1^*\\ Q_2=Q_2^*}} \\
&= \frac{\partial}{\partial T}\left(U_1(Q_1)+U_2(Q_2)+\lambda\left(Y-P_1\times\left(1+e^{-T}\right)Q_1-P_2Q_2\right)\right)\Big|_{\substack{\lambda=\lambda^*\\ Q_1=Q_1^*\\ Q_2=Q_2^*}} \\
&= \lambda e^{-T}P_1Q_1\Big|_{\substack{\lambda=\lambda^*\\ Q_1=Q_1^*\\ Q_2=Q_2^*}} = \lambda^* e^{-T}P_1Q_1^*.
\end{aligned}$$

Since $\lambda^* = \frac{U_1'}{P_1} = \frac{U_2'}{P_2} > 0$ and since $Q_1^* > 0$ and $e^{-T} > 0$ we have $\frac{\partial U^*(P_1,P_2,Y,T)}{\partial T} > 0$ and so increasing T makes the household better off.

Problem 3.32 *Suppose that a household maximizes utility $U(Q_1, Q_2)$ subject to the budget constraint $Y = P_1Q_1 + P_2Q_2$. The resulting clever (or indirect) utility function turns out to be*

$$U^*(P_1, P_2, Y) = Y^{\frac{1}{4}}\left(P_1^{-\frac{1}{2}} + P_2^{-\frac{1}{2}}\right)^{\frac{1}{2}}.$$

*[D***] Show that $U^*(P_1, P_2, Y)$ is homogeneous of degree k and determine k. Determine the Lagrange multiplier $\lambda^*(P_1, P_2, Y)$ and the demand curve $Q_1^*(P_1, P_2, Y)$. What is the income elasticity for Q_1?*

Answer: We have

$$\begin{aligned} U^*(\lambda P_1, \lambda P_2, \lambda Y) &= (\lambda Y)^{\frac{1}{4}}\left((\lambda P_1)^{-\frac{1}{2}} + (\lambda P_2)^{-\frac{1}{2}}\right)^{\frac{1}{2}} = \lambda^{\frac{1}{4}} Y^{\frac{1}{4}}\left(\lambda^{-\frac{1}{2}}\left(P_1^{-\frac{1}{2}} + P_2^{-\frac{1}{2}}\right)\right)^{\frac{1}{2}} \\ &= \lambda^{\frac{1}{4}}\left(\lambda^{-\frac{1}{2}}\right)^{\frac{1}{2}} Y^{\frac{1}{4}}\left(P_1^{-\frac{1}{2}} + P_2^{-\frac{1}{2}}\right)^{\frac{1}{2}} = \lambda^{\frac{1}{4}-\frac{1}{4}} Y^{\frac{1}{4}}\left(P_1^{-\frac{1}{2}} + P_2^{-\frac{1}{2}}\right)^{\frac{1}{2}} \\ &= \lambda^0 Y^{\frac{1}{4}}\left(P_1^{-\frac{1}{2}} + P_2^{-\frac{1}{2}}\right)^{\frac{1}{2}} = \lambda^0 U^*(P_1, P_2, Y) \end{aligned}$$

so that $U^*(P_1, P_2, Y)$ is homogeneous of degree $k = 0$.

Using the envelope theorem we have

$$\begin{aligned} \lambda^*(P_1, P_2, Y) &= \frac{\partial U^*(P_1, P_2, Y)}{\partial Y} \\ &= \frac{\partial}{\partial Y}\left(Y^{\frac{1}{4}}\left(P_1^{-\frac{1}{2}} + P_2^{-\frac{1}{2}}\right)^{\frac{1}{2}}\right) = \frac{1}{4}Y^{-\frac{3}{4}}\left(P_1^{-\frac{1}{2}} + P_2^{-\frac{1}{2}}\right)^{\frac{1}{2}} \end{aligned}$$

and

$$\begin{aligned} -\lambda^*(P_1, P_2, Y)\, Q_1^*(P_1, P_2, Y) &= \frac{\partial U^*(P_1, P_2, Y)}{\partial P_1} \\ &= -\frac{1}{4}Y^{\frac{1}{4}}\left(P_1^{-\frac{1}{2}} + P_2^{-\frac{1}{2}}\right)^{-\frac{1}{2}} P_1^{-\frac{3}{2}} \end{aligned}$$

so that

$$\begin{aligned} Q_1^*(P_1, P_2, Y) &= -\frac{\frac{\partial U^*(P_1, P_2, Y)}{\partial P_1}}{\frac{\partial U^*(P_1, P_2, Y)}{\partial Y}} \\ &= -\frac{-\frac{1}{4}Y^{\frac{1}{4}}\left(P_1^{-\frac{1}{2}} + P_2^{-\frac{1}{2}}\right)^{-\frac{1}{2}} P_1^{-\frac{3}{2}}}{\frac{1}{4}Y^{-\frac{3}{4}}\left(P_1^{-\frac{1}{2}} + P_2^{-\frac{1}{2}}\right)^{\frac{1}{2}}} = \frac{YP_1^{-\frac{3}{2}}}{P_1^{-\frac{1}{2}} + P_2^{-\frac{1}{2}}}. \end{aligned}$$

Since the exponent on Y is 1 the income elasticity of demand is 1.

Problem 3.33 *Suppose that a household maximizes utility $U(Q_1, Q_2)$ subject to the budget constraint $Y = P_1Q_1 + P_2Q_2$. The resulting clever (or indirect) utility function turns out to be*

$$U^*(P_1, P_2, Y) = 10YP_1^{-\frac{3}{4}}P_2^{-\frac{1}{4}}.$$

*[D***] Show that $U^*(P_1, P_2, Y)$ is homogeneous of degree k and determine k. Determine the Lagrange multiplier $\lambda^*(P_1, P_2, Y)$ and the demand curve $Q_1^*(P_1, P_2, Y)$. What is the income elasticity for Q_1? In the old days some economists argued that the marginal utility of income was a constant. Samuelson showed, using Euler's theorem, that the marginal utility of income cannot be a constant. How can this be done?*

Answer: The clever utility function has the Cobb-Douglas functional form so that it is homogenous of degree k, with k being the sum of the exponents or

$$k = 1 + -\frac{3}{4} + -\frac{1}{4} = 0.$$

Alternatively

$$\begin{aligned} U^*(\lambda P_1, \lambda P_2, \lambda Y) &= 10(\lambda Y)(\lambda P_1)^{-\frac{3}{4}}(\lambda P_2)^{-\frac{1}{4}} = \lambda^{1-\frac{3}{4}-\frac{1}{4}} \times 10 Y P_1^{-\frac{3}{4}} P_2^{-\frac{1}{4}} \\ &= \lambda^0 10 Y P_1^{-\frac{3}{4}} P_2^{-\frac{1}{4}} = \lambda^0 U^*(P_1, P_2, Y) \end{aligned}$$

so that $U^*(P_1, P_2, Y)$ is homogeneous of degree $k = 0$.

Using the envelope theorem we have

$$\lambda^*(P_1, P_2, Y) = \frac{\partial U^*(P_1, P_2, Y)}{\partial Y} = \frac{\partial}{\partial Y}\left(10 Y P_1^{-\frac{3}{4}} P_2^{-\frac{1}{4}}\right) = 10 P_1^{-\frac{3}{4}} P_2^{-\frac{1}{4}}$$

and

$$-\lambda^*(P_1, P_2, Y) Q_1^*(P_1, P_2, Y) = \frac{\partial U^*(P_1, P_2, Y)}{\partial P_1} = -\frac{3}{4} 10 Y P_1^{-\frac{7}{4}} P_2^{-\frac{1}{4}}$$

so that

$$Q_1^*(P_1, P_2, Y) = -\frac{\frac{\partial U^*(P_1, P_2, Y)}{\partial P_1}}{\frac{\partial U^*(P_1, P_2, Y)}{\partial Y}} = -\frac{-\frac{3}{4} 10 Y P_1^{-\frac{7}{4}} P_2^{-\frac{1}{4}}}{10 P_1^{-\frac{3}{4}} P_2^{-\frac{1}{4}}} = \frac{\frac{3}{4} Y}{P_1}.$$

Since the exponent on Y is 1 the income elasticity of demand is 1. Since $U^*(P_1, P_2, Y)$ is homogeneous of degree 0 and since $\lambda^* = \frac{\partial U^*(P_1, P_2, Y)}{\partial Y}$ it follows that λ^* is homogeneous of degree $k = -1$. Therefore by Euler's theorem

$$-1 \times \lambda^* = \frac{\partial \lambda^*}{\partial Y} Y + \frac{\partial \lambda^*}{\partial P_1} P_1 + \frac{\partial \lambda^*}{\partial P_2} P_2.$$

If λ^* is a constant then $\frac{\partial \lambda^*}{\partial Y} = \frac{\partial \lambda^*}{\partial P_1} = \frac{\partial \lambda^*}{\partial P_2} = 0$ which would then imply that $-\lambda^* = 0$, a contradiction (to be precise $\lambda^* \neq 0$ follows since from the first-order conditions $\frac{\partial U(Q_1^*, Q_2^*)}{\partial Q_1} = \lambda^* P_1 \neq 0$ and $\frac{\partial U(Q_1^*, Q_2^*)}{\partial Q_1} > 0$ and $P_1 > 0$).

Problem 3.34 *[B***] Write down the Lagrangian for the problem of maximizing utility $U(Q_1, Q_2)$ subject to the budget constraint $Y = P_1 Q_1 + P_2 Q_2$. Let $U^*(P_1, P_2, Y)$ be the clever (or indirect utility function). Using the envelope* *theorem find* $\frac{\partial U^*(P_1,P_2,Y)}{\partial Y}$ *and* $\frac{\partial U^*(P_1,P_2,Y)}{\partial P_1}$ *and show how these can be used to calculate* Q_1^*. *Using Young's theorem show that*

$$\lambda^* \frac{\partial Q_1^*}{\partial P_2} + Q_1^* \frac{\partial \lambda^*}{\partial P_2} = \lambda^* \frac{\partial Q_2^*}{\partial P_1} + Q_2^* \frac{\partial \lambda^*}{\partial P_1}.$$

Now suppose that the household has to pay a tax on income of T, *and a tax on* Q_1 *at a rate of* t_1 *so that the budget constraint becomes*

$$Y - T = (1 + t_1) \times P_1 Q_1 + P_2 Q_2.$$

Let $U^*(P_1, P_2, Y, T, t_1)$ *be the resulting clever utility function. Use the envelope theorem to calculate* $\frac{\partial U^*}{\partial T}$ *and* $\frac{\partial U^*}{\partial t_1}$.

Answer: Since

$$\begin{aligned}
\frac{\partial U^*}{\partial P_1} &= -\lambda^* Q_1^*, \Longrightarrow \frac{\partial^2 U^*}{\partial P_2 \partial P_1} = -\lambda^* \frac{\partial Q_1^*}{\partial P_2} - Q_1^* \frac{\partial \lambda^*}{\partial P_2} \\
\frac{\partial U^*}{\partial P_2} &= -\lambda^* Q_2^* \Longrightarrow \frac{\partial^2 U^*}{\partial P_1 \partial P_2} = -\lambda^* \frac{\partial Q_2^*}{\partial P_1} - Q_2^* \frac{\partial \lambda^*}{\partial P_1}
\end{aligned}$$

and so by Young's theorem

$$-\frac{\partial^2 U^*}{\partial P_2 \partial P_1} = -\frac{\partial^2 U^*}{\partial P_1 \partial P_2} = \lambda^* \frac{\partial Q_1^*}{\partial P_2} + Q_1^* \frac{\partial \lambda^*}{\partial P_2} = \lambda^* \frac{\partial Q_2^*}{\partial P_1} + Q_2^* \frac{\partial \lambda^*}{\partial P_1}.$$

With the taxes the Lagrangian is given by

$$\mathcal{L}(\lambda, Q_1, Q_2, P_1, P_2, Y, T, t_1) = U(Q_1, Q_2) + \lambda (Y - T - (1 + t_1) \times P_1 Q_1 - P_2 Q_2).$$

Thus

$$\begin{aligned}
\frac{\partial U^*}{\partial T} &= \frac{\partial \mathcal{L}(\lambda, Q_1, Q_2, P_1, P_2, Y, T, t_1)}{\partial T}\Big|_{\substack{\lambda=\lambda^* \\ Q_1=Q_1^* \\ Q_2=Q_2^*}} \\
&= (-\lambda)\Big|_{\substack{\lambda=\lambda^* \\ Q_1=Q_1^* \\ Q_2=Q_2^*}} = -\lambda^*
\end{aligned}$$

and

$$\begin{aligned}
\frac{\partial U^*}{\partial t_1} &= \frac{\partial \mathcal{L}(\lambda, Q_1, Q_2, P_1, P_2, Y, T, t_1)}{\partial t_1}\Big|_{\substack{\lambda=\lambda^* \\ Q_1=Q_1^* \\ Q_2=Q_2^*}} \\
&= (-\lambda P_1 Q_1)\Big|_{\substack{\lambda=\lambda^* \\ Q_1=Q_1^* \\ Q_2=Q_2^*}} = -\lambda^* P_1 Q_1^*.
\end{aligned}$$

Problem 3.35 *Suppose that the utility function takes the form*

$$U(Q_1, Q_2) = 2Q_1^{\frac{1}{2}} + 2Q_2^{\frac{1}{2}}.$$

*[D***] Write down the first-order conditions for utility maximization and solve for Q_1^* and Q_2^*. Show that the clever utility function then is*

$$U^*(P_1, P_2, Y) = 2Y^{\frac{1}{2}}\left(P_1^{-1} + P_2^{-1}\right)^{\frac{1}{2}}.$$

From $U^*(P_1, P_2, Y)$ use the envelope theorem to calculate Q_1^* and λ^*.

Answer: We have

$$\mathcal{L} = 2Q_1^{\frac{1}{2}} + 2Q_2^{\frac{1}{2}} + \lambda(Y - P_1Q_1 - P_2Q_2)$$

so that the first-order conditions yield

$$\begin{aligned}
Y - P_1Q_1^* - P_2Q_2^* &= 0 \\
&\implies Y = P_1Q_1^* + P_2Q_2^* \\
(Q_1^*)^{-\frac{1}{2}} - \lambda^* P_1 &= 0 \\
&\implies Q_1^* = (\lambda^*)^{-2}(P_1)^{-2} \\
(Q_2^*)^{-\frac{1}{2}} - \lambda^* P_2 &= 0 \\
&\implies Q_2^* = (\lambda^*)^{-2}(P_2)^{-2}.
\end{aligned}$$

From the second and third results we have

$$P_1Q_1^* = (\lambda^*)^{-2}P_1^{-1},\ P_2Q_2^* = (\lambda^*)^{-2}P_2^{-1}$$

and hence putting this in the budget constraint we have

$$\begin{aligned}
Y &= P_1Q_1^* + P_2Q_2^* = (\lambda^*)^{-2}P_1^{-1} + (\lambda^*)^{-2}P_2^{-1} \\
&\implies (\lambda^*)^{-2} = \frac{Y}{P_1^{-1} + P_2^{-1}}.
\end{aligned}$$

Thus

$$\begin{aligned}
Q_1^* &= (\lambda^*)^{-2}(P_1)^{-2} = \frac{YP_1^{-2}}{P_1^{-1} + P_2^{-1}} \\
Q_2^* &= (\lambda^*)^{-2}(P_2)^{-2} = \frac{YP_2^{-2}}{P_1^{-1} + P_2^{-1}}
\end{aligned}$$

so that

$$\begin{aligned}
U^*(P_1, P_2, Y) &= 2(Q_1^*)^{\frac{1}{2}} + 2(Q_2^*)^{\frac{1}{2}} \\
&= 2\left(\frac{YP_1^{-2}}{P_1^{-1} + P_2^{-1}}\right)^{\frac{1}{2}} + 2\left(\frac{YP_2^{-2}}{P_1^{-1} + P_2^{-1}}\right)^{\frac{1}{2}} \\
&= 2Y^{\frac{1}{2}}\left(\frac{P_1^{-1}}{\left(P_1^{-1} + P_2^{-1}\right)^{\frac{1}{2}}}\right) + 2Y^{\frac{1}{2}}\left(\frac{P_2^{-1}}{\left(P_1^{-1} + P_2^{-1}\right)^{\frac{1}{2}}}\right) \\
&= 2Y^{\frac{1}{2}}\frac{\left(P_1^{-1} + P_2^{-1}\right)^1}{\left(P_1^{-1} + P_2^{-1}\right)^{\frac{1}{2}}} = 2Y^{\frac{1}{2}}\left(P_1^{-1} + P_2^{-1}\right)^{\frac{1}{2}}.
\end{aligned}$$

Thus

$$\begin{aligned}\lambda^* &= \frac{\partial U^*(P_1,P_2,Y)}{\partial Y} = Y^{-\frac{1}{2}}\left(P_1^{-1}+P_2^{-1}\right)^{\frac{1}{2}} \\ -\lambda^* Q_1^* &= \frac{\partial U^*(P_1,P_2,Y)}{\partial P_1} = -Y^{\frac{1}{2}}\left(P_1^{-1}+P_2^{-1}\right)^{-\frac{1}{2}} P_1^{-2}\end{aligned}$$

so that

$$Q_1^* = -\frac{\frac{\partial U^*(P_1,P_2,Y)}{\partial P_1}}{\frac{\partial U^*(P_1,P_2,Y)}{\partial Y}} = \frac{Y^{\frac{1}{2}}\left(P_1^{-1}+P_2^{-1}\right)^{-\frac{1}{2}} P_1^{-2}}{Y^{-\frac{1}{2}}\left(P_1^{-1}+P_2^{-1}\right)^{\frac{1}{2}}} = \frac{YP_1^{-2}}{P_1^{-1}+P_2^{-1}}.$$

3.3 Mark Maximization

Problem 3.36 *[A**] Consider the following problem: a student has to write an exam with n questions. He has T minutes to write the exam. Let t_i be the number of minutes he spends on question i. Suppose the marks he gets for question i is determined by the function $M_i(t_i)$ where $M_i'(t_i) > 0$ and $M_i''(t_i) < 0$. The Lagrangian for this problem is then given by*

$$\mathcal{L}(\lambda, t_1, t_2, \ldots, t_n, T) = \sum_{i=1}^{n} M_i(t_i) + \lambda\left(T - \sum_{i=1}^{n} t_i\right).$$

What in every day language is the objective and constraint in this problem? What are the first-order conditions for this problem? Using economic terminology, what do these first-order conditions express?

Answer: The first-order conditions are

$$T - \sum_{i=1}^{n} t_i^* = 0 \text{ and } M_i'(t_i^*) - \lambda^* = 0 \text{ for } i = 1, 2, \ldots, n.$$

The first first-order condition $T - \sum_{i=1}^{n} t_i^* = 0$ states that the optimal solution obeys the time constraint, while the latter conditions imply that $M_i'(t_i^*) = \lambda^*$, or that the marginal return on each question must be the same.

Problem 3.37 *[A**] Will a solution to the first-order conditions corresponds to a global maximum? What are the second-order conditions using the bordered Hessian for this problem for the case where $n = 2$? Show that these second-order conditions will be satisfied.*

Answer: Since $\lambda^* = M_i'(t_i^*) > 0$ and since the objective function is concave while the constraint function is linear, it follows that a solution to the first-order conditions will correspond to a global maximum.

We have

$$\det[H] = \det\begin{bmatrix} 0 & -1 & -1 \\ -1 & M_1''(t_1^*) & 0 \\ -1 & 0 & M_2''(t_2^*) \end{bmatrix} = -M_1''(t_1^*) - M_2''(t_2^*) > 0.$$

Problem 3.38 *[A **] From the first-order conditions show that $\lambda^*(T) > 0$. Let $t_i^*(T)$ be the value of t_i which solves the maximization problem. Consider the function $M^*(T) \equiv \sum_{i=1}^{n} M_i(t_i^*(T))$. Use the envelope theorem to show that $\frac{\partial M^*(T)}{\partial T} = \lambda^*(T)$. What then is the appropriate interpretation of $\lambda^*(T)$?*

Answer: Since $M_i'(t_i) > 0$ it follows that $\lambda^*(T) = M_i'(t_i^*) > 0$. Using the envelope theorem we have

$$\frac{\partial M^*(T)}{\partial T} = \frac{\partial}{\partial T}\mathcal{L}(\lambda, t_1, t_2, \ldots, t_n, T)\,|_{\lambda=\lambda^*, t_i=t_i^*} = \lambda|_{\lambda=\lambda^*, t_i=t_i^*} = \lambda^*(T).$$

Thus $\lambda^*(T)$ is the marginal mark.

Problem 3.39 *[A **] For the case where $n = 2$ and using the total differential of the first-order conditions, show that there is a diminishing marginal return to extra time $\left(\frac{\partial \lambda^*}{\partial T} < 0\right)$ and that an increase in T always leads to an increase in the amount of time spend on each question.*

Answer: For $n = 2$ we have the total differential

$$\begin{bmatrix} 0 & -1 & -1 \\ -1 & M_1''(t_1^*) & 0 \\ -1 & 0 & M_2''(t_2^*) \end{bmatrix}\begin{bmatrix} d\lambda^* \\ dt_1^* \\ dt_2^* \end{bmatrix} + \begin{bmatrix} dT \\ 0 \\ 0 \end{bmatrix} = \begin{bmatrix} 0 \\ 0 \\ 0 \end{bmatrix}$$

where as usual the square matrix is the Hessian and has a positive determinant since

$$\begin{aligned} \det[H] &= \det\begin{bmatrix} 0 & -1 & -1 \\ -1 & M_1''(t_1^*) & 0 \\ -1 & 0 & M_2''(t_2^*) \end{bmatrix} \\ &= -M_1''(t_1^*) \cdot M_2''(t_2^*) > 0. \end{aligned}$$

It follows then that

$$\begin{bmatrix} 0 & -1 & -1 \\ -1 & M_1''(t_1^*) & 0 \\ -1 & 0 & M_2''(t_2^*) \end{bmatrix}\begin{bmatrix} \frac{\partial \lambda^*}{\partial T} \\ \frac{\partial t_1^*}{\partial T} \\ \frac{\partial t_2^*}{\partial T} \end{bmatrix} = \begin{bmatrix} -1 \\ 0 \\ 0 \end{bmatrix}$$

so that using Cramer's rule

$$\frac{\partial \lambda^*}{\partial T} = \frac{\det \begin{bmatrix} -1 & -1 & -1 \\ 0 & M_1''(t_1^*) & 0 \\ 0 & 0 & M_2''(t_2^*) \end{bmatrix}}{\det[H]} = -\frac{M_1''(t_1^*)\,M_2''(t_2^*)}{\det[H]} < 0$$

$$\frac{\partial t_1^*}{\partial T} = \frac{\det \begin{bmatrix} 0 & -1 & -1 \\ -1 & 0 & 0 \\ -1 & 0 & M_2''(t_2^*) \end{bmatrix}}{\det[H]} = -\frac{M_2''(t_2^*)}{\det[H]} > 0$$

$$\frac{\partial t_2^*}{\partial T} = \frac{\det \begin{bmatrix} 0 & -1 & -1 \\ -1 & M_1''(t_1^*) & 0 \\ -1 & 0 & 0 \end{bmatrix}}{\det[H]} = -\frac{M_1''(t_1^*)}{\det[H]} > 0.$$

Problem 3.40 *[A **] Now prove the same result for the general case by showing that $M^*(T)$ is a concave function of T.*

Answer: From the constraint we have

$$T = \sum_{i=1}^{n} t_i^*(T) \Longrightarrow 1 = \sum_{i=1}^{n} \frac{\partial t_i^*(T)}{\partial T}$$

so that at least one $\frac{\partial t_i^*(T)}{\partial T}$ must be positive. For that i we have

$$\begin{aligned} M_i'(t_i^*(T)) &= \lambda^*(T) \\ &\Longrightarrow \underbrace{M_i''(t_i^*(T))}_{-}\underbrace{\frac{\partial t_i^*(T)}{\partial T}}_{+} = \frac{\partial \lambda^*(T)}{\partial T} \\ &\Longrightarrow \frac{\partial \lambda^*(T)}{\partial T} < 0. \end{aligned}$$

It follows then that $M^*(T)$ is a concave function of T since from the envelope theorem $\lambda^*(T) = \frac{\partial M^*(T)}{\partial T}$ so that

$$\frac{\partial \lambda^*(T)}{\partial T} = \frac{\partial^2 M^*(T)}{\partial T^2} < 0.$$

From the first-order conditions and differentiating with respect to T we have now for **any** i

$$\begin{aligned} M_i(t_i^*(T)) &= \lambda^*(T) \\ &\Longrightarrow M_i''(t_i^*(T))\frac{\partial t_i^*(T)}{\partial T} = \frac{\partial \lambda^*(T)}{\partial T} \\ &\Longrightarrow \frac{\partial t_i^*(T)}{\partial T} = \frac{\frac{\partial \lambda^*(T)}{\partial T}}{M_i''(t_i^*(T))} > 0 \end{aligned}$$

since $M_i''(t_i^*(T)) < 0$ and $\frac{\partial \lambda^*(T)}{\partial T} < 0$ from above.

Problem 3.41 *[A **] Suppose that the function $M_i(t_i)$ is given by*

$$M_i(t_i) = m \times \left(1 - e^{-\delta t_i}\right), \ \delta > 0.$$

Show that m is the maximum obtainable mark on question i, that this form satisfies $M_i'(t_i) > 0$, $M_i''(t_i) < 0$, and that better students will have a higher value of δ.

Answer: Since $e^{-\delta t_i} \to 0$ as $t_i \to \infty$ and $e^{-\delta t_i} > 0$ we have

$$M_i(t_i) < m \text{ and } M_i(t_i) \to m \text{ as } t_i \to \infty.$$

Furthermore

$$M_i'(t_i) = \delta m e^{-\delta t_i} > 0, M_i''(t_i) = -\delta^2 m e^{-\delta t_i} < 0$$

and

$$\frac{\partial M_i(t_i)}{\partial \delta} = t_i m e^{-\delta t_i} > 0$$

so the higher δ is the higher the grade for a given amount of time t_i.

Problem 3.42 *Using the information in the previous question show that it is optimal to spend an equal amount of time on each question (i.e., that $t_i^*(T) = \frac{T}{n}$) and*

$$\begin{aligned} \lambda^*(T) &= \delta \times m \times e^{-\frac{\delta}{n}T} \\ M^*(T) &= n \times m \times (1 - e^{-\frac{\delta}{n}T}). \end{aligned}$$

If $T = 60$ and $n = 3$, what must δ be in order for the student to receive a 70% on the exam?

Answer: From the first-order conditions

$$M_i'(t_i^*) = \lambda^* \Longrightarrow \delta m e^{-\delta t_i^*} = \lambda^* \Longrightarrow t_i^* = -\frac{\ln\left(\frac{\lambda^*}{\delta m}\right)}{\delta}$$

so that

$$\begin{aligned} T - \sum_{i=1}^{n} t_i^* &= 0 \Longrightarrow -\frac{n \ln\left(\frac{\lambda^*}{\delta m}\right)}{\delta} = T \Longrightarrow \lambda^*(T) = \delta \times m \times e^{-\frac{\delta}{n}T} \\ &\Longrightarrow t_i^*(T) = -\frac{\ln\left(\frac{\delta \times m \times e^{-\frac{\delta}{n}T}}{\delta m}\right)}{\delta} = \frac{T}{n} \\ &\Longrightarrow M^*(T) = \sum_{i=1}^{n} m \times \left(1 - e^{-\delta t_i}\right) = n \times m \times (1 - e^{-\frac{\delta}{n}T}). \end{aligned}$$

If $T = 60$ and $n = 3$, then

$$\begin{aligned} M^*(60) &= 3 \times m \times (1 - e^{-\frac{\delta}{3}60}) = 3 \times m \times 0.7 \\ &\Longrightarrow 1 - e^{-\frac{\delta}{3}60} = 0.7 \Longrightarrow e^{-\frac{\delta}{3}60} = 0.3 \\ &\Longrightarrow -\delta \times 20 = \ln(0.3) \Longrightarrow \delta = -\frac{1}{20}\ln(0.3) \approx 0.0602. \end{aligned}$$

3.4 Monopolistic Price Discrimination

Problem 3.43 *[A**] Consider an airline having a total of Q seats to offer the travelling public. Let Q_1 be the number of seats devoted to business class travellers and let Q_2 be the number of seats offered to tourist class travellers. The revenue from business travellers is $R_1(Q_1)$ with marginal revenue $MR_1(Q_1) = R_1'(Q_1) > 0$ and $R_1''(Q_1) < 0$, and the revenue from tourist class travellers is $R_2(Q_2)$ with $MR_2(Q_2) = R_2'(Q_2) > 0$ and $R_2''(Q_2) < 0$. The airline wishes to allocate the given Q seats, which is equal to $Q_1 + Q_2$, so as to maximize the total revenue it gets*

$$R(Q_1, Q_2) = R_1(Q_1) + R_2(Q_2).$$

Set up the Lagrangian for this problem and find the first-order conditions. Show that the firm will choose Q_1^* and Q_2^* so as to equalize the marginal revenue from business and tourist class travellers.

Answer: To maximizing profits the airline must maximize the revenue from selling Q seats. Consider then maximizing $R(Q_1, Q_2)$ subject to the constraint $Q = Q_1 + Q_2$ which leads to the Lagrangian

$$\mathcal{L}(\lambda, Q_1, Q_2, Q) = R_1(Q_1) + R_2(Q_2) + \lambda(Q - Q_1 - Q_2)$$

so that the first-order conditions are

$$\begin{aligned} Q - Q_1^* - Q_2^* &= 0 \\ R_1'(Q_1^*) - \lambda^* &= 0 \\ R_2'(Q_2^*) - \lambda^* &= 0. \end{aligned}$$

It follows then that $MR_1(Q_1^*) = MR_2(Q_2^*) = \lambda^*$ so that the firm equates the marginal revenue from the two markets.

Problem 3.44 *[A**] The airline's revenue as a function of Q will be given by*

$$R^*(Q) = R_1(Q_1^*(Q)) + R_2(Q_2^*(Q)).$$

Use the envelope theorem to find $\frac{\partial R^*(Q)}{\partial Q}$ and show that marginal revenue in the two markets will be equal to $\frac{\partial R^*(Q)}{\partial Q}$.

Answer: We have

$$\begin{aligned} \frac{\partial R^*(Q)}{\partial Q} &= \frac{\partial \mathcal{L}(\lambda, Q_1, Q_2, Q)}{\partial Q}\Big|_{Q_1=Q_1^*, Q_2=Q_2^*, \lambda=\lambda^*} \\ &= (\lambda)\,|_{Q_1=Q_1^*, Q_2=Q_2^*, \lambda=\lambda^*} = \lambda^*. \end{aligned}$$

From the first-order conditions we then have

$$\frac{\partial R^*(Q)}{\partial Q} = \lambda^* = R_1'(Q_1^*) = R_2'(Q_2^*).$$

Problem 3.45 *[A **] Revenue in the two markets is given by*

$$R_1(Q_1) = P_1(Q_1)Q_1, \; R_2(Q_2) = P_2(Q_2)Q_2$$

where $P_1(Q_1)$ and $P_2(Q_2)$ are the inverse demand functions satisfying $P_1'(Q_1) < 0$ and $P_2'(Q_2) < 0$. Show that it is the market with the least elastic inverse demand curve that will have the higher price.

Answer: We have

$$\begin{aligned} R_1'(Q_1) &= P_1(Q_1) + P_1'(Q_1)Q_1 = P_1(Q_1)(1+\eta_1(Q_1)) \\ R_2'(Q_2) &= P_2(Q_2) + P_2'(Q_2)Q_2 = P_2(Q_2)(1+\eta_2(Q_2)) \end{aligned}$$

where the elasticities of the inverse demand curves are

$$\eta_1(Q_1) = \frac{P_1'(Q_1)Q_1}{P_1(Q_1)}, \; \eta_2(Q_2) = \frac{P_2'(Q_2)Q_2}{P_2(Q_2)}.$$

Since $R_1'(Q_1^*) = R_2'(Q_2^*)$ we have

$$P_1(Q_1^*)(1+\eta_1(Q_1^*)) = P_2(Q_2^*)(1+\eta_2(Q_2^*))$$

we have

$$\frac{P_1(Q_1^*)}{P_2(Q_2^*)} = \frac{1+\eta_2(Q_2^*)}{1+\eta_1(Q_1^*)}$$

so that

$$\eta_2(Q_2^*) > \eta_1(Q_1^*) \Longrightarrow P_1(Q_1^*) > P_2(Q_2^*).$$

3.5 Trade

Problem 3.46 *[A **] Consider a (small) country that produces two goods Q_1 and Q_2 with prices P_1 and P_2 and production possibilities curve given as an implicit function $\phi(Q_1, Q_2, T) = 0$ where*

$$\begin{aligned} \phi(Q_1, Q_2, T) &= \frac{T}{32} - \frac{5}{2}Q_1^2 - \frac{5}{2}Q_2^2 + 3Q_1Q_2 \\ &= \frac{T}{32} - \frac{1}{2}\begin{bmatrix} Q_1 & Q_2 \end{bmatrix}\begin{bmatrix} 5 & -3 \\ -3 & 5 \end{bmatrix}\begin{bmatrix} Q_1 \\ Q_2 \end{bmatrix} \end{aligned}$$

where T represents the level of technology. Write down the Lagrangian for the problem of maximizing Gross National Product R (R for Revenue)

$$R = P_1Q_1 + P_2Q_2 = \begin{bmatrix} P_1 & P_2 \end{bmatrix}\begin{bmatrix} Q_1 \\ Q_2 \end{bmatrix}$$

subject to the production possibilities curve constraint. Find the first-order conditions and solve. (Hint: The calculations are tedious unless you use matrix algebra.)

Answer: We have

$$\mathcal{L} = P_1Q_1 + P_2Q_2 + \lambda\left(\frac{T}{32} - \frac{5}{2}Q_1^2 - \frac{5}{2}Q_2^2 + 3Q_1Q_2\right)$$

with first-order conditions

$$\begin{aligned}
\frac{T}{32} - \frac{5}{2}(Q_1^*)^2 - \frac{5}{2}(Q_2^*)^2 + 3Q_1^*Q_2^* &= \frac{T}{32} - \frac{1}{2}\begin{bmatrix} Q_1^* & Q_2^* \end{bmatrix}\begin{bmatrix} 5 & -3 \\ -3 & 5 \end{bmatrix}\begin{bmatrix} Q_1^* \\ Q_2^* \end{bmatrix} = 0 \\
P_1 - \lambda^* \times (5Q_1^* - 3Q_2^*) &= 0 \\
P_2 - \lambda^* \times (5Q_2^* - 3Q_1^*) &= 0.
\end{aligned}$$

From the second and third equations we have

$$\begin{aligned}
5Q_1^* - 3Q_2^* &= P_1(\lambda^*)^{-1} \\
-3Q_1^* + 5Q_2^* &= P_2(\lambda^*)^{-1}
\end{aligned}$$

or in matrix notation

$$\begin{bmatrix} 5 & -3 \\ -3 & 5 \end{bmatrix}\begin{bmatrix} Q_1^* \\ Q_2^* \end{bmatrix} = \begin{bmatrix} P_1 \\ P_2 \end{bmatrix}(\lambda^*)^{-1}.$$

Solving we have

$$\begin{bmatrix} Q_1^* \\ Q_2^* \end{bmatrix} = \begin{bmatrix} 5 & -3 \\ -3 & 5 \end{bmatrix}^{-1}\begin{bmatrix} P_1 \\ P_2 \end{bmatrix}(\lambda^*)^{-1}.$$

Now putting Q_1^* and Q_2^* into the first first-order condition and using matrix algebra to simplify we have

$$\begin{aligned}
\frac{T}{32} &= \frac{1}{2}\begin{bmatrix} Q_1^* & Q_2^* \end{bmatrix}\begin{bmatrix} 5 & -3 \\ -3 & 5 \end{bmatrix}\begin{bmatrix} Q_1^* \\ Q_2^* \end{bmatrix} \\
&= \frac{1}{2}(\lambda^*)^{-1}\begin{bmatrix} P_1 & P_2 \end{bmatrix}\left(\begin{bmatrix} 5 & -3 \\ -3 & 5 \end{bmatrix}^{-1}\right)^T\begin{bmatrix} 5 & -3 \\ -3 & 5 \end{bmatrix} \\
&\quad \times \begin{bmatrix} 5 & -3 \\ -3 & 5 \end{bmatrix}^{-1}\begin{bmatrix} P_1 \\ P_2 \end{bmatrix}(\lambda^*)^{-1} \\
&= \frac{1}{2}(\lambda^*)^{-1}\begin{bmatrix} P_1 & P_2 \end{bmatrix}\begin{bmatrix} 5 & -3 \\ -3 & 5 \end{bmatrix}^{-1}\begin{bmatrix} 5 & -3 \\ -3 & 5 \end{bmatrix} \\
&\quad \times \begin{bmatrix} 5 & -3 \\ -3 & 5 \end{bmatrix}^{-1}\begin{bmatrix} P_1 \\ P_2 \end{bmatrix}(\lambda^*)^{-1} \\
&= \frac{1}{2}(\lambda^*)^{-2}\begin{bmatrix} P_1 & P_2 \end{bmatrix}\begin{bmatrix} 5 & -3 \\ -3 & 5 \end{bmatrix}^{-1}\begin{bmatrix} P_1 \\ P_2 \end{bmatrix} \\
&= \frac{1}{2}(\lambda^*)^{-2}\frac{1}{16}\begin{bmatrix} P_1 & P_2 \end{bmatrix}\begin{bmatrix} 5 & 3 \\ 3 & 5 \end{bmatrix}\begin{bmatrix} P_1 \\ P_2 \end{bmatrix} \\
&= \frac{(\lambda^*)^{-2}}{32}\left(5P_1^2 + 6P_1P_2 + 5P_2^2\right).
\end{aligned}$$

Thus

$$\lambda^* = \left(5P_1^2 + 6P_1P_2 + 5P_2^2\right)^{\frac{1}{2}} T^{-\frac{1}{2}}.$$

Now

$$\begin{aligned}
\begin{bmatrix} Q_1^* \\ Q_2^* \end{bmatrix} &= \begin{bmatrix} 5 & -3 \\ -3 & 5 \end{bmatrix}^{-1} \begin{bmatrix} P_1 \\ P_2 \end{bmatrix} (\lambda^*)^{-1} \\
&= \frac{1}{16} \begin{bmatrix} 5 & 3 \\ 3 & 5 \end{bmatrix} \begin{bmatrix} P_1 \\ P_2 \end{bmatrix} (\lambda^*)^{-1} \\
&= \frac{1}{16} \begin{bmatrix} 5P_1 + 3P_2 \\ 3P_1 + 5P_2 \end{bmatrix} \left(5P_1^2 + 6P_1P_2 + 5P_2^2\right)^{-\frac{1}{2}} T^{\frac{1}{2}}
\end{aligned}$$

or

$$\begin{aligned}
Q_1^*(P_1, P_2, T) &= \frac{5P_1 + 3P_2}{(5P_1^2 + 6P_1P_2 + 5P_2^2)^{\frac{1}{2}}} \frac{T^{\frac{1}{2}}}{16} \\
Q_2^*(P_1, P_2, T) &= \frac{3P_1 + 5P_2}{(5P_1^2 + 6P_1P_2 + 5P_2^2)^{\frac{1}{2}}} \frac{T^{\frac{1}{2}}}{16}.
\end{aligned}$$

Problem 3.47 *[A **] From the previous problem find the clever revenue function $R^*(P_1, P_2, T)$. Use the envelope theorem to show that $\frac{\partial R^*}{\partial P_1} = Q_1^*$ and $\frac{\partial R^*}{\partial T} = \frac{\lambda^*}{32}$ and verify that both these relationships are correct by directly calculating $\frac{\partial R^*}{\partial P_1}$ and $\frac{\partial R^*}{\partial T}$.*

Answer: To find $R^*(P_1, P_2, T)$ use

$$\begin{aligned}
R^*(P_1, P_2, T) &= \begin{bmatrix} P_1 & P_2 \end{bmatrix} \begin{bmatrix} Q_1^* \\ Q_2^* \end{bmatrix} \\
&= \begin{bmatrix} P_1 & P_2 \end{bmatrix} \begin{bmatrix} 5 & -3 \\ -3 & 5 \end{bmatrix}^{-1} \begin{bmatrix} P_1 \\ P_2 \end{bmatrix} (\lambda^*)^{-1} \\
&= \begin{bmatrix} P_1 & P_2 \end{bmatrix} \frac{1}{16} \begin{bmatrix} 5 & 3 \\ 3 & 5 \end{bmatrix} \begin{bmatrix} P_1 \\ P_2 \end{bmatrix} (\lambda^*)^{-1} \\
&= \frac{\frac{1}{16} \begin{bmatrix} P_1 & P_2 \end{bmatrix} \begin{bmatrix} 5 & 3 \\ 3 & 5 \end{bmatrix} \begin{bmatrix} P_1 \\ P_2 \end{bmatrix}}{(5P_1^2 + 6P_1P_2 + 5P_2^2)^{\frac{1}{2}} T^{-\frac{1}{2}}} \\
&= \frac{T^{\frac{1}{2}}}{16} \frac{5P_1^2 + 6P_1P_2 + 5P_2^2}{(5P_1^2 + 6P_1P_2 + 5P_2^2)^{\frac{1}{2}}} \\
&= \frac{T^{\frac{1}{2}}}{16} \left(5P_1^2 + 6P_1P_2 + 5P_2^2\right)^{\frac{1}{2}}.
\end{aligned}$$

Using the envelope theorem we have

$$\begin{aligned}
\frac{\partial R^*(P_1,P_2,T)}{\partial P_1} &= \frac{\partial}{\partial P_1}\left(P_1Q_1+P_2Q_2+\lambda\left(\frac{T}{32}-\frac{5}{2}Q_1^2-\frac{5}{2}Q_2^2+3Q_1Q_2\right)\right)\Big|_{\substack{\lambda=\lambda^*\\ Q_1^*=Q_1^*\\ Q_2^*=Q_2^*}} \\
&= (Q_1)\Big|_{\substack{\lambda=\lambda^*\\ Q_1^*=Q_1^*\\ Q_2^*=Q_2^*}} = Q_1^* \\
\frac{\partial R^*(P_1,P_2,T)}{\partial T} &= \frac{\partial}{\partial T}\left(P_1Q_1+P_2Q_2+\lambda\left(\frac{T}{32}-\frac{5}{2}Q_1^2-\frac{5}{2}Q_2^2+3Q_1Q_2\right)\right)\Big|_{\substack{\lambda=\lambda^*\\ Q_1^*=Q_1^*\\ Q_2^*=Q_2^*}} \\
&= \left(\frac{\lambda}{32}\right)\Big|_{\substack{\lambda=\lambda^*\\ Q_1^*=Q_1^*\\ Q_2^*=Q_2^*}} = \frac{\lambda^*}{32}.
\end{aligned}$$

Calculating these directly we have

$$\begin{aligned}
\frac{\partial R^*(P_1,P_2,T)}{\partial P_1} &= \frac{\partial}{\partial P_1}\left(\frac{T^{\frac{1}{2}}}{16}\left(5P_1^2+6P_1P_2+5P_2^2\right)^{\frac{1}{2}}\right) \\
&= \frac{T^{\frac{1}{2}}}{2\times 16}\left(5P_1^2+6P_1P_2+5P_2^2\right)^{-\frac{1}{2}}(2\times 5P_1+6P_2) \\
&= \frac{T^{\frac{1}{2}}}{2\times 16}\left(5P_1^2+6P_1P_2+5P_2^2\right)^{-\frac{1}{2}}(5P_1+3P_2)\times 2 \\
&= \frac{T^{\frac{1}{2}}}{16}\frac{5P_1+3P_2}{\left(5P_1^2+6P_1P_2+5P_2^2\right)^{\frac{1}{2}}} = Q_1^*
\end{aligned}$$

and

$$\begin{aligned}
\frac{\partial R^*(P_1,P_2,T)}{\partial T} &= \frac{\partial}{\partial T}\left(\frac{T^{\frac{1}{2}}}{16}\left(5P_1^2+6P_1P_2+5P_2^2\right)^{\frac{1}{2}}\right) \\
&= \frac{T^{-\frac{1}{2}}}{32}\left(5P_1^2+6P_1P_2+5P_2^2\right)^{\frac{1}{2}} \\
&= \frac{1}{32}\underbrace{T^{-\frac{1}{2}}\left(5P_1^2+6P_1P_2+5P_2^2\right)^{\frac{1}{2}}}_{\lambda^*} = \frac{\lambda^*}{32}.
\end{aligned}$$

Problem 3.48 *[A**] Consider a (small) country that produces two goods Q_1 and Q_2 with prices P_1 and P_2 and production possibilities curve given as an implicit function $\phi(Q_1,Q_2,T)=0$ where*

$$\phi(Q_1,Q_2,T)=\frac{T}{2}-\frac{1}{2}Q_1^2-\frac{1}{2}Q_2^2.$$

Here T represents the (exogenous) level of technology. Write down the Lagrangian for the problem of maximizing revenue $R=P_1Q_1+P_2Q_2$ subject to the production possibilities curve constraint. Find the first-order conditions and *solve. Find the clever revenue function* $R^*(P_1, P_2, T)$ *and from* $R^*(P_1, P_2, T)$ *show that* $\frac{\partial R^*}{\partial P_1} = Q_1^*$ *and* $\frac{\partial R^*}{\partial T} = \frac{\lambda^*}{2}$. *Now use the envelope theorem to show that* $\frac{\partial R^*}{\partial P_1} = Q_1^*$ *and* $\frac{\partial R^*}{\partial T} = \frac{\lambda^*}{2}$.

Answer: We have

$$\mathcal{L} = P_1 Q_1 + P_2 Q_2 + \lambda\left(\frac{T}{2} - \frac{1}{2}Q_1^2 - \frac{1}{2}Q_2^2\right)$$

so that the first-order conditions yield

$$\begin{aligned} \frac{T}{2} - \frac{1}{2}(Q_1^*)^2 - \frac{1}{2}(Q_2^*)^2 &= 0 \Longrightarrow T = (Q_1^*)^2 + (Q_2^*)^2 \\ P_1 - \lambda^* Q_1^* &= 0 \Longrightarrow Q_1^* = P_1(\lambda^*)^{-1} \\ P_2 - \lambda^* Q_2^* &= 0 \Longrightarrow Q_2^* = P_2(\lambda^*)^{-1}. \end{aligned}$$

Thus

$$\begin{aligned} T &= (Q_1^*)^2 + (Q_2^*)^2 = \left(P_1(\lambda^*)^{-1}\right)^2 + \left(P_2(\lambda^*)^{-1}\right)^2 \\ &\Longrightarrow (\lambda^*)^{-2}\left(P_1^2 + P_2^2\right) = T \\ &\Longrightarrow (\lambda^*)^{-1} = T^{\frac{1}{2}}\left(P_1^2 + P_2^2\right)^{-\frac{1}{2}} \\ &\Longrightarrow \lambda^* = T^{-\frac{1}{2}}\left(P_1^2 + P_2^2\right)^{\frac{1}{2}}. \end{aligned}$$

Using the result for $(\lambda^*)^{-1}$ we have

$$\begin{aligned} Q_1^* &= P_1(\lambda^*)^{-1} = T^{\frac{1}{2}} P_1\left(P_1^2 + P_2^2\right)^{-\frac{1}{2}} \\ Q_2^* &= P_2(\lambda^*)^{-1} = T^{\frac{1}{2}} P_2\left(P_1^2 + P_2^2\right)^{-\frac{1}{2}}. \end{aligned}$$

Now

$$\begin{aligned} R^*(P_1, P_2, T) &= P_1 Q_1^* + P_2 Q_2^* = P_1 T^{\frac{1}{2}} P_1\left(P_1^2 + P_2^2\right)^{-\frac{1}{2}} + P_2 T^{\frac{1}{2}} P_2\left(P_1^2 + P_2^2\right)^{-\frac{1}{2}} \\ &= T^{\frac{1}{2}}\left(P_1^2 + P_2^2\right)^{-\frac{1}{2}}\left(P_1^2 + P_2^2\right) = T^{\frac{1}{2}}\left(P_1^2 + P_2^2\right)^{\frac{1}{2}}. \end{aligned}$$

From $R^*(P_1, P_2, T) = T^{\frac{1}{2}}\left(P_1^2 + P_2^2\right)^{\frac{1}{2}}$ we have

$$\begin{aligned} \frac{\partial R^*(P_1, P_2, T)}{\partial P_1} &= \frac{\partial}{\partial P_1}\left(T^{\frac{1}{2}}\left(P_1^2 + P_2^2\right)^{\frac{1}{2}}\right) = \frac{T^{\frac{1}{2}}}{2}\left(P_1^2 + P_2^2\right)^{-\frac{1}{2}} \times 2P_1 \\ &= T^{\frac{1}{2}} P_1\left(P_1^2 + P_2^2\right)^{-\frac{1}{2}} = Q_1^* \end{aligned}$$

and

$$\begin{aligned} \frac{\partial R^*(P_1, P_2, T)}{\partial T} &= \frac{\partial}{\partial T}\left(T^{\frac{1}{2}}\left(P_1^2 + P_2^2\right)^{\frac{1}{2}}\right) = \frac{T^{-\frac{1}{2}}\left(P_1^2 + P_2^2\right)^{\frac{1}{2}}}{2} \\ &= \frac{\lambda^*}{2}. \end{aligned}$$

Using the envelope theorem we have

$$\begin{aligned}
\frac{\partial R^*(P_1,P_2,T)}{\partial P_1} &= \frac{\partial}{\partial P_1}\left(P_1Q_1+P_2Q_2+\lambda\left(\frac{T}{2}-\frac{1}{2}Q_1^2-\frac{1}{2}Q_2^2\right)\right)|_{\substack{\lambda=\lambda^*\\ Q_1^*=Q_1^*\\ Q_2^*=Q_2^*}}\\
&= (Q_1)|_{\substack{\lambda=\lambda^*\\ Q_1^*=Q_1^*\\ Q_2^*=Q_2^*}} = Q_1^*\\
\frac{\partial R^*(P_1,P_2,T)}{\partial T} &= \frac{\partial}{\partial T}\left(P_1Q_1+P_2Q_2+\lambda\left(\frac{T}{2}-\frac{1}{2}Q_1^2-\frac{1}{2}Q_2^2\right)\right)|_{\substack{\lambda=\lambda^*\\ Q_1^*=Q_1^*\\ Q_2^*=Q_2^*}}\\
&= \left(\frac{\lambda}{2}\right)|_{\substack{\lambda=\lambda^*\\ Q_1^*=Q_1^*\\ Q_2^*=Q_2^*}} = \frac{\lambda^*}{2}.
\end{aligned}$$

3.6 Cost Minimization

Problem 3.49 *[D***] Consider the problem of choosing the values of two factors of production (say labour and capital) L and K which minimize the cost $C \equiv WL+RK$ subject to the constraint that Q units are produced or that $Q = L^{\frac{1}{3}}K^{\frac{1}{3}}$. Find the first-order conditions for this problem and solve. Find the cost function $C^*(Q,W,R)$ and verify that*

$$\begin{aligned}
\frac{\partial C^*(Q,W,R)}{\partial Q} &= \lambda^*(Q,W,R)\\
\frac{\partial C^*(Q,W,R)}{\partial W} &= L^*(Q,W,R)\\
\frac{\partial C^*(Q,W,R)}{\partial R} &= K^*(Q,W,R).
\end{aligned}$$

Answer: From the Lagrangian

$$\mathcal{L} = WL + RK + \lambda\left(Q - L^{\frac{1}{3}}K^{\frac{1}{3}}\right)$$

the first-order conditions yield

$$\begin{aligned}
Q-(L^*)^{\frac{1}{3}}(K^*)^{\frac{1}{3}} &= 0 \Longrightarrow Q=(L^*)^{\frac{1}{3}}(K^*)^{\frac{1}{3}}\\
W-\frac{1}{3}\lambda^*(L^*)^{-\frac{2}{3}}(K^*)^{\frac{1}{3}} &= 0 \Longrightarrow \lambda^*(L^*)^{-\frac{2}{3}}(K^*)^{\frac{1}{3}} = 3W\\
R-\frac{1}{3}\lambda^*(L^*)^{\frac{1}{3}}(K^*)^{-\frac{2}{3}} &= 0 \Longrightarrow \lambda^*(L^*)^{\frac{1}{3}}(K^*)^{-\frac{2}{3}} = 3R.
\end{aligned}$$

Dividing the last two results yields

$$\frac{\lambda^*(L^*)^{-\frac{2}{3}}(K^*)^{\frac{1}{3}}}{\lambda^*(L^*)^{\frac{1}{3}}(K^*)^{-\frac{2}{3}}} = \frac{K^*}{L^*} = \frac{3W}{3R} = WR^{-1} \Longrightarrow K^* = WR^{-1}L^*.$$

Now

$$\begin{aligned} Q-(L^*)^{\frac{1}{3}}(K^*)^{\frac{1}{3}} &= 0 \Longrightarrow Q=(L^*)^{\frac{1}{3}}(K^*)^{\frac{1}{3}} \Longrightarrow Q=(L^*)^{\frac{1}{3}}\left(WR^{-1}L^*\right)^{\frac{1}{3}} \\ &\Longrightarrow (L^*)^{\frac{2}{3}}W^{\frac{1}{3}}R^{-\frac{1}{3}}=Q \Longrightarrow L^*=Q^{\frac{3}{2}}W^{-\frac{1}{2}}R^{\frac{1}{2}} \end{aligned}$$

and

$$K^*=WR^{-1}L^*=WR^{-1}Q^{\frac{3}{2}}W^{-\frac{1}{2}}R^{\frac{1}{2}}=Q^{\frac{3}{2}}W^{\frac{1}{2}}R^{-\frac{1}{2}}.$$

Now to find λ^* we use

$$\begin{aligned} W-\frac{1}{3}\lambda^*(L^*)^{-\frac{2}{3}}(K^*)^{\frac{1}{3}} &= 0 \Longrightarrow W=\frac{\frac{1}{3}\lambda^*\overbrace{(L^*)^{\frac{1}{3}}(K^*)^{\frac{1}{3}}}^{Q}}{L^*} \\ &\Longrightarrow W=\frac{\frac{1}{3}\lambda^*Q}{L^*}\Longrightarrow \lambda^*=3\frac{WL^*}{Q} \\ &\Longrightarrow \lambda^*=3\frac{W\left(Q^{\frac{3}{2}}W^{-\frac{1}{2}}R^{\frac{1}{2}}\right)}{Q}=3Q^{\frac{1}{2}}W^{\frac{1}{2}}R^{\frac{1}{2}}. \end{aligned}$$

We have

$$C^*(Q,W,R)=W\times Q^{\frac{3}{2}}W^{-\frac{1}{2}}R^{\frac{1}{2}}+R\times Q^{\frac{3}{2}}W^{\frac{1}{2}}R^{-\frac{1}{2}}=2Q^{\frac{3}{2}}W^{\frac{1}{2}}R^{\frac{1}{2}}$$

We then have

$$\begin{aligned} \frac{\partial C^*(Q,W,R)}{\partial Q} &= \frac{\partial}{\partial Q}\left(2Q^{\frac{3}{2}}W^{\frac{1}{2}}R^{\frac{1}{2}}\right)=3Q^{\frac{1}{2}}W^{\frac{1}{2}}R^{\frac{1}{2}}=\lambda^*(Q,W,R) \\ \frac{\partial C^*(Q,W,R)}{\partial W} &= \frac{\partial}{\partial W}\left(2Q^{\frac{3}{2}}W^{\frac{1}{2}}R^{\frac{1}{2}}\right)=Q^{\frac{3}{2}}W^{-\frac{1}{2}}R^{\frac{1}{2}}=L^*(Q,W,R) \\ \frac{\partial C^*(Q,W,R)}{\partial R} &= \frac{\partial}{\partial R}\left(2Q^{\frac{3}{2}}W^{\frac{1}{2}}R^{\frac{1}{2}}\right)=Q^{\frac{3}{2}}W^{\frac{1}{2}}R^{-\frac{1}{2}}=K^*(Q,W,R). \end{aligned}$$

Problem 3.50 *[A**] Consider the problem of minimizing costs*

$$C\equiv WL+RK$$

with a Cobb-Douglas production function $Q=L^{\alpha}K^{\beta}$. Find the first-order conditions and solve. Show that the cost function is

$$C^*(Q,W,R)=(\alpha+\beta)\,\alpha^{-\frac{\alpha}{\alpha+\beta}}\beta^{-\frac{\beta}{\alpha+\beta}}Q^{\frac{1}{\alpha+\beta}}W^{\frac{\alpha}{\alpha+\beta}}R^{\frac{\beta}{\alpha+\beta}}$$

and verify that

$$\frac{\partial C^*(Q,W,R)}{\partial Q}=\lambda^*(Q,W,R)\,,\ \frac{\partial C^*(Q,W,R)}{\partial W}=L^*(Q,W,R)\,,\ \frac{\partial C^*(Q,W,R)}{\partial R}=K^*(Q,W,R)\,.$$

Answer: From the Lagrangian

$$\mathcal{L}=WL+RK+\lambda\left(Q-L^{\alpha}K^{\beta}\right)$$

the first-order conditions yield

$$\begin{aligned}
Q-(L^*)^{\alpha}(K^*)^{\beta} &= 0 \Longrightarrow Q=(L^*)^{\alpha}(K^*)^{\beta} \\
W-\alpha\lambda^*(L^*)^{\alpha-1}(K^*)^{\beta} &= 0 \Longrightarrow \lambda^*(L^*)^{\alpha-1}(K^*)^{\beta}=\alpha^{-1}W \\
R-\beta\lambda^*(L^*)^{\alpha}(K^*)^{\beta-1} &= 0 \Longrightarrow \lambda^*(L^*)^{\alpha}(K^*)^{\beta-1}=\beta^{-1}R.
\end{aligned}$$

Dividing the two results yields

$$\frac{\lambda^*(L^*)^{\alpha-1}(K^*)^{\beta}}{\lambda^*(L^*)^{\alpha}(K^*)^{\beta-1}}=\frac{K^*}{L^*}=\frac{\alpha^{-1}W}{\beta^{-1}R}=\alpha^{-1}\beta WR^{-1}\Longrightarrow K^*=\alpha^{-1}\beta WR^{-1}L^*.$$

Now

$$\begin{aligned}
Q-L^{*\alpha}K^{*\beta} &= 0 \Longrightarrow Q=(L^*)^{\alpha}(K^*)^{\beta}\Longrightarrow Q=(L^*)^{\alpha}\left(\alpha^{-1}\beta WR^{-1}L^*\right)^{\beta} \\
&\Longrightarrow (L^*)^{\alpha+\beta}\alpha^{-\beta}\beta^{\beta}W^{\beta}R^{-\beta}=Q\Longrightarrow L^*=\alpha^{\frac{\beta}{\alpha+\beta}}\beta^{-\frac{\beta}{\alpha+\beta}}Q^{\frac{1}{\alpha+\beta}}W^{-\frac{\beta}{\alpha+\beta}}R^{\frac{\beta}{\alpha+\beta}}
\end{aligned}$$

and

$$\begin{aligned}
K^* &= \alpha^{-1}\beta WR^{-1}L^*=\alpha^{-1}\beta WR^{-1}\times\alpha^{\frac{\beta}{\alpha+\beta}}\beta^{-\frac{\beta}{\alpha+\beta}}Q^{\frac{1}{\alpha+\beta}}W^{-\frac{\beta}{\alpha+\beta}}R^{\frac{\beta}{\alpha+\beta}} \\
&= \alpha^{-\frac{\alpha}{\alpha+\beta}}\beta^{\frac{\alpha}{\alpha+\beta}}Q^{\frac{1}{\alpha+\beta}}W^{\frac{\alpha}{\alpha+\beta}}R^{-\frac{\alpha}{\alpha+\beta}}.
\end{aligned}$$

To find λ^* we use

$$\begin{aligned}
W-\alpha\lambda^*(L^*)^{\alpha-1}(K^*)^{\beta} &= 0 \\
&\Longrightarrow W=\frac{\alpha\lambda^*\overbrace{(L^*)^{\alpha}(K^*)^{\beta}}^{Q}}{L^*} \\
&\Longrightarrow W=\frac{\alpha\lambda^*Q}{L^*}\Longrightarrow\lambda^*=\frac{1}{\alpha}\frac{WL^*}{Q} \\
&\Longrightarrow \lambda^*=\frac{1}{\alpha}\frac{W\left(\alpha^{\frac{\beta}{\alpha+\beta}}\beta^{-\frac{\beta}{\alpha+\beta}}Q^{\frac{1}{\alpha+\beta}}W^{-\frac{\beta}{\alpha+\beta}}R^{\frac{\beta}{\alpha+\beta}}\right)}{Q} \\
&= \alpha^{-\frac{\alpha}{\alpha+\beta}}\beta^{-\frac{\beta}{\alpha+\beta}}Q^{\frac{1}{\alpha+\beta}-1}W^{\frac{\alpha}{\alpha+\beta}}R^{\frac{\beta}{\alpha+\beta}}.
\end{aligned}$$

Here is another method:

$$\begin{aligned}
Q-L^{*\alpha}K^{*\beta} &= 0\Longrightarrow\alpha\ln(L^*)+\beta\ln(K^*)=\ln(Q) \\
W-\alpha\lambda^*L^{*\alpha-1}K^{*\beta} &= 0\Longrightarrow\ln(\lambda^*)+(\alpha-1)\ln(L^*)+\beta\ln(K^*)=\ln\left(\frac{W}{\alpha}\right) \\
R-\beta\lambda^*L^{*\alpha}K^{*\beta-1} &= 0\Longrightarrow\ln(\lambda^*)+\alpha\ln(L^*)+(\beta-1)\ln(K^*)=\ln\left(\frac{R}{\beta}\right).
\end{aligned}$$

so that in matrix notation we have

$$\begin{bmatrix} 0 & \alpha & \beta \\ 1 & \alpha-1 & \beta \\ 1 & \alpha & \beta-1 \end{bmatrix}\begin{bmatrix}\ln(\lambda^*)\\ \ln(L^*)\\ \ln(K^*)\end{bmatrix}=\begin{bmatrix}\ln(Q)\\ \ln\left(\frac{W}{\alpha}\right)\\ \ln\left(\frac{R}{\beta}\right)\end{bmatrix}.$$

The adjoint and determinant are

$$\begin{bmatrix} 0 & \alpha & \beta \\ 1 & \alpha-1 & \beta \\ 1 & \alpha & \beta-1 \end{bmatrix}^{-1} = \frac{1}{\alpha+\beta}\begin{bmatrix} 1-\alpha-\beta & \alpha & \beta \\ 1 & -\beta & \beta \\ 1 & \alpha & -\alpha \end{bmatrix}$$

so that

$$\begin{bmatrix} \ln(\lambda^*) \\ \ln(L^*) \\ \ln(K^*) \end{bmatrix} = \frac{1}{\alpha+\beta}\begin{bmatrix} 1-\alpha-\beta & \alpha & \beta \\ 1 & -\beta & \beta \\ 1 & \alpha & -\alpha \end{bmatrix}\begin{bmatrix} \ln(Q) \\ \ln\left(\frac{W}{\alpha}\right) \\ \ln\left(\frac{R}{\beta}\right) \end{bmatrix}.$$

The cost function is given by

$$\begin{aligned} C^*(Q,W,R) &= WL^* + RK^* \\ &= W \times \alpha^{\frac{\beta}{\alpha+\beta}} \beta^{-\frac{\beta}{\alpha+\beta}} Q^{\frac{1}{\alpha+\beta}} W^{-\frac{\beta}{\alpha+\beta}} R^{\frac{\beta}{\alpha+\beta}} \\ &\quad + R \times \alpha^{-\frac{\alpha}{\alpha+\beta}} \beta^{\frac{\alpha}{\alpha+\beta}} Q^{\frac{1}{\alpha+\beta}} W^{\frac{\alpha}{\alpha+\beta}} R^{-\frac{\alpha}{\alpha+\beta}} \\ &= \alpha^{\frac{\beta}{\alpha+\beta}} \beta^{-\frac{\beta}{\alpha+\beta}} Q^{\frac{1}{\alpha+\beta}} W^{\frac{\alpha}{\alpha+\beta}} R^{\frac{\beta}{\alpha+\beta}} \\ &\quad + \alpha^{-\frac{\alpha}{\alpha+\beta}} \beta^{\frac{\alpha}{\alpha+\beta}} Q^{\frac{1}{\alpha+\beta}} W^{\frac{\alpha}{\alpha+\beta}} R^{\frac{\beta}{\alpha+\beta}} \\ &= \alpha^{\frac{\beta}{\alpha+\beta}} \beta^{\frac{\alpha}{\alpha+\beta}} Q^{\frac{1}{\alpha+\beta}} W^{\frac{\alpha}{\alpha+\beta}} R^{\frac{\beta}{\alpha+\beta}} \left(\beta^{-1} + \alpha^{-1}\right) \\ &= \frac{\alpha^{\frac{\beta}{\alpha+\beta}}}{\alpha} \frac{\beta^{\frac{\alpha}{\alpha+\beta}}}{\beta} Q^{\frac{1}{\alpha+\beta}} W^{\frac{\alpha}{\alpha+\beta}} R^{\frac{\beta}{\alpha+\beta}} (\alpha+\beta) \\ &= (\alpha+\beta)\, \alpha^{-\frac{\alpha}{\alpha+\beta}} \beta^{\frac{-\beta}{\alpha+\beta}} Q^{\frac{1}{\alpha+\beta}} W^{\frac{\alpha}{\alpha+\beta}} R^{\frac{\beta}{\alpha+\beta}}. \end{aligned}$$

Thus

$$\begin{aligned} \frac{\partial C^*(Q,W,R)}{\partial Q} &= \frac{\partial}{\partial Q}\left((\alpha+\beta)\, \alpha^{-\frac{\alpha}{\alpha+\beta}} \beta^{\frac{-\beta}{\alpha+\beta}} Q^{\frac{1}{\alpha+\beta}} W^{\frac{\alpha}{\alpha+\beta}} R^{\frac{\beta}{\alpha+\beta}}\right) \\ &= \alpha^{-\frac{\alpha}{\alpha+\beta}} \beta^{\frac{-\beta}{\alpha+\beta}} Q^{\frac{1}{\alpha+\beta}-1} W^{\frac{\alpha}{\alpha+\beta}} R^{\frac{\beta}{\alpha+\beta}} = \lambda^*(Q,W,R) \\ \frac{\partial C^*(Q,W,R)}{\partial W} &= \frac{\partial}{\partial W}\left((\alpha+\beta)\, \alpha^{-\frac{\alpha}{\alpha+\beta}} \beta^{\frac{-\beta}{\alpha+\beta}} Q^{\frac{1}{\alpha+\beta}} W^{\frac{\alpha}{\alpha+\beta}} R^{\frac{\beta}{\alpha+\beta}}\right) \\ &= \alpha \times \alpha^{-\frac{\alpha}{\alpha+\beta}} \beta^{\frac{-\beta}{\alpha+\beta}} Q^{\frac{1}{\alpha+\beta}} W^{\frac{\alpha}{\alpha+\beta}-1} R^{\frac{\beta}{\alpha+\beta}} \\ &= \alpha^{\frac{\beta}{\alpha+\beta}} \beta^{\frac{-\beta}{\alpha+\beta}} Q^{\frac{1}{\alpha+\beta}} W^{\frac{-\beta}{\alpha+\beta}} R^{\frac{\beta}{\alpha+\beta}} = L^*(Q,W,R) \\ \frac{\partial C^*(Q,W,R)}{\partial R} &= \frac{\partial}{\partial R}\left((\alpha+\beta)\, \alpha^{-\frac{\alpha}{\alpha+\beta}} \beta^{\frac{-\beta}{\alpha+\beta}} Q^{\frac{1}{\alpha+\beta}} W^{\frac{\alpha}{\alpha+\beta}} R^{\frac{\beta}{\alpha+\beta}}\right) \\ &= \beta\alpha^{-\frac{\alpha}{\alpha+\beta}} \beta^{\frac{-\beta}{\alpha+\beta}} Q^{\frac{1}{\alpha+\beta}} W^{\frac{\alpha}{\alpha+\beta}} R^{\frac{\beta}{\alpha+\beta}-1} \\ &= \alpha^{-\frac{\alpha}{\alpha+\beta}} \beta^{\frac{\alpha}{\alpha+\beta}} Q^{\frac{1}{\alpha+\beta}} W^{\frac{\alpha}{\alpha+\beta}} R^{-\frac{\alpha}{\alpha+\beta}} = K^*(Q,W,R). \end{aligned}$$

Problem 3.51 *[D***] Find λ^*, L^*, K^* when the production function is given by*

$$Q = F(L,K) = 2L^{\frac{1}{2}} + 2K^{\frac{1}{2}}.$$

Use L^* and K^* to find the cost function $C^*(Q,W,R) = WL^* + RK^*$ and verify that $\frac{\partial C^*}{\partial Q} = \lambda^*$, $\frac{\partial C^*}{\partial W} = L^*$ and $\frac{\partial C^*}{\partial R} = K^*$.

Answer: We have

$$\mathcal{L}(\lambda, L, K, Q, W, R) = WL + RK + \lambda\left(Q - 2L^{\frac{1}{2}} - 2K^{\frac{1}{2}}\right)$$

so that the first-order conditions yield

$$\begin{aligned}
Q - 2L^{*\frac{1}{2}} - 2K^{*\frac{1}{2}} &\implies Q = 2L^{*\frac{1}{2}} + 2K^{*\frac{1}{2}} \\
W - \lambda^*(L^*)^{-\frac{1}{2}} &\implies (L^*)^{\frac{1}{2}} = \frac{\lambda^*}{W} \\
R - \lambda^*(K^*)^{-\frac{1}{2}} &\implies (K^*)^{\frac{1}{2}} = \frac{\lambda^*}{R}.
\end{aligned}$$

Putting the second and third results into the first we have

$$\begin{aligned}
Q &= 2L^{*\frac{1}{2}} + 2K^{*\frac{1}{2}} = 2\frac{\lambda^*}{W} + 2\frac{\lambda^*}{R} \\
&\implies \lambda^*(Q, W, R) = \frac{1}{2}\frac{Q}{W^{-1} + R^{-1}}.
\end{aligned}$$

Now

$$\begin{aligned}
(L^*)^{\frac{1}{2}} &= \frac{\lambda^*}{W} \\
&\implies L^* = (\lambda^*)^2 W^{-2} = \left(\frac{1}{2}\frac{Q}{W^{-1} + R^{-1}}\right)^2 W^{-2} \\
&\implies L^*(Q, W, R) = \frac{1}{4}Q^2\frac{W^{-2}}{(W^{-1} + R^{-1})^2}.
\end{aligned}$$

Similarly

$$\begin{aligned}
(K^*)^{\frac{1}{2}} &= \frac{\lambda^*}{R} \\
&\implies K^* = (\lambda^*)^2 R^{-2} = \left(\frac{1}{2}\frac{Q}{W^{-1} + R^{-1}}\right)^2 R^{-2} \\
&\implies K^*(Q, W, R) = \frac{1}{4}Q^2\frac{R^{-2}}{(W^{-1} + R^{-1})^2}.
\end{aligned}$$

Thus

$$\begin{aligned}
C^*(Q, W, R) &= WL^* + RK^* \\
&= W\left(\frac{1}{4}Q^2\frac{W^{-2}}{(W^{-1} + R^{-1})^2}\right) + R\left(\frac{1}{4}Q^2\frac{R^{-2}}{(W^{-1} + R^{-1})^2}\right) \\
&= \frac{1}{4}\frac{Q^2}{(W^{-1} + R^{-1})}.
\end{aligned}$$

Now

$$\begin{aligned}
\frac{\partial C^*}{\partial Q} &= \frac{\partial}{\partial Q}\left(\frac{1}{4}\frac{Q^2}{(W^{-1}+R^{-1})}\right) = \frac{1}{2}\frac{Q}{(W^{-1}+R^{-1})} = \lambda^* \\
\frac{\partial C^*}{\partial W} &= \frac{\partial}{\partial W}\left(\frac{1}{4}\frac{Q^2}{(W^{-1}+R^{-1})}\right) = \frac{1}{4}Q^2\frac{W^{-2}}{(W^{-1}+R^{-1})^2} = L^* \\
\frac{\partial C^*}{\partial R} &= \frac{\partial}{\partial R}\left(\frac{1}{4}\frac{Q^2}{(W^{-1}+R^{-1})}\right) = \frac{1}{4}Q^2\frac{R^{-2}}{(W^{-1}+R^{-1})^2} = K^*.
\end{aligned}$$

Problem 3.52 *[A**] Suppose we are given a function*

$$y = f(x_1, x_2, \ldots, x_n)$$

where the ith elasticity is

$$\eta_i = \frac{\partial y}{\partial x_i}\frac{x_i}{y} = \frac{\partial f(x_1, x_2, \ldots, x_n)}{\partial x_i}\frac{x_i}{f(x_1, x_2, \ldots, x_n)}.$$

We can rewrite the function as

$$\tilde{y} = \ln\left(f\left(e^{\tilde{x}_1}, e^{\tilde{x}_2}, \ldots, e^{\tilde{x}_n}\right)\right)$$

where $\tilde{y} = \ln(y)$ and $\tilde{x}_i = \ln(x_i)$. Show that

$$\eta_i = \frac{\partial \tilde{y}}{\partial \tilde{x}_i} = \frac{\partial \ln(y)}{\partial \ln(x_i)}.$$

Answer: We have

$$\begin{aligned}
\frac{\partial \tilde{y}}{\partial \tilde{x}_i} &= \frac{\partial}{\partial \tilde{x}_i}\ln\left(f\left(e^{\tilde{x}_1}, e^{\tilde{x}_2}, \ldots, e^{\tilde{x}_n}\right)\right) \\
&= \frac{1}{f\left(e^{\tilde{x}_1}, e^{\tilde{x}_n}, \ldots, e^{\tilde{x}_n}\right)}\frac{\partial f\left(e^{\tilde{x}_1}, e^{\tilde{x}_2}, \ldots, e^{\tilde{x}_n}\right)}{\partial x_i}\frac{\partial e^{\tilde{x}_i}}{\partial x_i} \\
&= \frac{1}{f(x_1, x_2, \ldots, x_n)}\frac{\partial f(x_1, x_2, \ldots, x_n)}{\partial x_i}e^{\tilde{x}_i} \\
&= \frac{1}{f(x_1, x_2, \ldots, x_n)}\frac{\partial f(x_1, x_2, \ldots, x_n)}{\partial x_i}x_i = \eta_i.
\end{aligned}$$

Problem 3.53 *[B***] Given the firm's cost function $C^*(Q, W, R)$ what are the elasticities with respect to Q, W and R?*

Answer: We have the elasticity with respect to Q of

$$\frac{\partial \ln(C^*(Q, W, R))}{\partial \ln(Q)} = \frac{\partial C^*(Q, W, R)}{\partial Q}\frac{Q}{C^*(Q, W, R)} = \frac{\text{Marginal Cost}}{\text{Average Cost}}.$$

Note that if there are increasing returns to scale then marginal cost is less than average cost and the elasticity will be less than 1. If there are constant returns to scale marginal cost equals average cost so the elasticity will be 1. If there are decreasing returns to scale then marginal cost is greater than average cost and the elasticity will be greater than 1.

We have the elasticity with respect to W of

$$\begin{aligned}\frac{\partial \ln (C^*(Q,W,R))}{\partial \ln (W)} &= \frac{\partial C^*(Q,W,R)}{\partial W}\frac{W}{C^*(Q,W,R)} \\ &= \frac{WL^*(Q,W,R)}{C^*(Q,W,R)} = \text{labour's cost share.}\end{aligned}$$

We have the elasticity with respect to R of

$$\begin{aligned}\frac{\partial \ln (C^*(Q,W,R))}{\partial \ln (R)} &= \frac{\partial C^*(Q,W,R)}{\partial R}\frac{R}{C^*(Q,W,R)} \\ &= \frac{RK^*(Q,W,R)}{C^*(Q,W,R)} = \text{capital's cost share.}\end{aligned}$$

Problem 3.54 *[D***] Suppose the firm's cost function is given by*

$$C^*(Q,W,R) = Q^{\frac{1}{3}}W^{\frac{1}{4}}R^{\delta}.$$

What is the exponent on R (i.e., δ) equal to? What do the exponents on Q and W tell you? Calculate the conditional factor demands L^* and K^* as well as λ^*.

Answer: Since $C^*(Q,W,R)$ is homogeneous of degree 1 in W and R the exponents on W and R must sum to 1 here or $\delta = 1 - \frac{1}{4} = \frac{3}{4}$. Since

$$\frac{\partial \ln (C^*(Q,W,R))}{\partial \ln (Q)} = \frac{\frac{\partial C^*(Q,W,R)}{\partial Q}}{\frac{C^*(Q,W,R)}{Q}} = \frac{\text{Marginal Cost}}{\text{Average Cost}} = \frac{1}{3}$$

it follows that marginal cost is less than average cost so there are increasing returns to scale. In fact the production function is homogeneous of degree 3.

Similarly

$$\frac{\partial \ln (C^*(Q,W,R))}{\partial \ln (W)} = \frac{\frac{\partial C^*(Q,W,R)}{\partial W}}{\frac{C^*(Q,W,R)}{W}} = \frac{WL^*}{C^*} = \text{labour's cost share} = \frac{1}{4}$$

it follows that labour's cost share is $\frac{1}{4}$ or 25%.

Problem 3.55 *Calculate L^*, K^*, λ^* when*

$$C^*(Q,W,R) = Q^2\left(W^{\frac{1}{3}} + R^{\frac{1}{3}}\right)^3.$$

Answer: We have

$$\begin{aligned}
L^* &= \frac{\partial C^*(Q,W,R)}{\partial W} = Q^2\frac{\partial}{\partial W}\left(W^{\frac{1}{3}}+R^{\frac{1}{3}}\right)^3 \\
&= Q^2\times 3\left(W^{\frac{1}{3}}+R^{\frac{1}{3}}\right)^2\left(\frac{1}{3}W^{-\frac{2}{3}}\right) \\
&= Q^2\left(W^{\frac{1}{3}}+R^{\frac{1}{3}}\right)^2 W^{-\frac{2}{3}} \\
K^* &= \frac{\partial C^*(Q,W,R)}{\partial R} = Q^2\frac{\partial}{\partial R}\left(W^{\frac{1}{3}}+R^{\frac{1}{3}}\right)^3 \\
&= Q^2\times 3\left(W^{\frac{1}{3}}+R^{\frac{1}{3}}\right)^2\left(\frac{1}{3}R^{-\frac{2}{3}}\right) \\
&= Q^2\left(W^{\frac{1}{3}}+R^{\frac{1}{3}}\right)^2 R^{-\frac{2}{3}} \\
\lambda^* &= \frac{\partial C^*(Q,W,R)}{\partial Q} = \frac{\partial}{\partial Q}Q^2\left(W^{\frac{1}{3}}+R^{\frac{1}{3}}\right)^3 \\
&= 2Q\left(W^{\frac{1}{3}}+R^{\frac{1}{3}}\right)^3.
\end{aligned}$$

Problem 3.56 *[D***] Suppose that a cost function is given by*

$$C^*(Q,W,R) = Q\left(W^{\frac{1}{2}}+R^{\frac{1}{2}}\right)^2$$

Show that $C^*(Q,W,R)$ is homogeneous of degree 1 with respect to W and R. Find $L^*(Q,W,R)$ and $K^*(Q,W,R)$ using Shephard's lemma and show that they are homogeneous of degree 0 *with respect to W and R. Show that* $\frac{\partial L^*(Q,W,R)}{\partial W} < 0$.

Answer: We have

$$\begin{aligned}
C^*(Q,\lambda W,\lambda R) &= Q\left((\lambda W)^{\frac{1}{2}}+(\lambda R)^{\frac{1}{2}}\right)^2 = Q\left(\lambda^{\frac{1}{2}}\left((W)^{\frac{1}{2}}+(R)^{\frac{1}{2}}\right)\right)^2 \\
&= \lambda\left(QW^{\frac{1}{2}}+R^{\frac{1}{2}}\right)^2 = \lambda^1 C^*(Q,W,R).
\end{aligned}$$

We have

$$\begin{aligned}
L^*(Q,W,R) &= \frac{\partial C^*(Q,W,R)}{\partial W} = \frac{\partial}{\partial W}Q\left(W^{\frac{1}{2}}+R^{\frac{1}{2}}\right)^2 \\
&= Q\frac{\left(W^{\frac{1}{2}}+R^{\frac{1}{2}}\right)}{W^{\frac{1}{2}}} = Q\left(1+R^{\frac{1}{2}}W^{-\frac{1}{2}}\right) \\
K^*(Q,W,R) &= \frac{\partial C^*(Q,W,R)}{\partial R} = \frac{\partial}{\partial R}Q\left(W^{\frac{1}{2}}+R^{\frac{1}{2}}\right)^2 \\
&= Q\frac{\left((W)^{\frac{1}{2}}+(R)^{\frac{1}{2}}\right)}{(R)^{\frac{1}{2}}} = Q\left(1+W^{\frac{1}{2}}R^{-\frac{1}{2}}\right).
\end{aligned}$$

We have

$$\frac{\partial L^*(Q,W,R)}{\partial W} = \frac{\partial}{\partial W}Q\left(1+R^{\frac{1}{2}}W^{-\frac{1}{2}}\right) = -\frac{1}{2}QR^{\frac{1}{2}}W^{-\frac{3}{2}} < 0.$$

Problem 3.57 *[B***] Suppose that a cost function is given by*

$$C^*(Q,W,R) = (\delta_1 W^\alpha + \delta_2 R^\alpha)^{\frac{1}{\alpha}} Q^\beta$$

where W and R are the prices of inputs L and K, $0 < \alpha < 1$, and $\beta > 0$. Show that $C^*(Q,W,R)$ is homogeneous of degree 1 with respect to W and R. Find $L^*(Q,W,R)$ and $K^*(Q,W,R)$ using Shephard's lemma. Find the marginal cost function and determine the Lagrange multiplier $\lambda^*(Q,W,R)$ for the cost minimization problem. Find the average cost functions. Suppose there are decreasing returns to scale. What can you say about β then? Is there increasing returns to scale?

Answer: We have

$$\begin{aligned} C^*(Q,\tau W,\tau R) &= \left(\delta_1 (\tau W)^\alpha + \delta_2 (\tau R)^\alpha\right)^{\frac{1}{\alpha}} Q^\beta = \left(\tau^\alpha \left(\delta_1 W^\alpha + \delta_2 R^\alpha\right)\right)^{\frac{1}{\alpha}} Q^\beta \\ &= \tau^1 \left(\delta_1 W^\alpha + \delta_2 R^\alpha\right)^{\frac{1}{\alpha}} Q^\beta = \tau^1 C^*(Q,W,R). \end{aligned}$$

We have

$$\begin{aligned} L^*(Q,W,R) &= \frac{\partial C^*(Q,W,R)}{\partial W} = (\delta_1 W^\alpha + \delta_2 R^\alpha)^{\frac{1}{\alpha}-1} Q^\beta \times \delta_1 W^{\alpha-1} \\ K^*(Q,W,R) &= \frac{\partial C^*(Q,W,R)}{\partial R} = (\delta_1 W^\alpha + \delta_2 R^\alpha)^{\frac{1}{\alpha}-1} Q^\beta \times \delta_2 R^{\alpha-1} \\ \lambda^*(Q,W,R) &= \frac{\partial C^*(Q,W,R)}{\partial Q} = \beta\,(\delta_1 W^\alpha + \delta_2 R^\alpha)^{\frac{1}{\alpha}} Q^{\beta-1}. \end{aligned}$$

If there are decreasing returns to scale then $\beta > 1$ since

$$AC^*(Q,W,R) = \frac{(\delta_1 W^\alpha + \delta_2 R^\alpha)^{\frac{1}{\alpha}} Q^\beta}{Q} = (\delta_1 W^\alpha + \delta_2 R^\alpha)^{\frac{1}{\alpha}} Q^{\beta-1}$$

so that to have average costs increasing in Q we require the exponent on Q be positive or $\beta > 1$. Similarly if there are increasing returns to scale then average costs must fall with Q so that the exponent on Q must be negative or $\beta < 1$.

Problem 3.58 *[B***] Suppose that an econometrician using data for a firm finds that the cost function is given by*

$$\ln\left(C^*(Q,W,R)\right) = 3 + 0.87\ln(Q) + 0.63\ln(W) + \delta\ln(R).$$

What does this tell you about the firm? The econometrician has forgotten to give you the coefficient δ. Can you figure it out on your own or would you need to ask him to find it for you?

Answer: There are increasing returns to scale since the coefficient on $\ln(Q)$ is less than 1 or

$$\frac{\partial \ln\left(C^*(Q,W,R)\right)}{\partial \ln(Q)} = \frac{MC}{AC} = 0.87 < 1.$$

Labour's share of costs is

$$\frac{\partial \ln (C^* (Q, W, R))}{\partial \ln (W)} = \frac{WL^*}{C^*} = 0.63$$

while capital's share of costs is

$$\delta = \frac{\partial \ln (C^* (Q, W, R))}{\partial \ln (R)} = \frac{RK^*}{C^*} = 1 - 0.63 = 0.37.$$

Problem 3.59 *[B***] Suppose we modify the production function so that $Q = F(L, K, T)$ where T is the temperature, which is* ***exogenous****. Suppose that $\frac{\partial F(L,K,T)}{\partial T} > 0$; that is an increase in temperature always increases output. If $C^* (Q, W, R, T)$ is the resulting clever cost function, use the envelope theorem to show that increasing temperature reduces costs or that $\frac{\partial C^*(Q,W,R,T)}{\partial T} < 0$.*

Answer: The Lagrangian is

$$\mathcal{L}(\lambda, L, K, Q, W, R, T) = WL + RK + \lambda (Q - F(L, K, T)).$$

From the first-order conditions we have $W = \lambda^* \frac{\partial F(L^*,K^*,T)}{\partial L}$. Since $W > 0$ and $\frac{\partial F(L^*,K^*,T)}{\partial L} > 0$ it follows that $\lambda^* > 0$. Now

$$\begin{aligned} \frac{\partial C^* (Q, W, R, T)}{\partial T} &= \frac{\partial \mathcal{L}(\lambda, L, K, Q, W, R, T)}{\partial T} |_{\lambda=\lambda^*, L=L^*, K=K^*} \\ &= \left(-\lambda \frac{\partial F(L, K, T)}{\partial T} \right) |_{\lambda=\lambda^*, L=L^*, K=K^*} = -\lambda^* \frac{\partial F(L^*, K^*, T)}{\partial T} < 0 \end{aligned}$$

since $\lambda^* > 0$ and $\frac{\partial F(L^*,K^*,T)}{\partial T} > 0$.

Problem 3.60 *[A**] Suppose we have a production function with n inputs $L_1, L_2, \ldots, L_n$ with nominal prices $W_1, W_2, \ldots, W_n$ and a Cobb-Douglas production function*

$$Q = F(L_1, L_2, \ldots, L_n) = (L_1^{\alpha_1} L_2^{\alpha_2} \times \cdots \times L_n^{\alpha_n})^k$$

where

$$\alpha_1 + \alpha_2 + \cdots + \alpha_n = 1$$

and $k > 0$. Costs are therefore given by

$$W_1 L_1 + W_2 L_2 + \cdots + W_n L_n.$$

Let L be the $n \times 1$ vector of the L_i 's and let W be the $n \times 1$ vector of the W_i's. Show that k determines the degree of homogeneity of the production function. Find the first-order conditions for cost minimization and solve for $L_i^* (Q, W)$ and $\lambda^* (Q, W)$. Show that $\lambda^* (Q, W)$ is homogeneous of degree 1 in W while $L_i^* (Q, W)$ is homogeneous of degree 0. Show that the firm's cost function is $C^* (Q, W) = \lambda^* (Q, W)$. Why isn't λ^* ***marginal*** *cost here? Verify Shephard's lemma that $\frac{\partial C^*(Q,W)}{\partial W_i} = L_i^* (Q, W)$.*

Answer: We have

$$\begin{aligned} F(\tau L_1, \tau L_2, \ldots, \tau L_n) &= \left((\tau L_1)^{\alpha_1} (\tau L_2)^{\alpha_2} \times \cdots \times (\tau L_n)^{\alpha_n}\right)^k \\ &= \left(\tau^{\overbrace{\alpha_1 + \alpha_2 + \cdots + \alpha_n}^{1}} \left(L_1^{\alpha_1} L_2^{\alpha_2} \times \cdots \times L_n^{\alpha_n}\right)\right)^k \\ &= \tau^k \left(L_1^{\alpha_1} L_2^{\alpha_2} \times \cdots \times L_n^{\alpha_n}\right)^k. \end{aligned}$$

By using the $\ln()$ function we can re-write the production constraint

$$Q = \left(L_1^{\alpha_1} L_2^{\alpha_2} \times \cdots \times L_n^{\alpha_n}\right)^k \Longrightarrow \frac{1}{k} \ln(Q) - \sum_{i=1}^{n} \alpha_i \ln(L_i) = 0.$$

The Lagrangian is therefore

$$\mathcal{L}(\lambda, L, Q, W) = W_1 L_1 + W_2 L_2 + \cdots + W_n L_n + \lambda \left(\frac{1}{k} \ln(Q) - \sum_{i=1}^{n} \alpha_i \ln(L_i)\right).$$

The first-order conditions then yield

$$\begin{aligned} \frac{1}{k} \ln(Q) - \sum_{i=1}^{n} \alpha_i \ln(L_i^*) &= 0 \\ &\Longrightarrow \frac{1}{k} \ln(Q) = \sum_{i=1}^{n} \alpha_i \ln(L_i^*) \\ W_i - \frac{\lambda^* \alpha_i}{L_i^*} &= 0 \\ &\Longrightarrow L_i^* = \frac{\lambda^* \alpha_i}{W_i}. \end{aligned}$$

Substituting the last result into the first yields

$$\frac{1}{k} \ln(Q) = \sum_{i=1}^{n} \alpha_i \ln\left(\frac{\lambda^* \alpha_i}{W_i}\right) \Longrightarrow \frac{1}{k} \ln(Q) = \ln(\lambda^*) + \sum_{i=1}^{n} \alpha_i \ln(\alpha_i) - \sum_{i=1}^{n} \alpha_i \ln(W_i)$$

so that solving for λ^* yields

$$\lambda^*(Q, W) = \frac{Q^{\frac{1}{k}} W_1^{\alpha_1} W_2^{\alpha_2} \times \cdots \times W_n^{\alpha_n}}{\alpha_1^{\alpha_1} \alpha_2^{\alpha_2} \times \cdots \times \alpha_n^{\alpha_n}}.$$

Now substituting this into the expression for L_i^* yields

$$L_i^*(Q, W) = \frac{\lambda^* \alpha_i}{W_i} = \left(\frac{Q^{\frac{1}{k}} W_1^{\alpha_1} W_2^{\alpha_2} \times \cdots \times W_n^{\alpha_n}}{\alpha_1^{\alpha_1} \alpha_2^{\alpha_2} \times \cdots \times \alpha_n^{\alpha_n}}\right) \frac{\alpha_i}{W_i}.$$

We have

$$\begin{aligned} C^*(Q,W) &= W_1L_1^*(Q,W)+W_2L_2^*(Q,W)+\cdots+W_nL_n^*(Q,W) \\ &= W_1\frac{\lambda^*\alpha_1}{W_1}+W_2\frac{\lambda^*\alpha_2}{W_2}+\cdots+W_n\frac{\lambda^*\alpha_n}{W_n}=\lambda^*\left(\underbrace{\alpha_1+\alpha_2+\cdots+\alpha_n}_{1}\right) \\ &= \lambda^*(Q,W)=\frac{Q^{\frac{1}{k}}W_1^{\alpha_1}W_2^{\alpha_2}\times\cdots\times W_n^{\alpha_n}}{\alpha_1^{\alpha_1}\alpha_2^{\alpha_2}\times\cdots\times\alpha_n^{\alpha_n}}. \end{aligned}$$

The Cobb-Douglas production function has a cost function that has the Cobb-Douglas functional form!

It is a simple matter then to verify that

$$\begin{aligned} \frac{\partial C^*(Q,W)}{\partial W_i} &= \frac{\partial\lambda}{\partial W_i}\left(\frac{Q^{\frac{1}{k}}W_1^{\alpha_1}W_2^{\alpha_2}\times\cdots\times W_n^{\alpha_n}}{\alpha_1^{\alpha_1}\alpha_2^{\alpha_2}\times\cdots\times\alpha_n^{\alpha_n}}\right) \\ &= \frac{Q^{\frac{1}{k}}W_1^{\alpha_1}W_2^{\alpha_2}\times\cdots\times W_n^{\alpha_n}}{\alpha_1^{\alpha_1}\alpha_2^{\alpha_2}\times\cdots\times\alpha_n^{\alpha_n}}\frac{\alpha_i}{W_i}=L_i^*(Q,W). \end{aligned}$$

Problem 3.61 *[A **] Suppose we have a production function with n inputs $L_1,L_2,\ldots,L_n$ with nominal prices $W_1,W_2,\ldots,W_n$ and a Constant Elasticity of Substitution (CES) production function*

$$Q=F(L_1,L_2,\ldots,L_n)=(\alpha_1L_1^\rho+\alpha_2L_2^\rho+\cdots+\alpha_nL_n^\rho)^{\frac{k}{\rho}}$$

where

$$\alpha_1+\alpha_2+\cdots+\alpha_n=1$$

and $k>0$. Costs are given by

$$W_1L_1+W_2L_2+\cdots+W_nL_n.$$

Let L be the $n\times$ 1 vector of the L_i 's and let W be the $n\times$ 1 vector of the W_i 's. Let L be the $n\times$ 1 vector of the L_i 's and let W be the $n\times$ 1 vector of the W_i 's.Show that k determines the degree of homogeneity of the production function and that as $k\to0$ the CES becomes a Cobb-Douglas production function. Find the first-order conditions for cost minimization and solve for $L_i^*(Q,W)$ and $\lambda^*(Q,W)$. Find the firm's cost function $C^*(Q,W)$. Verify Shephard's lemma that $\frac{\partial C^*(Q,W)}{\partial W_i}=L_i^*(Q,W)$.

Answer: Verify that we can rewrite the production relation as

$$\frac{Q^{\frac{\rho}{k}}-1}{\rho}=\sum_{i=1}^{n}\alpha_i\frac{L_i^\rho-1}{\rho}.$$

From L'Hôpital's rule we have

$$\lim_{\rho\to0}\frac{x^\rho-1}{\rho}=\lim_{\rho\to0}\frac{\frac{d}{d\rho}\left(e^{\rho\ln(x)}-1\right)}{\frac{d}{d\rho}(\rho)}=\lim_{\rho\to0}\frac{\ln(x)\,e^{\rho\ln(x)}}{1}=\ln(x)$$

so that as $\rho \to 0$ we have

$$\frac{1}{k}\ln(Q) = \sum_{i=1}^{n} \alpha_i \ln(L_i) \Longrightarrow Q = \left(L_1^{\alpha_1} L_2^{\alpha_2} \times \cdots \times L_n^{\alpha_n}\right)^k.$$

We can re-write the production constraint as

$$Q^{\frac{\rho}{k}} - \sum_{i=1}^{n} \alpha_i L_i^{\rho} = 0.$$

The Lagrangian is therefore

$$\mathcal{L}(\lambda, L, Q, W) = W_1 L_1 + W_2 L_2 + \cdots + W_n L_n + \lambda\left(Q^{\frac{\rho}{k}} - \sum_{i=1}^{n} \alpha_i L_i^{\rho}\right).$$

The first-order conditions then yield

$$\begin{aligned}
Q^{\frac{\rho}{k}} - \sum_{i=1}^{n} \alpha_i (L_i^*)^{\rho} &= 0 \\
&\Longrightarrow Q^{\frac{\rho}{k}} = \sum_{i=1}^{n} \alpha_i (L_i^*)^{\rho} \\
W_i - \lambda^* \alpha_i \rho (L_i^*)^{\rho-1} &= 0 \\
&\Longrightarrow W_i = \lambda^* \alpha_i \rho (L_i^*)^{\rho-1} \text{ for } i = 1, 2, \ldots, n.
\end{aligned}$$

Solving for $(L_i^*)^{\rho}$ yields

$$\begin{aligned}
(L_i^*)^{\rho} &= W_i^{\frac{\rho}{\rho-1}} \alpha_i^{-\frac{\rho}{\rho-1}} \rho^{-\frac{\rho}{\rho-1}} (\lambda^*)^{-\frac{\rho}{\rho-1}} \\
&\Longrightarrow \alpha_i (L_i^*)^{\rho} = (\lambda^*)^{-\frac{\rho}{\rho-1}} \alpha_i^{-\frac{1}{\rho-1}} W_i^{\frac{\rho}{\rho-1}} \rho^{-\frac{\rho}{\rho-1}} \\
&\Longrightarrow Q^{\frac{\rho}{k}} = \sum_{i=1}^{n} \alpha_i (L_i^*)^{\rho} = (\lambda^*)^{-\frac{\rho}{\rho-1}} \rho^{-\frac{\rho}{\rho-1}} \sum_{i=1}^{n} \alpha_i^{-\frac{1}{\rho-1}} W_i^{\frac{\rho}{\rho-1}} \\
&\Longrightarrow (\lambda^*)^{-\frac{\rho}{\rho-1}} = \frac{Q^{\frac{\rho}{k}} \rho^{\frac{\rho}{\rho-1}}}{\sum_{i=1}^{n} \alpha_i^{-\frac{1}{\rho-1}} W_i^{\frac{\rho}{\rho-1}}}
\end{aligned}$$

and so

$$(L_i^*)^{\rho} = \alpha_i^{-\frac{\rho}{\rho-1}} W_i^{\frac{\rho}{\rho-1}} \rho^{-\frac{\rho}{\rho-1}} (\lambda^*)^{-\frac{\rho}{\rho-1}} = Q^{\frac{\rho}{k}} \frac{\alpha_i^{-\frac{\rho}{\rho-1}} W_i^{\frac{\rho}{\rho-1}}}{\sum_{i=1}^{n} \alpha_i^{-\frac{1}{\rho-1}} W_i^{\frac{\rho}{\rho-1}}}$$

so that

$$L_i^*(Q, W) = Q^{\frac{1}{k}} \frac{\left(\alpha_i^{-\frac{\rho}{\rho-1}} W_i^{\frac{\rho}{\rho-1}}\right)^{\frac{1}{\rho}}}{\left(\sum_{i=1}^{n} \alpha_i^{-\frac{1}{\rho-1}} W_i^{\frac{\rho}{\rho-1}}\right)^{\frac{1}{\rho}}} = Q^{\frac{1}{k}} \frac{\alpha_i^{-\frac{1}{\rho-1}} W_i^{\frac{1}{\rho-1}}}{\left(\sum_{i=1}^{n} \alpha_i^{-\frac{1}{\rho-1}} W_i^{\frac{\rho}{\rho-1}}\right)^{\frac{1}{\rho}}}.$$

We have

$$\begin{aligned}
C^*(Q,W) &= \sum_{i=1}^{n} W_i L_i^*(Q,W) = \sum_{i=1}^{n} W_i \frac{Q^{\frac{1}{k}} \alpha_i^{-\frac{1}{\rho-1}} W_i^{\frac{1}{\rho-1}}}{\left(\sum_{i=1}^{n} \alpha_i^{-\frac{1}{\rho-1}} W_i^{\frac{\rho}{\rho-1}}\right)^{\frac{1}{\rho}}} \\
&= \frac{Q^{\frac{1}{k}}}{\left(\sum_{i=1}^{n} \alpha_i^{-\frac{1}{\rho-1}} W_i^{\frac{\rho}{\rho-1}}\right)^{\frac{1}{\rho}}} \left(\sum_{i=1}^{n} \alpha_i^{-\frac{1}{\rho-1}} W_i^{\frac{\rho}{\rho-1}}\right)^{1} \\
&= Q^{\frac{1}{k}} \left(\sum_{i=1}^{n} \alpha_i^{-\frac{1}{\rho-1}} W_i^{\frac{\rho}{\rho-1}}\right)^{\frac{\rho-1}{\rho}}.
\end{aligned}$$

It is an interesting fact that the CES production function has a cost function which is of the CES form.

Problem 3.62 *[A **] Suppose that a firm has a Leontief production function given by*

$$Q = F(L,K) = \min\left[\sigma^{\alpha} L^{\alpha}, K^{\alpha}\right]$$

where $\min[X,Y]$ *is the minimum of* X *or* Y *(for example* $\min[5,4] = 4$ *while* $\min[2,7] = 2$*) and* $\alpha > 0, A > 0$ *and* $\sigma > 0$*. Prove that the firm always hires labour and capital so that* $K = \sigma L$*. Show that the production function is homogeneous of degree* α*. Show without calculus that the firm's cost function is given by*

$$C^*(Q,W,R) = \left(\frac{W}{\sigma} + R\right) Q^{\frac{1}{\alpha}}$$

and verify that Shephard's lemma holds for this cost function.

Answer: The production function is homogeneous of degree α since

$$F(\tau L, \tau K) = \min\left[\sigma^{\alpha}(\tau L)^{\alpha}, (\tau K)^{\alpha}\right] = \min\left[\tau^{\alpha}\sigma^{\alpha}L^{\alpha}, \tau^{\alpha}K^{\alpha}\right] = \tau^{\alpha}\min\left[\sigma^{\alpha}L^{\alpha}, K^{\alpha}\right] = \tau^{\alpha}F(L,K)$$

where we use

$$\min[aX, aY] = a\min[X,Y] \text{ for } a > 0.$$

Suppose that $K > \sigma L$. Then $K^{\alpha} > \sigma^{\alpha}L^{\alpha}$ and so the firm would produce

$$Q = \min\left[\sigma^{\alpha}L^{\alpha}, K^{\alpha}\right] = \sigma^{\alpha}L^{\alpha}.$$

Now suppose the firm reduced the amount of capital it hires to $K_0 = \sigma L < K$. Then $K_0^{\alpha} = \sigma^{\alpha}L^{\alpha}$ and so

$$Q = \min\left[\sigma^{\alpha}L^{\alpha}, K_0^{\alpha}\right] = A\sigma^{\alpha}L^{\alpha}$$

and so the firm produces the same amount of output but hires less capital. Thus $K > \sigma L$ cannot be profit maximizing or cost minimizing and so we conclude that $K \leq \sigma L$.

Now suppose that $K < \sigma L$. Then $K^\alpha < \sigma^\alpha L^\alpha$ and so the firm would produce

$$Q = \min\left[\sigma^\alpha L^\alpha, K^\alpha\right] = AK^\alpha.$$

If the firm reduced the amount of labour it hires to $L_0 = \frac{1}{\sigma}K < L$ then $\sigma^\alpha L_0^\alpha = K^\alpha$ and so

$$Q = \min\left[\sigma^\alpha L_0^\alpha, K^\alpha\right] = AK^\alpha$$

and so the firm produces the same amount of output but hires less labour. Thus $K < \sigma L$ cannot be profit maximizing or cost minimizing and so we conclude that $K \geq \sigma L$.

Combining the two results $K \geq \sigma L$ and $K \leq \sigma L$ we conclude that $K = \sigma L$. Since $K = \sigma L$ we have

$$Q = \min\left[\sigma^\alpha L^\alpha, K^\alpha\right] = \sigma^\alpha L^\alpha = AK^\alpha.$$

We solve for the cost minimizing L^*, K^*, which are independent of W, R, as

$$\begin{aligned} Q &= \sigma^\alpha L^\alpha \Longrightarrow L^*(Q) = \frac{1}{\sigma}Q^{\frac{1}{\alpha}} \\ Q &= K^\alpha \Longrightarrow K^*(Q) = Q^{\frac{1}{\alpha}} \end{aligned}$$

so that

$$\begin{aligned} C^*(Q, W, R) &= WL^*(Q) + RK^*(Q) \\ &= W \times \frac{1}{\sigma}Q^{\frac{1}{\alpha}} + R \times Q^{\frac{1}{\alpha}} = \left(\frac{W}{\sigma} + R\right)Q^{\frac{1}{\alpha}}. \end{aligned}$$

Shephard's lemma holds as

$$\begin{aligned} \frac{\partial C^*(Q, W, R)}{\partial W} &= \frac{\partial}{\partial W}\left(\left(\frac{W}{\sigma} + R\right)Q^{\frac{1}{\alpha}}\right) = \frac{1}{\sigma}Q^{\frac{1}{\alpha}} = L^*(Q) \\ \frac{\partial C^*(Q, W, R)}{\partial R} &= \frac{\partial}{\partial R}\left(\left(\frac{W}{\sigma} + R\right)Q^{\frac{1}{\alpha}}\right) = Q^{\frac{1}{\alpha}} = K^*(Q). \end{aligned}$$

Problem 3.63 *[A***] Use Euler's theorem and the first-order conditions for cost minimization to prove that if the production function $Q = F(L, K)$ is homogeneous of degree k, then the cost function takes the form*

$$C^*(Q, W, R) = A(W, R)\, Q^{\frac{1}{k}}.$$

Average costs are then

$$AC^*(Q, W, R) \equiv \frac{C^*(Q, W, R)}{Q}.$$

Use the result to show that if $0 < k < 1$ then average costs will increase with Q (i.e., decreasing returns to scale).

Answer: The Lagrangian is

$$\mathcal{L}(\lambda, L, K, Q, W, R) = WL + RK + \lambda (Q - F(L, K))$$

so that the first-order conditions yield

$$\begin{aligned} Q - F(L^*, K^*) &= 0 \Longrightarrow Q = F(L^*, K^*) \\ W - \lambda^* \frac{\partial F(L^*, K^*)}{\partial L} &= 0 \Longrightarrow \frac{\partial F(L^*, K^*)}{\partial L} = \frac{W}{\lambda^*} \\ R - \lambda^* \frac{\partial F(L^*, K^*)}{\partial K} &= 0 \Longrightarrow \frac{\partial F(L^*, K^*)}{\partial K} = \frac{R}{\lambda^*}. \end{aligned}$$

Since the production function is homogeneous Euler's theorem yields

$$kF(L^*, K^*) = kQ = \frac{\partial F(L^*, K^*)}{\partial L} L^* + \frac{\partial F(L^*, K^*)}{\partial K} K^*.$$

Since $\frac{\partial F}{\partial L} = \frac{W}{\lambda^*}, \frac{\partial F}{\partial K} = \frac{R}{\lambda^*}$ and $C^* \equiv WL^* + RK^*$ we have

$$kQ = \frac{W}{\lambda^*} L^* + \frac{R}{\lambda^*} K^* = \frac{WL^* + RK^*}{\lambda^*} = \frac{C^*}{\lambda^*}.$$

Since $\lambda^* = \frac{\partial C^*}{\partial Q}$ by the envelope theorem we have

$$\frac{C^*(Q, W, R)}{\frac{\partial C^*(Q,W,R)}{\partial Q}} = kQ \Longrightarrow \frac{\partial C^*(Q, W, R)}{\partial Q} \frac{Q}{C^*(Q, W, R)} = \frac{1}{k}.$$

The left-hand side is the *elasticity* of the cost function with respect to Q. The right-hand side is a constant $\frac{1}{k}$ that does not depend on Q. Thus the elasticity is constant for all Q and so the cost function must take the form Ax^b where A does not depend on Q and $b = \frac{1}{k}$ is the elasticity. Since A can depend on W and R we write $A = A(W, R)$. Thus

$$C^*(Q, W, R) = A(W, R) Q^{\frac{1}{k}}.$$

Now

$$\begin{aligned} AC^*(Q, W, R) &= \frac{C^*(Q, W, R)}{Q} = \frac{A(W, R) Q^{\frac{1}{k}}}{Q} \\ &= A(W, R) Q^{\frac{1}{k} - 1} = A(W, R) Q^{\frac{1-k}{k}} \end{aligned}$$

so that

$$\frac{\partial AC^*(Q, W, R)}{\partial Q} = \left(\frac{1-k}{k}\right) A(W, R) Q^{\frac{1-k}{k} - 1} > 0$$

since

$$0 < k < 1 \Longrightarrow \frac{1-k}{k} > 0.$$

Problem 3.64 *[B***] Suppose a firm has a production function*

$$Q = e^{-X}\left(2L^{\frac{1}{2}} + 2K^{\frac{1}{2}}\right)$$

where X is an exogenous externality (say pollution from another plant). Write down the Lagrangian for cost minimization, find the first-order conditions and solve. Find the cost function $C^*(Q, W, R, X)$. Using the envelope theorem show that $\frac{\partial C^*}{\partial X} = \lambda^* Q$ and verify this result by directly calculating $\frac{\partial C^*}{\partial X}$.

Answer: The (naive) Lagrangian is

$$\mathcal{L}(\lambda, L, K, Q, W, R) = WL + RK + \lambda\left(Q - e^{-X}\left(2L^{\frac{1}{2}} + 2K^{\frac{1}{2}}\right)\right).$$

The first-order conditions yield

$$\begin{aligned}
Q - e^{-X}\left(2(L^*)^{\frac{1}{2}} + 2(K^*)^{\frac{1}{2}}\right) &= 0 \\
&\Longrightarrow Q = e^{-X}\left(2(L^*)^{\frac{1}{2}} + 2(K^*)^{\frac{1}{2}}\right). \\
W - \lambda^* e^{-X}(L^*)^{-\frac{1}{2}} &= 0 \\
&\Longrightarrow (L^*)^{\frac{1}{2}} = e^{-X}\lambda^* W^{-1} \\
&\Longrightarrow L^* = e^{-2X}(\lambda^*)^2 W^{-2} \\
R - \lambda^* e^{-X}(K^*)^{-\frac{1}{2}} &= 0 \\
&\Longrightarrow (K^*)^{\frac{1}{2}} = e^{-X}\lambda^* R^{-1} \\
&\Longrightarrow K^* = e^{-2X}(\lambda^*)^2 R^{-2}.
\end{aligned}$$

Putting the second and third results into the first yields

$$\begin{aligned}
Q &= e^{-X}\left(2(L^*)^{\frac{1}{2}} + 2(K^*)^{\frac{1}{2}}\right) \\
&= e^{-X}\left(2\lambda^* e^{-X}W^{-1} + 2\lambda^* e^{-X}R^{-1}\right) \\
&= 2e^{-2X}\lambda^*\left(W^{-1} + R^{-1}\right) \\
\Longrightarrow \lambda^* &= \lambda^*(Q, W, R, X) = \frac{Q}{2}e^{2X}\left(W^{-1} + R^{-1}\right)^{-1}.
\end{aligned}$$

Now

$$\begin{aligned} L^*(Q,W,R,X) &= e^{-2X}(\lambda^*)^2 W^{-2} \\ &= e^{-2X}\left(\frac{Q}{2}e^{2X}\left(W^{-1}+R^{-1}\right)^{-1}\right)^2 W^{-2} \\ &= \frac{Q^2}{4}e^{-2X}e^{4X}W^{-2}\left(W^{-1}+R^{-1}\right)^{-2} \\ &= \frac{Q^2}{4}e^{2X}W^{-2}\left(W^{-1}+R^{-1}\right)^{-2} \\ K^*(Q,W,R) &= e^{-2X}(\lambda^*)^2 R^{-2} \\ &= e^{-2X}\left(\frac{Q}{2}e^{2X}\left(W^{-1}+R^{-1}\right)^{-1}\right)^2 R^{-2} \\ &= \frac{Q^2}{4}e^{-2X}e^{4X}R^{-2}\left(W^{-1}+R^{-1}\right)^{-2} \\ &= \frac{Q^2}{4}e^{2X}R^{-2}\left(W^{-1}+R^{-1}\right)^{-2}. \end{aligned}$$

The cost function then is

$$\begin{aligned} C^*(Q,W,R,X) &= WL^* + RK^* \\ &= W\frac{Q^2}{4}e^{2X}W^{-2}\left(W^{-1}+R^{-1}\right)^{-2} + R\frac{Q^2}{4}e^{2X}R^{-2}\left(W^{-1}+R^{-1}\right)^{-2} \\ &= \frac{Q^2}{4}e^{2X}\left(W^{-1}+R^{-1}\right)^{-2}\left(W^{-1}+R^{-1}\right) = \frac{Q^2}{4}e^{2X}\left(W^{-1}+R^{-1}\right)^{-1} \\ &= \frac{Q^2}{4}e^{2X}\left(W^{-1}+R^{-1}\right)^{-1}. \end{aligned}$$

To find $\frac{\partial C^*}{\partial X}$ use the envelope theorem as

$$\begin{aligned} \frac{\partial C^*}{\partial X} &= \frac{\partial}{\partial X}\left(WL + RK + \lambda\left(Q - e^{-X}F(L,K)\right)\right)\Big|_{\substack{\lambda=\lambda^* \\ L=L^* \\ K=K^*}} \\ &= \left(\lambda e^{-X}F(L,K)\right)\Big|_{\substack{\lambda=\lambda^* \\ L=L^* \\ K=K^*}} = \lambda^* e^{-X}F(L^*,K^*) = \lambda^* Q \end{aligned}$$

since $Q = e^{-X}F(L^*,K^*)$. Calculating $\frac{\partial C^*}{\partial X}$ directly we have

$$\begin{aligned} \frac{\partial C^*}{\partial X} &= \frac{\partial}{\partial X}\left(\frac{Q^2}{4}e^{2X}\left(W^{-1}+R^{-1}\right)^{-1}\right) = \frac{Q^2}{2}e^{2X}\left(W^{-1}+R^{-1}\right)^{-1} \\ &= Q\underbrace{\frac{Q}{2}e^{2X}\left(W^{-1}+R^{-1}\right)^{-1}}_{\lambda^*} = Q\lambda^*. \end{aligned}$$

Problem 3.65 *[A ***] A function $f(x)$ with a Taylor series*

$$\bar{f}(x) = f\left(x^0\right) + \nabla f\left(x^0\right)^T\left(x - x^0\right)$$

is concave if and only if for all x^0

$$f(x) \le \bar{f}(x) = f(x^0) + \nabla f(x^0)^T (x - x^0).$$

Use this result and Shephard's lemma to prove that the cost function $C^*(Q, W, R)$ *is a concave function of* W, R.

Answer: Think of W_0, R_0 as Period 0 prices and W, R as Period 1 prices. Define C^*, L^*, K^* for both periods as

$$\begin{aligned} C_0^* &\equiv C^*(Q, W_0, R_0), C_1^* \equiv C^*(Q, W, R) \\ L_0^* &\equiv L^*(Q, W_0, R_0), L_1^* \equiv L^*(Q, W, R) \\ K_0^* &\equiv K^*(Q, W_0, R_0), K_1^* \equiv K^*(Q, W, R). \end{aligned}$$

The Taylor series for $C^*(Q, W, R)$ at W_0, R_0 then is

$$\bar{C}^*(Q, W, R) = C^*(Q, W_0, R_0) + \frac{\partial C^*(Q, W_0, R_0)}{\partial W}(W - W_0) + \frac{\partial C^*(Q, W_0, R_0)}{\partial R}(R - R_0)$$

or

$$\bar{C}^*(Q, W, R) = C_0^* + \frac{\partial C^*(Q, W_0, R_0)}{\partial W}(W - W_0) + \frac{\partial C^*(Q, W_0, R_0)}{\partial R}(R - R_0).$$

By Shephard's lemma

$$\frac{\partial C^*(Q, W_0, R_0)}{\partial W} = L^*(Q, W_0, R_0) \equiv L_0^* \text{ and } \frac{\partial C^*(Q, W_0, R_0)}{\partial R} = -K^*(Q, W_0, R_0) \equiv K_0^*$$

we have

$$\begin{aligned} \bar{C}^*(Q, W, R) &= C_0^* + L_0^* \times (W - W_0) + K_0^* \times (R - R_0) \\ &= C_0^* - (W_0 \times L_0^* + R_0 \times K_0^*) + W \times L_0^* + R \times K_0^* \\ &= W \times L_0^* + R \times K_0^* \end{aligned}$$

since $C_0^* \equiv W_0 \times L_0^* + R_0 \times K_0^*$. Now

$$\bar{C}^*(Q, W, R) = W \times L_0^* + R \times K_0^*$$

is costs when there are Period 1 prices W, R but L_0^*, K_0^* are optimal for Period 0 prices W_0, R_0. Picking labour and capital that are optimal for the wrong period will generally increase costs, and certainly cannot lower costs and so

$$C^*(Q, W, R) \equiv W \times L_1^* + R \times K_1^* \le \bar{C}^*(Q, W, R) = W \times L_0^* + R \times K_0^*$$

and so the cost function is concave.

Problem 3.66 *[A***] Suppose that the naive function* $f(y, x)$ *can be written as*

$$f(y, x) = \sum_{i=1}^{n} x_i \phi_i(y) = x^T \phi(y)$$

where $\phi(y) = [\phi_i(y)]$ is an $m \times 1$ vector, and suppose that the constraint function $h(y,x) = 0$ does not depend on x as $h(y,x) = h(y)$. Show that if $f(y,x)$ is to be minimized then the clever function $f^*(x)$ is homogeneous of degree 1, and $y^*(x)$ is homogeneous of degree 0. Show that $f^*(x)$ is ***concave****. Use this result to show that when Q is held fixed that $C^*(Q,W,R)$ is concave and homogeneous of degree 1 in W,R.*

Answer: Suppose that $y^*(x)$ minimizes $f(y,x)$ subject to $h(y) = 0$ so that $h(y^*(x)) = 0$. The naive function is homogeneous of degree 1 in x since

$$f(y,\tau x) = (\tau x)^T \phi(y) = \tau\left(x^T \phi(y)\right) = \tau f(y,x).$$

Since $\tau > 0$ and since $h(y^*(x)) = 0$, it follows that $y^*(x)$ minimizes $\tau f(y,x)$ subject to $h(y) = 0$ and so

$$y^*(\tau x) = y^*(x).$$

Thus $y^*(x)$ is homogeneous of degree 0. Furthermore

$$f^*(\tau x) = (\tau x)^T \phi(y^*(\tau x)) = (\tau x)^T \phi(y^*(x)) = \tau x^T \phi(y^*(x)) = \tau f^*(x)$$

where the second result follows from the homogeneity of $y^*(x)$. This proves that $f^*(x)$ is homogeneous of degree 1.

Let x_1 and x_2 (which are $m \times 1$ vectors) be two values of x and let $x_3 = \lambda x_1 + (1-\lambda) x_2$ where $0 \leq \lambda \leq 1$ be a convex combination of x_1 and x_2. We need to show that

$$f^*(x_3) \leq \lambda f^*(x_1) + (1-\lambda) f^*(x_2).$$

We have

$$f^*(x_3) = (\lambda x_1 + (1-\lambda) x_2)^T \phi(y^*(x_3)) = \lambda x_1^T \phi(y^*(x_3)) + (1-\lambda) x_2^T \phi(y^*(x_3)).$$

But

$$x_1^T \phi(y^*(x_3)) = f(y^*(x_3), x_1) \geq f(y^*(x_1), x_1) = x_1^T \phi(y^*(x_1)) = f^*(x_1)$$

since $y^*(x_1)$ minimizes $f(y,x_1)$ subject to $h(y) = 0$ and $h(y^*(x_3)) = 0$. Similarly

$$x_2^T \phi(y^*(x_3)) = f(y^*(x_3), x_2) \geq f(y^*(x_2), x_2) = x_2^T \phi(y^*(x_2)) = f^*(x_2)$$

since $y^*(x_2)$ maximizes $f(y,x_2)$ subject to $h(y) = 0$ and $h(y^*(x_3)) = 0$. We therefore conclude that

$$f^*(x_3) = \lambda x_1^T \phi(y^*(x_3)) + (1-\lambda) x_2^T \phi(y^*(x_3)) \geq \lambda f^*(x_1) + (1-\lambda) f^*(x_2)$$

so that $f^*(x)$ is concave.

The naive cost function is a linear function of the exogenous variables W, R when Q is held fixed as

$$C(L, K, W, R) \equiv WL + RK = x^T \phi(y) = f(y, x)$$

where

$$x = \begin{bmatrix} W \\ R \end{bmatrix}, \; y = \begin{bmatrix} L \\ K \end{bmatrix}, \; \phi(y) = y = \begin{bmatrix} L \\ K \end{bmatrix}.$$

From this and the fact that the constraint does not depend on W or R, it follows that $C^*(Q, W, R)$ is convex and homogeneous of degree 1 in W and R.

Chapter 4

Summation and Random Variables

4.1 Summation

Problem 4.1 *[D***] Suppose that* $n = 25$ *and*

$$\sum_{i=1}^{n} X_i = 100, \sum_{i=1}^{n} X_i^2 = 500, \sum_{i=1}^{n} Y_i = 125, \sum_{i=1}^{n} Y_i^2 = 850, \sum_{i=1}^{n} X_i Y_i = 700.$$

Based on this information calculate

$$\sum_{i=1}^{n} (3X_i + 4Y_i + 3), \sum_{i=1}^{n} (3X_i + 4)^2, \sum_{i=1}^{n} (3X_i + 4Y_i)^2, \sum_{i=1}^{n} \left(X_i - \bar{X}\right)^2, \sum_{i=1}^{n} \left(X_i - \bar{X}\right)\left(Y_i - \bar{Y}\right).$$

Now calculate $\hat{\beta}$ *for* $Y_i = \alpha + \beta X_i + e_i$.

Answer: We have

$$\begin{aligned}
\sum_{i=1}^{n}(3X_i+4Y_i+3) &= 3\sum_{i=1}^{n}X_i+4\sum_{i=1}^{n}Y_i+3n \\
&= 3\times 100+4\times 125+3\times 25=875 \\
\sum_{i=1}^{n}(3X_i+4)^2 &= \sum_{i=1}^{n}\left(9X_i^2+24X_i+16\right) \\
&= 9\sum_{i=1}^{n}X_i^2+24\sum_{i=1}^{n}X_i+16n \\
&= 9\times 500+24\times 100+16\times 25=7300 \\
\sum_{i=1}^{n}(3X_i+4Y_i)^2 &= \sum_{i=1}^{n}\left(9X_i^2+24X_iY_i+16Y_i^2\right) \\
&= 9\sum_{i=1}^{n}X_i^2+24\sum_{i=1}^{n}X_iY_i+16\sum_{i=1}^{n}Y_i^2 \\
&= 9\times 500+24\times 700+16\times 850=34900 \\
\sum_{i=1}^{n}\left(X_i-\overline{X}\right)^2 &= \sum_{i=1}^{n}\left(X_i^2-2\bar{X}X_i+\bar{X}^2\right)=\sum_{i=1}^{n}X_i^2-n\bar{X}^2 \\
&= 500-25\left(\frac{100}{25}\right)^2=100 \\
\sum_{i=1}^{n}\left(X_i-\bar{X}\right)\left(Y_i-\bar{Y}\right) &= \sum_{i=1}^{n}X_iY_i-n\bar{X}\bar{Y} \\
&= 700-25\left(\frac{100}{25}\right)\left(\frac{125}{25}\right)=200.
\end{aligned}$$

Thus

$$\hat{\beta}=\frac{\sum_{i=1}^{n}\left(X_i-\bar{X}\right)\left(Y_i-\bar{Y}\right)}{\sum_{i=1}^{n}\left(X_i-\bar{X}\right)^2}=\frac{200}{100}=2.$$

Problem 4.2 *[D***] If $\sum_{i=1}^{7}X_i=35$ then what is $\bar{X}$ and what is $\sum_{i=1}^{7}(4X_i+2)$? If $\sum_{i=1}^{7}X_i^2=255$ then what is $\sum_{i=1}^{7}(4X_i+2)^2$ and what is $\sum_{i=1}^{7}\left(X_i-\bar{X}\right)^2$?*

Answer: We have

$$\begin{aligned}
\bar{X} &= \frac{1}{n}\sum_{i=1}^{n}X_i=\frac{1}{7}\times 35=5, \sum_{i=1}^{7}(4X_i+2) \\
&= 4\sum_{i=1}^{7}X_i+\sum_{i=1}^{7}2=4\times 35+7\times 2=154.
\end{aligned}$$

If $\sum_{i=1}^{7} X_i^2 = 255$ then

$$\begin{aligned}\sum_{i=1}^{7}(4X_i+2)^2 &= \sum_{i=1}^{7}\left(16X_i^2+16X_i+4\right) = 16\sum_{i=1}^{7}X_i^2+16\sum_{i=1}^{7}X_i+\sum_{i=1}^{7}4 \\ &= 16\times 255+16\times 35+7\times 4 = 4668.\end{aligned}$$

Also

$$\sum_{i=1}^{7}\left(X_i-\bar{X}\right)^2 = \sum_{i=1}^{7}X_i^2 - 7\bar{X}^2 = 255-7\times 25 = 80.$$

Problem 4.3 *[D***] If $n=10$ and $\sum_{i=1}^{n} X_i = 50$ then what is $\bar{X}$ and what is $\sum_{i=1}^{n}(3X_i+2)$? If $\sum_{i=1}^{n} X_i^2 = 275$ then what are $\sum_{i=1}^{n}(3X_i+2)^2$ and $\sum_{i=1}^{n}\left(X_i-\bar{X}\right)^2$?*

Answer: We have

$$\begin{aligned}\bar{X} &= \frac{1}{n}\sum_{i=1}^{n}X_i = \frac{1}{10}\times 50 = 5 \\ \sum_{i=1}^{n}(3X_i+2) &= 3\sum_{i=1}^{n}X_i+\sum_{i=1}^{n}2 = 3\times 50+10\times 2 = 170.\end{aligned}$$

If $\sum_{i=1}^{n} X_i^2 = 275$ then

$$\begin{aligned}\sum_{i=1}^{n}(3X_i+2)^2 &= \sum_{i=1}^{n}\left(9X_i^2+12X_i+4\right) = 9\sum_{i=1}^{n}X_i^2+12\sum_{i=1}^{n}X_i+\sum_{i=1}^{n}4 \\ &= 9\times 275+12\times 50+10\times 4 = 3115\end{aligned}$$

and

$$\sum_{i=1}^{n}\left(X_i-\bar{X}\right)^2 = \sum_{i=1}^{n}X_i^2 - n\bar{X}^2 = 275-10\times 5^2 = 25.$$

Problem 4.4 *[D***] If $n=100$ and*

$$\sum_{i=1}^{n}X_i = 600 \text{ and } \sum_{i=1}^{n}X_i^2 = 5000$$

then what is $\bar{X}$ and what is $\sum_{i=1}^{n}(4X_i+2)$? What is $\sum_{i=1}^{n}(4X_i-3)^2$, and what is $\hat{\sigma}^2 = \frac{1}{n}\sum_{i=1}^{n}\left(X_i-\bar{X}\right)^2$?

Answer: We have

$$\begin{aligned}\bar{X} &= \frac{1}{n}\sum_{i=1}^{n}X_i = \frac{1}{100}\times 600 = 6 \\ \sum_{i=1}^{n}(4X_i+2) &= 4\sum_{i=1}^{n}X_i+\sum_{i=1}^{n}2 = 4\times 600+100\times 2 = 2600.\end{aligned}$$

Also

$$\begin{aligned}\sum_{i=1}^{n}(4X_i-3)^2 &= \sum_{i=1}^{n}(16X_i^2-24X_i+9)=16\sum_{i=1}^{n}X_i^2-24\sum_{i=1}^{n}X_i+\sum_{i=1}^{n}9\\ &= 16\times 5000-24\times 600+100\times 9=66500\end{aligned}$$

and

$$\sum_{i=1}^{n}\left(X_i-\bar{X}\right)^2=\sum_{i=1}^{n}X_i^2-n\bar{X}^2=5000-100\times 6^2=1400$$

so that

$$\hat{\sigma}^2=\frac{1}{n}\sum_{i=1}^{n}\left(X_i-\bar{X}\right)^2=\frac{1}{100}\times 1400=14.$$

Problem 4.5 *[B***] Prove that*

$$\sum_{i=1}^{n}(X_i-m)^2=\sum_{i=1}^{n}\left(X_i-\bar{X}\right)^2+n\left(\bar{X}-m\right)^2$$

(hint: use $X_i-m=\left(X_i-\bar{X}\right)+\left(\bar{X}-m\right)$*).*

Answer: For $\sum_{i=1}^{n}(X_i-m)^2$ we have

$$\begin{aligned}\sum_{i=1}^{n}(X_i-m)^2 &= \sum_{i=1}^{n}\left(\left(X_i-\bar{X}\right)+\left(\bar{X}-m\right)\right)^2\\ &= \sum_{i=1}^{n}\left(\left(X_i-\bar{X}\right)^2+2\left(\bar{X}-m\right)\left(X_i-\bar{X}\right)+\left(\bar{X}-m\right)^2\right)\\ &= \sum_{i=1}^{n}\left(X_i-\bar{X}\right)^2+\sum_{i=1}^{n}2\underbrace{\left(\bar{X}-m\right)}_{\text{no } i}\left(X_i-\bar{X}\right)+\sum_{i=1}^{n}\underbrace{\left(\bar{X}-m\right)^2}_{\text{no } i}\\ &= \sum_{i=1}^{n}\left(X_i-\bar{X}\right)^2+2\left(\bar{X}-m\right)\underbrace{\sum_{i=1}^{n}\left(X_i-\bar{X}\right)}_{0}+n\left(\bar{X}-m\right)^2\end{aligned}$$

where $\sum_{i=1}^{n}\left(X_i-\bar{X}\right)=0$ since

$$\sum_{i=1}^{n}\left(X_i-\bar{X}\right)=\sum_{i=1}^{n}X_i-\sum_{i=1}^{n}\bar{X}=\sum_{i=1}^{n}X_i-n\bar{X}=\sum_{i=1}^{n}X_i-\sum_{i=1}^{n}X_i=0.$$

Problem 4.6 *[D***] If*

$$\sum_{i=1}^{10}X_i=70 \text{ and } \sum_{i=1}^{10}X_i^2=500$$

then what is $\sum_{i=1}^{10}(3X_i-2)$, $\sum_{i=1}^{10}\left(3X_i^2-2\right)$ *and* $\sum_{i=1}^{10}(3X_i-2)^2$*?*

Answer: We have

$$
\begin{aligned}
\sum_{i=1}^{10}(3X_i-2) &= 3\sum_{i=1}^{10}X_i-\sum_{i=1}^{10}2=3\times 70-2\times 10=190\\
\sum_{i=1}^{10}(3X_i^2-2) &= 3\sum_{i=1}^{10}X_i^2-\sum_{i=1}^{10}2=3\times 500-2\times 10=1480\\
\sum_{i=1}^{10}(3X_i-2)^2 &= \sum_{i=1}^{10}(9X_i^2-12X_i+4)=9\sum_{i=1}^{10}X_i^2-12\sum_{i=1}^{10}X_i+\sum_{i=1}^{10}4\\
&= 9\times 500-12\times 70+4\times 10=3700.
\end{aligned}
$$

Problem 4.7 *[C***] If*

$$\hat{\sigma}^2=\frac{1}{n}\sum_{i=1}^{n}\left(X_i-\bar{X}\right)^2,\bar{X}=\frac{1}{n}\sum_{i=1}^{n}X_i$$

prove that

$$\hat{\sigma}^2=\frac{1}{n}\sum_{i=1}^{n}X_i^2-\bar{X}^2.$$

*[B**] Prove that*

$$\frac{1}{n}\sum_{i=1}^{n}(X_i-m)^2=\hat{\sigma}^2+\left(\bar{X}-m\right)^2.$$

Hint use $X_i-m=\left(X_i-\bar{X}\right)+\left(\bar{X}-m\right)$ *and hold on to those brackets! Use this result to prove without calculus that the sum of squares function*

$$S(m)=\frac{1}{n}\sum_{i=1}^{n}(X_i-m)^2$$

has a global minimum at $m^*=\bar{X}$.

Answer: We have

$$
\begin{aligned}
\hat{\sigma}^2 &= \frac{1}{n}\sum_{i=1}^{n}\left(X_i-\bar{X}\right)^2=\frac{1}{n}\sum_{i=1}^{n}\left(X_i^2-2\bar{X}X_i+\bar{X}^2\right)\\
&= \frac{1}{n}\left(\sum_{i=1}^{n}X_i^2-2\bar{X}\sum_{i=1}^{n}X_i+\sum_{i=1}^{n}\bar{X}^2\right)=\frac{1}{n}\left(\sum_{i=1}^{n}X_i^2-2\bar{X}\left(n\bar{X}\right)+n\bar{X}^2\right)\\
&= \frac{1}{n}\left(\sum_{i=1}^{n}X_i^2-2n\bar{X}^2+n\bar{X}^2\right)=\frac{1}{n}\left(\sum_{i=1}^{n}X_i^2-n\bar{X}^2\right)\\
&= \frac{1}{n}\sum_{i=1}^{n}X_i^2-\bar{X}^2.
\end{aligned}
$$

As well

$$\begin{aligned}
\frac{1}{n}\sum_{i=1}^{n}(X_i-m)^2 &= \frac{1}{n}\sum_{i=1}^{n}\left((X_i-\bar{X})+(\bar{X}-m)\right)^2 \\
&= \frac{1}{n}\sum_{i=1}^{n}\left((X_i-\bar{X})^2+2(\bar{X}-m)(X_i-\bar{X})+(\bar{X}-m)^2\right) \\
&= \frac{1}{n}\sum_{i=1}^{n}(X_i-\bar{X})^2+\frac{1}{n}\sum_{i=1}^{n}\underbrace{2(\bar{X}-m)}_{\text{no } i}(X_i-\bar{X})+\frac{1}{n}\sum_{i=1}^{n}\underbrace{(\bar{X}-m)^2}_{\text{no } i} \\
&= \frac{1}{n}\sum_{i=1}^{n}(X_i-\bar{X})^2+2(\bar{X}-m)\underbrace{\frac{1}{n}\sum_{i=1}^{n}(X_i-\bar{X})}_{0}+\frac{1}{n}n(\bar{X}-m)^2 \\
&= \hat{\sigma}^2+(\bar{X}-m)^2
\end{aligned}$$

where $\sum_{i=1}^{n}(X_i-\bar{X})=0$ since

$$\sum_{i=1}^{n}(X_i-\bar{X})=\sum_{i=1}^{n}X_i-\sum_{i=1}^{n}\bar{X}=\sum_{i=1}^{n}X_i-n\bar{X}=\sum_{i=1}^{n}X_i-\sum_{i=1}^{n}X_i=0.$$

Since

$$S(m)=\frac{1}{n}\sum_{i=1}^{n}(X_i-m)^2=\hat{\sigma}^2+(\bar{X}-m)^2$$

and since $(\bar{X}-m)^2\geq 0$ with equality only if $\bar{X}-m=0\Longrightarrow m=\bar{X}$, it follows that $m^*=\bar{X}$ is a global minimizer for $S(m)$.

Problem 4.8 *[D***] If* $n=10$ *and*

$$\sum_{i=1}^{n}X_i=60\ ,\sum_{i=1}^{n}X_i^2=520,\ \sum_{i=1}^{n}X_iY_i=360,\sum_{i=1}^{n}Y_i=120,\sum_{i=1}^{n}Y_i^2=1690$$

then what is $\bar{X},\bar{Y},\sum_{i=1}^{n}(4X_i+3Y_i)$ *and* $\sum_{i=1}^{n}(4X_i+3Y_i)^2$. *If* $\hat{\sigma}^2=\frac{1}{10}\sum_{i=1}^{10}(X_i-\bar{X})^2$ *then what is* $\hat{\sigma}^2$*? If*

$$\hat{\beta}=\frac{\sum_{i=1}^{n}(X_i-\bar{X})(Y_i-\bar{Y})}{\sum_{i=1}^{n}(X_i-\bar{X})^2}$$

then what is $\hat{\beta}$*? Find the value of* β *which minimizes* $S(\beta)=\sum_{i=1}^{n}(Y_i-\beta X_i)^2$.

Answer: We have

$$\begin{aligned}
\bar{X} &= \frac{1}{n}\sum_{i=1}^{n} X_i = \frac{1}{10}\times 60 = 6,\ \bar{Y} = \frac{1}{n}\sum_{i=1}^{n} Y_i = 12 \\
\sum_{i=1}^{n}(4X_i + 3Y_i) &= 4\sum_{i=1}^{n} X_i + 3\sum_{i=1}^{n} Y_i = 4\times 60 + 3\times 120 = 600 \\
\sum_{i=1}^{n}(4X_i + 3Y_i)^2 &= \sum_{i=1}^{n}\left(16X_i^2 + 24X_iY_i + 9Y_i^2\right) = 16\sum_{i=1}^{n} X_i^2 + 24\sum_{i=1}^{n} X_iY_i + 9\sum_{i=1}^{n} Y_i^2 \\
&= 16\times 520 + 24\times 360 + 9\times 1690 = 32170
\end{aligned}$$

and

$$\hat{\sigma}^2 = \frac{1}{n}\sum_{i=1}^{n}(X_i - \bar{X})^2 = \frac{1}{n}\left(\sum_{i=1}^{n} X_i^2 - n\bar{X}^2\right) = \frac{1}{n}\sum_{i=1}^{n} X_i^2 - \bar{X}^2 = \frac{1}{10}\times 520 - 6^2 = 16.$$

For $\hat{\beta}$ we have

$$\begin{aligned}
\hat{\beta} &= \frac{\sum_{i=1}^{n}(X_i - \bar{X})(Y_i - \bar{Y})}{\sum_{i=1}^{n}(X_i - \bar{X})^2} = \frac{\sum_{i=1}^{n} X_iY_i - n\times\bar{X}\bar{Y}}{\sum_{i=1}^{n} X_i^2 - n\times\bar{X}^2} \\
&= \frac{360 - 10\times 6\times 12}{520 - 10\times 6^2} = -\frac{9}{4}.
\end{aligned}$$

For $S(\beta)$ we have

$$\begin{aligned}
S(\beta) &= \sum_{i=1}^{n}(Y_i - \beta X_i)^2 = \sum_{i=1}^{n}\left(Y_i^2 - 2\beta X_iY_i + \beta^2X_i^2\right) \\
&= \sum_{i=1}^{n} Y_i^2 - 2\beta\sum_{i=1}^{n} X_iY_i + \beta^2\sum_{i=1}^{n} X_i^2 = 1690 - 720\beta + 520\beta^2
\end{aligned}$$

so that

$$S'(\beta^*) = 0 \Longrightarrow -720 + 1040\beta^* = 0 \Longrightarrow \beta^* = \frac{720}{1040} = \frac{9}{13}$$

with $S''(\beta) = 1040 > 0$ for all β so that $\beta^* = \frac{9}{13}$ is a global minimizer.

Problem 4.9 *[D***] If $n = 10$ and*

$$\sum_{i=1}^{n} X_i = 500, \sum_{i=1}^{n} X_i^2 = 27250, \sum_{i=1}^{n} X_iY_i = 30500, \sum_{i=1}^{n} Y_i = 700, \sum_{i=1}^{n} Y_i^2 = 55000$$

then calculate $\bar{X}, \hat{\sigma}^2 = \frac{1}{n}\sum_{i=1}^{n}(X_i - \bar{X})^2, \sum_{i=1}^{n}(2X_i - 3Y_i)^2$ *and*

$$\hat{\beta} = \frac{\sum_{i=1}^{n}(X_i - \bar{X})(Y_i - \bar{Y})}{\sum_{i=1}^{n}(X_i - \bar{X})^2}.$$

Answer: We have

$$\begin{aligned}
\bar{X} &= \frac{1}{10} \times 500 = 50,\ \hat{\sigma}^2 = \frac{1}{10} 27250 - 50^2 = 225 \\
\sum_{i=1}^{n} (2X_i - 3Y_i)^2 &= \sum_{i=1}^{n} (4X_i^2 - 12X_iY_i + 9Y_i^2) = 4\sum_{i=1}^{n} X_i^2 - 12\sum_{i=1}^{n} X_iY_i + 9\sum_{i=1}^{n} Y_i^2 \\
&= 4 \times 27250 - 12 \times 30500 + 9 \times 55000 = 238000 \\
\hat{\beta} &= \frac{\sum_{i=1}^{n} X_iY_i - n\bar{X}\bar{Y}}{\sum_{i=1}^{n} X_i^2 - n\bar{X}^2} = \frac{30500 - 10 \times 50 \times 70}{27250 - 10 \times 50^2} = -2.
\end{aligned}$$

4.2 Integration

Problem 4.10 *[D***] What is the anti-derivative of e^{-x}? Show that*

$$\int_0^T e^{-x} dx = 1 - e^{-T}.$$

Answer: The anti-derivative of e^{-x} is $-e^{-x}$ so that

$$\int_0^T e^{-x} dx = -e^{-x}|_{x=0}^{T} = 1 - e^{-T}.$$

Problem 4.11 *[A***] If $G_n = \int_o^T x^n e^{-x} dx$ use integration by parts to derive a relationship between G_n and G_{n-1} for $n > 0$. What then is G_2?*

Answer: Using integration by parts with

$$\begin{aligned}
u &= x^n & v' &= e^{-x} \\
u' &= nx^{n-1} & v &= -e^{-x}
\end{aligned}$$

we have for $n > 0$

$$G_n = \int_0^T x^n e^{-x} dx = -x^n e^{-x}|_{x=0}^{T} + n\int_0^T x^{n-1} e^{-x} dx = -T^n e^{-T} + nG_{n-1}.$$

Thus

$$G_2 = -T^2 e^{-T} + 2G_1 = -T^2 e^{-T} + 2\left(-Te^{-T} + G_0\right).$$

since

$$G_1 = -T^1 e^{-T} + 1G_0 = -Te^{-T} + G_0.$$

But

$$G_0 = \int_0^T e^{-x} dx = 1 - e^{-T}$$

so that

$$G_2 = -T^2 e^{-T} + 2\left(-Te^{-T} + \left(1 - e^{-T}\right)\right) = -e^{-T}\left(T^2 + 2T + 2\right) + 2.$$

Problem 4.12 *[B***] If*

$$\Gamma(n,\alpha) = \int_0^\infty x^{n-1} e^{-\alpha x} dx$$

for $\alpha > 0$ use integration by parts to derive a relationship between $\Gamma(n,\alpha)$ and $\Gamma(n-1,\alpha)$ for $n > 0$. Show that if n is a positive integer then $\Gamma(n,\alpha) = \frac{1}{\alpha^n}(n-1)!$.

Answer: Using integration by parts with

$$\begin{array}{ll} u = x^{n-1} & v' = e^{-\alpha x} \\ u' = (n-1)\, x^{n-2} & v = -\frac{1}{\alpha} e^{-\alpha x} \end{array}$$

we have for $n > 0$

$$\Gamma(n,\alpha) = \int_0^\infty x^{n-1} e^{-\alpha x} dx = -\frac{1}{\alpha} x^{n-1} e^{-\alpha x}|_{x=0}^{\infty} + \frac{n-1}{\alpha} \int_0^\infty x^{n-2} e^{-\alpha x} dx = \frac{n-1}{\alpha} \Gamma(n-1,\alpha).$$

Consider now a proof by induction. For $n = 1$ we have $\Gamma(n,\alpha) = \frac{1}{\alpha^n}(n-1)!$ since replacing n with 1

$$\Gamma(1,\alpha) = \int_0^\infty e^{-\alpha x} dx = \frac{1}{\alpha} = \frac{0!}{\alpha^1}.$$

Now assume the statement is true for $n-1$ so that $\Gamma(n-1,\alpha) = \frac{(n-2)!}{\alpha^{n-1}}$. Then

$$\Gamma(n,\alpha) = \frac{n-1}{\alpha} \Gamma(n-1,\alpha) = \frac{n-1}{\alpha} \frac{(n-2)!}{\alpha^{n-1}} = \frac{(n-1)!}{\alpha^n}$$

and so the result follows by induction.

Problem 4.13 *[C***] Calculate the following integrals*

$$\int_1^5 x^2 dx, \int_0^3 e^{-x} dx, \int_0^3 \left(e^{-x}+2\right)^2 dx, \int_0^3 x e^{-x} dx, \int_0^3 x^2 e^{-x} dx.$$

Answer: We have

$$\begin{aligned} \int_1^5 x^2 dx &= \frac{x^3}{3}|_{x=1}^5 = \frac{5^3}{3} - \frac{1^3}{3} = \frac{124}{3}, \int_0^3 e^{-x} dx = 1 - e^{-3} \\ \int_0^3 \left(e^{-x}+2\right)^2 dx &= \int_0^3 \left(e^{-2x} + 4e^{-x} + 4\right) dx = \int_0^3 e^{-2x} dx + 4\int_0^3 e^{-x} dx + 4\int_0^3 dx \\ &= -\frac{1}{2} e^{-6} - 4e^{-3} + \frac{33}{2} \\ \int_0^3 x e^{-x} dx &= -4e^{-3} + 1, \int_0^3 x^2 e^{-x} dx = -17e^{-3} + 2. \end{aligned}$$

Problem 4.14 *[C***] Calculate*

$$\int_0^T te^{-rt}dt$$

using integration by parts. From this find

$$\int_0^\infty te^{-rt}dt$$

assuming that $r > 0$.

Answer: Using integration by parts with

$$\begin{array}{ll} u = t & v' = e^{-rt} \\ u' = 1 & v = -\frac{e^{-rt}}{r} \end{array}$$

we have

$$\begin{aligned} \int_0^T te^{-rt}dt &= -\frac{e^{-rt}}{r} \times t|_{t=0}^T + \frac{1}{r}\int_0^T e^{-rt}dt = -\frac{e^{-Tr}}{r}T + \frac{1}{r}\int_0^T e^{-rt}dt \\ &= -\frac{e^{-Tr}}{r}T - \frac{1}{r}\frac{e^{-rt}}{r}|_{t=0}^T = -\frac{e^{-Tr}}{r}T + \frac{1}{r^2}\left(1 - e^{-rT}\right). \end{aligned}$$

Since $r > 0$ it follows that $e^{-rT} \to 0$ and $e^{-rT}T \to 0$ so that

$$\int_0^\infty te^{-rt}dt = \frac{1}{r^2}.$$

Problem 4.15 *Using integration by parts, calculate*

$$\int_0^T t^2e^{-rt}dt, \text{ where } r > 0$$

and find the limit of this expression as $T \to \infty$ *when* $r > 0$.

Answer: Letting $u = t^2$ and $v' = e^{-rt}$ we have

$$\begin{array}{ll} u = t^2 & v' = e^{-rt} \\ u' = 2t & v = -\frac{e^{-rt}}{r} \end{array}$$

so that using integration by parts we have

$$\int_0^T t^2e^{-rt}dt = -\frac{e^{-rt}}{r}t^2|_{t=0}^T + \frac{2}{r}\int_0^T te^{-rt}dt = -\frac{e^{-Tr}}{r}T^2 + \frac{2}{r}\int_0^T te^{-rt}dt.$$

The new integral was found in the last problem so that

$$\begin{aligned}\int_0^T t^2 e^{-rt}dt &= -\frac{e^{-Tr}}{r}T^2 + \frac{2}{r}\int_0^T te^{-rt}dt \\ &= -\frac{e^{-Tr}}{r}T^2 + \frac{2}{r}\left(\frac{1}{r^2}\left(1-e^{-rT}\right) - \frac{T}{r}e^{-Tr}.\right).\end{aligned}$$

Letting $T \to \infty$ we have

$$\int_0^\infty t^2 e^{-rt}dt = \frac{2}{r^3}.$$

Problem 4.16 *[D***] Using integration by parts find*

$$\int_0^5 xe^{-2x}dx.$$

*[A***] The incomplete gamma function is given by*

$$\Gamma(n,m) = \int_m^\infty x^{n-1}e^{-x}dx$$

for $m \geq 0$. *Use integration by parts to find the relationship between* $\Gamma(n,m)$ *and* $\Gamma(n-1,m)$.

Answer: Using integration by parts with

$$\begin{array}{ll} u = x & v' = e^{-2x} \\ u' = 1 & v = -\frac{e^{-2x}}{2} \end{array}$$

we have

$$\begin{aligned}\int_0^5 xe^{-2x}dx &= -\frac{xe^{-2x}}{2}|_{x=0}^5 + \frac{1}{2}\int_0^5 e^{-2x}dx = -\frac{5}{2}e^{-10} + \frac{1}{2}\left(-\frac{e^{-2x}}{2}|_{x=0}^5\right) \\ &= -\frac{5}{2}e^{-10} + \frac{1}{4}\left(1-e^{-10}\right) = \frac{1}{4} - \frac{11}{4}e^{-10}.\end{aligned}$$

Now given

$$\begin{array}{ll} u = x^{n-1} & v' = e^{-x} \\ u' = (n-1)x^{n-2} & v = -e^{-x} \end{array}$$

we have

$$\begin{aligned}\Gamma(n,m) &= \int_m^\infty x^{n-1}e^{-x}dx = -x^{n-1}e^{-x}|_{x=m}^\infty + (n-1)\int_m^\infty x^{n-2}e^{-x}dx \\ &= m^{n-1}e^{-m} + (n-1)\Gamma(n-1,m).\end{aligned}$$

Problem 4.17 *[A**] If* $L^*(Q,W,R)$ *is the conditional factor demand for labour, what is* $\int_{W_1}^{W_2} L^*(Q,W,R)\,dW$ *equal to?*

Answer: By Shephard's lemma we have

$$L^*(Q,W,R) = \frac{\partial C^*(Q,W,R)}{\partial W}$$

so that the anti-derivative of $L^*(Q,W,R)$ (with respect to W) is $C^*(Q,W,R)$. Thus

$$\begin{aligned}\int_{W_1}^{W_2} L^*(Q,W,R)\,dW &= \int_{W_1}^{W_2} \frac{\partial C^*(Q,W,R)}{\partial W}dW \\ &= C^*(Q,W,R)\,|_{W_1}^{W_2} \\ &= C^*(Q,W_2,R) - C^*(Q,W_1,R).\end{aligned}$$

Problem 4.18 *Consider the standard normal density*

$$p(x) = \frac{e^{-\frac{1}{2}x^2}}{\sqrt{2\pi}}, \quad -\infty < x < \infty$$

It is a fact that $\int_{-\infty}^{\infty} p(x)\,dx = 1$ *and that if the random variable* X *has this density then* $E[X] = 0$. *Show that* $Var[X] = 1$ *(hint: use integration by parts with* $u = x, v' = xp(x)$ *and show that* $v = -p(x)$. *Now show that* $E[X^4] = 3$ *again using integration by parts (hint use* $u = x^3$ *and the same* v' *as before). The value 3 is the* ***kurtosis*** *of the standard normal distribution, a fact commonly used in applied work to test if normality is consistent with a set of data.*

Answer: Using integration by parts with

$$\begin{aligned} u &= x & v' &= x\frac{e^{-\frac{x^2}{2}}}{\sqrt{2\pi}} \\ u' &= 1 & v &= -\frac{e^{-\frac{x^2}{2}}}{\sqrt{2\pi}} \end{aligned}$$

we have

$$\begin{aligned} Var[X] &= \int_{-\infty}^{\infty} (x - E[X])^2 p(x)\,dx = \int_{-\infty}^{\infty} x^2 \frac{e^{-\frac{x^2}{2}}}{\sqrt{2\pi}}dx \\ &= -x\frac{e^{-\frac{x^2}{2}}}{\sqrt{2\pi}}|_{x=-\infty}^{\infty} + \int_{-\infty}^{\infty} \frac{e^{-\frac{x^2}{2}}}{\sqrt{2\pi}}dx = 1 \end{aligned}$$

since

$$\lim_{x\to\pm\infty} xe^{-\frac{x^2}{2}} = 0$$

using L'Hôpital's rule as

$$\lim_{x\to\pm\infty} xe^{-\frac{x^2}{2}} = \lim_{x\to\pm\infty} \frac{x}{e^{\frac{x^2}{2}}} = \lim_{x\to\pm\infty} \frac{1}{xe^{\frac{x^2}{2}}} = 0$$

and

$$\int_{-\infty}^{\infty} \frac{e^{-\frac{x^2}{2}}}{\sqrt{2\pi}}dx = \int_{-\infty}^{\infty} p(x)\,dx = 1.$$

Now using integration by parts with

$$u = x^3 \qquad v' = x\frac{e^{-\frac{x^2}{2}}}{\sqrt{2\pi}}$$
$$u' = 3x^2 \qquad v = -\frac{e^{-\frac{x^2}{2}}}{\sqrt{2\pi}}$$

we have

$$\begin{aligned} E\left[X^4\right] &= \int_{-\infty}^{\infty} x^4 p(x)\,dx = \int_{-\infty}^{\infty} x^4\frac{e^{-\frac{x^2}{2}}}{\sqrt{2\pi}}dx \\ &= -x^3\frac{e^{-\frac{x^2}{2}}}{\sqrt{2\pi}}|_{-\infty}^{\infty} + 3\int_{-\infty}^{\infty} x^2\frac{e^{-\frac{x^2}{2}}}{\sqrt{2\pi}}dx = 3Var\left[X\right] = 3 \end{aligned}$$

since $Var\left[X\right] = 1$ and

$$\lim_{x\to\pm\infty} x^3 e^{-\frac{x^2}{2}} = 0$$

using L'Hôpital's rule as

$$\begin{aligned} \lim_{x\to\pm\infty} x^3 e^{-\frac{x^2}{2}} &= \lim_{x\to\pm\infty} \frac{x^3}{e^{\frac{x^2}{2}}} = \lim_{x\to\pm\infty} \frac{3x^2}{xe^{\frac{x^2}{2}}} \\ &= \lim_{x\to\pm\infty} \frac{6x}{(x^2+1)\,e^{\frac{x^2}{2}}} = \lim_{x\to\pm\infty} \frac{6}{(x^3+3x)\,e^{\frac{x^2}{2}}} = 0. \end{aligned}$$

Problem 4.19 *[A**] The nth uncentred moment is defined as $\mu_n \equiv E\left[X^n\right]$. For the standard normal distribution use proof by induction to show that $\mu_n = 0$ for n odd and that for n even*

$$\mu_n = (n-1)\times(n-3)\times\cdots\times 3\times 1.$$

Calculate μ_n for $n = 0, 1, 2, 3, 4, 5, 6$.

Answer: Using integration by parts with

$$u = x^{n-1} \qquad v' = x\frac{e^{-\frac{x^2}{2}}}{\sqrt{2\pi}}$$
$$u' = (n-1)\,x^{n-2} \qquad v = -\frac{e^{-\frac{x^2}{2}}}{\sqrt{2\pi}}$$

we have for $n > 0$

$$\mu_n = \int_{-\infty}^{\infty} x^n\frac{e^{-\frac{x^2}{2}}}{\sqrt{2\pi}}dx = -x^{n-1}\frac{e^{-\frac{x^2}{2}}}{\sqrt{2\pi}}|_{x=-\infty}^{\infty} + (n-1)\int_{-\infty}^{\infty} x^{n-2}\frac{e^{-\frac{x^2}{2}}}{\sqrt{2\pi}}dx = (n-1)\,\mu_{n-2}$$

since

$$\lim_{x\to\pm\infty} x^{n-1} e^{-\frac{x^2}{2}} = 0.$$

We now use a proof by induction to show that $\mu_n = 0$ for n odd. We know that $\mu_1 = E\left[X\right] = 0$. Now assume that $\mu_k = 0$ for $k < n$ and k odd. In particular if n is odd then so too is $n-2$ and so μ_{n-2} is also odd so that by the induction hypothesis $\mu_{n-2} = 0$ and so

$$\mu_n = (n-1)\,\mu_{n-2} = 0$$

so that $\mu_n = 0$ for all n odd.

For n even we have

$$\mu_0 = E\left[X^0\right] = 1.$$

Since if n is even so too is $n-2$ we have from the induction hypothesis that

$$\mu_{n-2} = (n-3) \times \cdots \times 3 \times 1$$

so that

$$\mu_n = (n-1) \times \mu_{n-2} = (n-1) \times (n-3) \times \cdots \times 3 \times 1.$$

We thus have

$n =$	0	1	2	3	4	5	6
$\mu_n =$	1	0	1	0	3	0	15

Problem 4.20 *Calculate $\int_1^3 x^3 dx$ and $\int_0^5 e^{-2x} dx$. Using integration by parts calculate $\int_0^5 x e^{-2x} dx$ and $\int_0^5 x^2 e^{-2x} dx$. If $\int_0^1 f(x)\,dx = 2$, $\int_0^1 f(x)^2\,dx = 4$ then what is $\int_0^1 (f(x) - 5)^2\,dx$? Find the value of c which minimizes $S(c) = \int_0^1 (f(x) - c)^2\,dx$.*

Answer: We have

$$\begin{aligned}
\int_1^3 x^3 dx &= \frac{x^4}{4}\Big|_1^3 = \frac{3^4}{4} - \frac{1^4}{4} = 20 \\
\int_0^5 e^{-2x} dx &= -\frac{e^{-2x}}{2}\Big|_0^5 = \frac{-e^{-2\times 5}}{2} - \left(-\frac{e^{-2\times 0}}{2}\right) = \frac{1}{2} - \frac{1}{2}e^{-10}.
\end{aligned}$$

Using integration by parts on

$$\int_0^5 x e^{-2x} dx$$

with

$$\begin{array}{ll} u = x & v' = e^{-2x} \\ u' = 1 & v = -\frac{e^{-2x}}{2} \end{array}$$

we have that

$$\int_0^5 x e^{-2x} dx = -x\frac{e^{-2x}}{2}\Big|_0^5 + \frac{1}{2}\int_0^5 e^{-2x} dx = -5\frac{e^{-10}}{2} + \frac{1}{2}\left(\frac{1}{2} - \frac{1}{2}e^{-10}\right) = \frac{1}{4} - \frac{11}{4}e^{-10}$$

and for

$$\int_0^5 x^2 e^{-2x} dx$$

with

$$\begin{array}{ll} u = x^2 & v' = e^{-2x} \\ u' = 2x & v = -\frac{e^{-2x}}{2} \end{array}$$

that

$$\begin{aligned} \int_0^5 x^2 e^{-2x} dx &= -x^2 \frac{e^{-2x}}{2} |_0^5 + \int_0^5 xe^{-2x} dx \\ &= -\frac{25}{2} e^{-10} + \frac{1}{4} - \frac{11}{4} e^{-10} = -\frac{61}{4} e^{-10} + \frac{1}{4}. \end{aligned}$$

We have

$$\begin{aligned} \int_0^1 (f(x) - 5)^2 dx &= \int_0^1 \left(f(x)^2 - 10f(x) + 25 \right) dx \\ &= \int_0^1 f(x)^2 dx - 10 \int_0^1 f(x) dx + \int_0^1 25dx \\ &= 4 - 10 \times 2 + 25 = 9. \end{aligned}$$

To minimize $S(c)$ we have

$$\begin{aligned} S(c) &= \int_0^1 (f(x) - c)^2 dx = \int_0^1 \left(f(x)^2 - 2cf(x) + c^2 \right) dx \\ &= \int_0^1 f(x)^2 dx - 2c \int_0^1 f(x) dx + \int_0^1 c^2 dx = 4 - 4c + c^2 \end{aligned}$$

so that the first-order condition $S'(c^*) = 0$ yields

$$2c^* - 4 = 0 \Longrightarrow c^* = 2$$

with $S''(c) = 2 > 0$ so that c^* is a global minimizer.

Problem 4.21 *[D***] Calculate $\int_0^5 e^{-x} dx$ and $\int_0^5 e^{-2x} dx$. Using integration by parts calculate $\int_0^5 xe^{-x} dx$. What then is $\int_0^5 (e^{-x} + x)^2 dx$?*

Answer: We have

$$\begin{aligned} \int_0^5 e^{-x} dx &= -e^{-x} |_0^5 = -e^{-5} - (-e^{-0}) = 1 - e^{-5} \\ \int_0^5 e^{-2x} dx &= -\frac{e^{-2x}}{2} |_0^5 = \frac{-e^{-2\times 5}}{2} - \left(-\frac{e^{-2\times 0}}{2} \right) = \frac{1 - e^{-10}}{2}. \end{aligned}$$

Using integration by parts on

$$\int_0^5 xe^{-x} dx$$

with

$$\begin{array}{ll} u = x & v' = e^{-x} \\ u' = 1 & v = -e^{-x} \end{array}$$

we have

$$\int_0^5 xe^{-x}dx = -xe^{-x}|_0^5 + \int_0^5 e^{-x}dx = -5e^{-5} + 1 - e^{-5} = 1 - 6e^{-5}.$$

Now

$$\begin{aligned} \int_0^5 \left(e^{-x} + x\right)^2 dx &= \int_0^5 \left(e^{-2x} + 2xe^{-x} + x^2\right) dx \\ &= \int_0^5 e^{-2x}dx + 2\int_0^5 xe^{-x}dx + \int_0^5 x^2 dx \\ &= \frac{1}{2} - \frac{1}{2}e^{-10} + 2 \times \left(1 - 6e^{-5}\right) + \frac{x^3}{3}|_0^5 \\ &= \frac{1}{2} - \frac{1}{2}e^{-10} + 2 \times \left(1 - 6e^{-5}\right) + \frac{125}{3}. \end{aligned}$$

Problem 4.22 *[A**] If $Q = 2L^{\frac{1}{2}} + 2K^{\frac{1}{2}}$ calculate the Lagrange multiplier λ^* from the first-order conditions without bothering to calculate L^* and K^*. Use $\frac{\partial C^*}{\partial Q} = \lambda^*$ to calculate C^* using the fundamental theorem of integral calculus. Show that the firm's cost function is*

$$C^*\left(Q, W, R\right) = \frac{Q^2}{4}\left(W^{-1} + R^{-1}\right)^{-1}.$$

From $C^*\left(Q, W, R\right)$ use Shephard's lemma to calculate $L^*\left(Q, W, R\right)$. Show that the cost share of labour $\frac{WL^*}{C^*}$ is $\frac{W^{-1}}{W^{-1}+R^{-1}}$.

Answer: The Lagrangian is

$$\mathcal{L}\left(\lambda, L, K, Q, W, R\right) = WL + RK + \lambda\left(Q - 2L^{\frac{1}{2}} - 2K^{\frac{1}{2}}\right)$$

with first-order conditions

$$\begin{aligned} Q - 2L^{*\frac{1}{2}} - 2K^{*\frac{1}{2}} &= 0 \Longrightarrow Q = 2L^{*\frac{1}{2}} + 2K^{*\frac{1}{2}} \\ W - \lambda^* L^{*-\frac{1}{2}} &= 0 \Longrightarrow L^* = \left(\lambda^*\right)^2 W^{-2} \\ R - \lambda^* K^{*-\frac{1}{2}} &= 0 \Longrightarrow K^* = \left(\lambda^*\right)^2 R^{-2}. \end{aligned}$$

Substituting the second and third into the production function then yields

$$\begin{aligned} Q &= 2L^{*\frac{1}{2}} + 2K^{*\frac{1}{2}} = 2\left(\left(\lambda^*\right)^2 W^{-2}\right)^{\frac{1}{2}} + 2\left(\left(\lambda^*\right)^2 R^{-2}\right)^{\frac{1}{2}} \\ &= 2\lambda^* W^{-1} + 2\lambda^* R^{-1} = 2\lambda^*\left(W^{-1} + R^{-1}\right) \\ &\Longrightarrow \lambda^* = \frac{Q}{2}\left(W^{-1} + R^{-1}\right)^{-1}. \end{aligned}$$

We know that

$$\frac{\partial C^*}{\partial Q} = \lambda^* = \frac{Q}{2}\left(W^{-1} + R^{-1}\right)^{-1}$$

so that C^* is the anti-derivative of λ^* with respect to Q (treating W and R as constants). The anti-derivative of $\frac{Q}{2}$ is $\frac{Q^2}{4}$ so that

$$C^* = \frac{Q^2}{4}\left(W^{-1}+R^{-1}\right)^{-1}+\phi$$

where ϕ may depend on W, R but not on Q. Since $Q=0 \Longrightarrow C^*=0$ we have $\phi=0$ so that $C^* = \frac{Q^2}{4}\left(W^{-1}+R^{-1}\right)^{-1}$.

Thus

$$\frac{\partial C^*}{\partial W} = \frac{\partial}{\partial W}\left(\frac{Q^2}{4}\left(W^{-1}+R^{-1}\right)^{-1}\right) = \frac{Q^2}{4}\left(W^{-1}+R^{-1}\right)^{-2}W^{-2} = L^*.$$

Now the cost share of labour $\frac{WL^*}{C^*}$ is

$$\frac{WL^*}{C^*} = \frac{W \times \frac{Q^2}{4}\left(W^{-1}+R^{-1}\right)^{-2}W^{-2}}{\frac{Q^2}{4}\left(W^{-1}+R^{-1}\right)^{-1}} = W^{-1}\left(W^{-1}+R^{-1}\right)^{-1} = \frac{W^{-1}}{W^{-1}+R^{-1}}.$$

4.3 Random Variables

Problem 4.23 *[C***] Consider an investor who can invest in three assets with returns R_0, R_1 and R_2 where*

$$\begin{aligned} E[R_0] &= 3, Var[R_0]=0,\ E[R_1]=10, Var[R_1]=25 \\ E[R_2] &= 7, Var[R_2]=9, Cov[R_1,R_2]=-9. \end{aligned}$$

What kind of asset does R_0 correspond to? Calculate $E\left[R_1^2\right]$, $E[R_1R_2]$ and the correlation coefficient ρ between R_1 and R_2. The return on a portfolio is $R = R_0\omega_o + \omega_1 R_1 + \omega_2 R_2$ where $1 = \omega_o + \omega_1 + \omega_2$. If $\omega_o = 0.3, \omega_1 = 0.5$ and $\omega_2 = 0.2$ calculate $E[R]$, $Var[R]$, and $E\left[R^2\right]$. If $\omega_0 = 0$ find the portfolio with the least risk. If the investor wants an expected return of 8, what portfolio does this with the minimum amount of risk? What is the variance of this portfolio?

Answer: R_0 has a variance of 0 and hence is a degenerate random variable; that is $R_0 = 3$ with probability 1. It is therefore a riskless asset (something like Canada savings bonds).

We have

$$\begin{aligned} E\left[R_1^2\right] &= E[R_1]^2 + Var[R_1] = 10^2+25 = 125 \\ E[R_1R_2] &= E[R_1]E[R_2] + Cov[R_1,R_2] = 7\times 10 + -9 = 61 \\ \rho &= \frac{Cov[R_1,R_2]}{\sqrt{Var[R_1]}\sqrt{Var[R_2]}} = \frac{-9}{\sqrt{25}\sqrt{9}} = -\frac{3}{5}. \end{aligned}$$

We have

$$E[R] = 0.3E[R_0] + 0.5E[R_1] + 0.2E[R_2] = 0.3\times 3 + 0.5\times 10 + 0.2\times 7 = 7.3$$

and using the fact that $R_0 = 3$ is a constant and therefore does not affect the variance we have

$$\begin{aligned} Var[R] &= Var[0.3 \times R_0 + 0.5R_1 + 0.2R_2] = Var[0.5R_1 + 0.2R_2] \\ &= (0.5)^2 Var[R_1] + (0.2)^2 Var[R_2] + 2 \times 0.5 \times 0.2 \times Cov[R_1, R_2] \\ &= (0.5)^2 \times 25 + (0.2)^2 \times 9 + 2 \times 0.5 \times 0.2 \times (-9) = 4.81 \\ E[R^2] &= E[R]^2 + Var[R] = 7.3^2 + 4.81 = 58.1. \end{aligned}$$

If $\omega_0 = 0$ then since the asset weights sum to 1 we have

$$1 = \omega_o + \omega_1 + \omega_2 \Longrightarrow \omega_2 = 1 - \omega_1$$

and so replacing ω_2 with $1 - \omega_1$ in

$$\begin{aligned} Var[R] &= Var[\omega_o R_0 + \omega_1 R_1 + \omega_2 R_2] = Var[\omega_1 R_1 + \omega_2 R_2] = Var[\omega_1 R_1 + (1 - \omega_1) R_2] \\ &= \omega_1^2 Var[R_1] + (1 - \omega_1)^2 Var[R_2] + 2\omega_1 (1 - \omega_1) Cov[R_1, R_2] \\ &= (\omega_1)^2 25 + (1 - \omega_1)^2 9 + 2 \times \omega_1 \times (1 - \omega_1) \times (-9) \end{aligned}$$

we have

$$\begin{aligned} f(\omega_1) &= Var[R] = (\omega_1)^2 25 + (1 - \omega_1)^2 9 + 2 \times \omega_1 \times (1 - \omega_1) \times (-9) \\ &\Longrightarrow f'(\omega_1) = 50\omega_1 - 18(1 - \omega_1) - 18(1 - 2\omega_1) \\ &\Longrightarrow f'(\omega_1^*) = 0 = 50\omega_1^* - 18(1 - \omega_1^*) - 18(1 - 2\omega_1^*) \\ &\Longrightarrow \omega_1^* = \frac{9}{26}, \omega_2^* = 1 - \frac{9}{26} = \frac{17}{26}. \end{aligned}$$

This is a global minimizer since

$$f''(\omega_1) = 50 + 18 + 36 = 104 > 0$$

so that $f(\omega_1)$ is globally convex.

Since the portfolio needs to have an expected return of 8 and the weights of the portfolio sum to 1 we have

$$\begin{aligned} E[R] &= 8 = \omega_o 3 + \omega_1 10 + \omega_2 7 \\ 1 &= \omega_o + \omega_1 + \omega_2 \end{aligned}$$

or

$$8 = (1 - \omega_1 - \omega_2) 3 + \omega_1 10 + \omega_2 7 \Longrightarrow 5 = \omega_1 7 + \omega_2 4.$$

This will act as the constraint. The variance of the portfolio is

$$Var[R] = 25\omega_1^2 + 9\omega_2^2 - 18\omega_1\omega_2.$$

Minimizing $\frac{Var[R]}{2}$ is equivalent to minimizing $Var[R]$ but is slightly more convenient. Given the constraint the Lagrangian is

$$\mathcal{L}(\lambda, \omega_1, \omega_2) = \frac{25\omega_1^2 + 9\omega_2^2 - 18\omega_1\omega_2}{2} + \lambda(5 - \omega_1 7 - \omega_2 4)$$

and the first-order conditions are

$$\begin{aligned} 5 - 7\omega_1^* - 4\omega_2^* &= 0 \\ 25\omega_1^* - 9\omega_2^* - 7\lambda^* &= 0 \\ -9\omega_1^* + 9\omega_2^* - 4\lambda^* &= 0. \end{aligned}$$

The first-order conditions in matrix notation are then

$$\begin{bmatrix} 0 & -7 & -4 \\ -7 & 25 & -9 \\ -4 & -9 & 9 \end{bmatrix} \begin{bmatrix} \lambda^* \\ \omega_1^* \\ \omega_2^* \end{bmatrix} = \begin{bmatrix} -5 \\ 0 \\ 0 \end{bmatrix}$$

so that using

$$\det \begin{bmatrix} 0 & -7 & -4 \\ -7 & 25 & -9 \\ -4 & -9 & 9 \end{bmatrix} = -1345$$

and solving by Cramer's rule yields

$$\begin{aligned} \omega_1^* &= \frac{\det \begin{bmatrix} 0 & -5 & -4 \\ -7 & 0 & -9 \\ -4 & 0 & 9 \end{bmatrix}}{-1345} = \frac{-495}{-1345} = \frac{99}{269} \approx 0.37 \\ \omega_2^* &= \frac{\det \begin{bmatrix} 0 & -7 & -5 \\ -7 & 25 & 0 \\ -4 & -9 & 0 \end{bmatrix}}{-1345} = \frac{-815}{-1345} = \frac{163}{269} \approx 0.61. \end{aligned}$$

The second-order conditions for a global minimum are satisfied since the objective function has a positive definite Hessian

$$H = \det \begin{bmatrix} 25 & -9 \\ -9 & 9 \end{bmatrix} \Longrightarrow M_1 = 25 > 0, M_2 = 144 > 0$$

and so is globally convex, and the constraint is $5 - \omega_1 7 - \omega_2 4$ linear.

Thus the optimal portfolio has

$$\omega_1^* = \frac{99}{269} \approx 0.37, \omega_2^* = \frac{163}{269} \approx 0.61$$

or 37% of the portfolio in asset 1, 61% in asset 2, and

$$\omega_o^* = 1 - \omega_1^* - \omega_2^* = \frac{7}{269} \approx 0.02$$

or 2% in the riskless asset.

The variance of the portfolio is then given by

$$Var\left[R\right] = 25\omega_1^2 + 9\omega_2^2 - 18\omega_1\omega_2 = 25\left(\frac{99}{269}\right)^2 + 9\left(\frac{163}{269}\right)^2 - 18 \times \frac{99}{269} \times \frac{163}{269} \approx 2.68.$$

Problem 4.24 *[C***] Let R_1, R_2, R_3 be three random variables having the following properties*

$$\begin{aligned} E[R_1] &= 7, E[R_2] = 10, E[R_3] = 3 \\ Var[R_1] &= 9, Var[R_2] = 16, Cov[R_1, R_2] = -6, Var[R_3] = 0. \end{aligned}$$

What is $E\left[R_1^2\right]$, $E[R_1R_2]$ and the correlation coefficient between R_1 and R_2? What kind of random variable is R_3? If

$$R = 0.3R_1 + 0.5R_2 + 0.2R_3,$$

what is $E[R]$ and $Var[R]$?

Answer: We have

$$\begin{aligned} E\left[R_1^2\right] &= E[R_1]^2 + Var[R_1] = 7^2 + 9 = 58 \\ E[R_1R_2] &= E[R_1]E[R_2] + Cov[R_1, R_2] = 7 \times 10 + -6 = 64 \\ \rho &= \frac{Cov[R_1, R_2]}{\sqrt{Var[R_1]}\sqrt{Var[R_2]}} = \frac{-6}{\sqrt{9}\sqrt{16}} = -\frac{1}{2} \end{aligned}$$

Since $Var[R_3] = 0$ it follows that R_3 is a degenerate random variable or $R_3 = 3$.
For R

$$E[R] = 0.3E[R_1] + 0.5E[R_2] + 0.2E[R_3] = 0.3 \times 7 + 0.5 \times 10 + 0.2 \times 3 = 7.7.$$

Since $R_3 = 3$ is a constant we have

$$\begin{aligned} Var[R] &= Var[0.3R_1 + 0.5R_2 + 0.2 \times 3] \\ &= Var[0.3R_1 + 0.5R_2] \\ &= (0.3)^2 Var[R_1] + (0.5)^2 Var[R_2] + 2 \times 0.5 \times 0.3 Cov[R_1, R_2] \\ &= (0.3)^2 \times 9 + (0.5)^2 \times 16 + 2 \times 0.5 \times 0.3 \times -6 = 3.01. \end{aligned}$$

Problem 4.25 *[B***] For any two random variables X and Y, what value of a would minimize $Var[Y - aX]$?*

Answer: We have

$$f(a) = Var[Y - aX] = Var[Y] + a^2 Var[X] - 2a Cov[X, Y]$$

so that

$$f'(a^*) = 0 \Longrightarrow 2a^* Var[X] - 2Cov[X, Y] = 0 \Longrightarrow a^* = \frac{Cov[X, Y]}{Var[X]}.$$

This is a global minimizer since $f''(a) = 2Var[X] > 0$ so that $f(a)$ is globally convex.

Problem 4.26 *[B***] Consider a barrel with 30,000 dice in it, let X_i be the outcome of the ith die and let S be the sum of all the dice. Each of the dice is fair so that the probabilities of $1, 2, 3, 4, 5$ and 6 are all $\frac{1}{6}$. Calculate $E[X_i]$, $E[X_i^2]$, $Var[X_i]$, $E[S]$ and $Var[S]$. It is a fact that 95% of the die rolls will fall in a range $E[S] \pm 1.96\sqrt{Var[S]}$. Calculate this range and compare this with the range of all possible values.*

Answer: We have

$$\begin{aligned} E[X_i] &= \frac{1}{6}\times 1+\frac{1}{6}\times 2+\frac{1}{6}\times 3+\frac{1}{6}\times 4+\frac{1}{6}\times 5+\frac{1}{6}\times 6=\frac{7}{2} \\ E[X_i^2] &= \frac{1}{6}\times 1^2+\frac{1}{6}\times 2^2+\frac{1}{6}\times 3^2+\frac{1}{6}\times 4^2+\frac{1}{6}\times 5^2+\frac{1}{6}\times 6^2=\frac{91}{6} \\ Var[X_i] &= E[X_i^2]-E[X_i]^2=\frac{91}{6}-\left(\frac{7}{2}\right)^2=\frac{35}{12}. \end{aligned}$$

If

$$S = X_1 + X_2 + \cdots + X_{30000}$$

then

$$\begin{aligned} E[S] &= E[X_1+X_2+\cdots+X_{30000}] = E[X_1]+E[X_2]+\cdots+E[X_{30000}] \\ &= \frac{7}{2}+\frac{7}{2}+\cdots+\frac{7}{2} = 30000\times\frac{7}{2} = 105000 \end{aligned}$$

and

$$\begin{aligned} Var[S] &= Var[X_1+X_2+\cdots+X_{30000}] \\ &= 1^2 Var[X_1]+1^2 Var[X_2]+\cdots+1^2 Var[X_{30000}] \\ &= \frac{35}{12}+\frac{35}{12}+\cdots+\frac{35}{12} = 30000\times\frac{35}{12} = 87500. \end{aligned}$$

We therefore have 95% of the outcomes of S falling in the range

$$E[S] \pm 1.96\sqrt{Var[S]} = 105000 \pm 1.96\sqrt{87500} = 105000 \pm 580.$$

The lowest possible value of S occurs when each die is a 1 which results in $S = 30000$, while the highest possible value of S occurs when each die is a 6 which results in $S = 180000$. Thus the total range of S is $180000 - 30000 = 150000$. Despite this large range, almost all values of S will occur in a narrow band around 105000; for example 95% of the outcomes will occur in the range 105000 ± 580. This is an illustration of the law of large numbers.

Problem 4.27 *[B***] Consider a barrel with 20,000 dice in it, let X_i be the outcome of the ith die and let S be the sum of all the dice. Each of the dice is crooked so that the probabilities of $1, 2, 3, 4, 5$ and 6 are respectively $\frac{1}{10}, \frac{1}{10}, \frac{1}{10}, \frac{1}{10}, \frac{1}{10}$ and $\frac{1}{2}$. Calculate $E[X_i]$, $E[X_i^2]$, $Var[X_i]$, $E[S]$ and $Var[S]$. It is a fact that 95% of the die rolls will fall in a range*

$$E[S] \pm 1.96\sqrt{Var[S]}.$$

Calculate this range and compare this with the range of all possible values.

Answer: We have

$$\begin{aligned} E[X_i] &= \frac{1}{10}\times 1+\frac{1}{10}\times 2+\frac{1}{10}\times 3+\frac{1}{10}\times 4+\frac{1}{10}\times 5+\frac{1}{2}\times 6=\frac{9}{2}. \\ E\left[X_i^2\right] &= \frac{1}{10}\times 1^2+\frac{1}{10}\times 2^2+\frac{1}{10}\times 3^2+\frac{1}{10}\times 4^2+\frac{1}{10}\times 5^2+\frac{1}{2}\times 6^2=\frac{47}{2} \\ Var[X_i] &= E\left[X_i^2\right]-E[X_i]^2=\frac{47}{2}-\left(\frac{9}{2}\right)^2=\frac{13}{4}. \end{aligned}$$

If $S = X_1 + X_2 + \cdots + X_{20000}$ then

$$\begin{aligned} E[S] &= E[X_1+X_2+\cdots+X_{20000}] = E[X_1]+E[X_2]+\cdots+E[X_{20000}] \\ &= \frac{9}{2}+\frac{9}{2}+\cdots+\frac{9}{2}=20000\times\frac{9}{2}=90000 \end{aligned}$$

and

$$\begin{aligned} Var[S] &= Var[X_1+X_2+\cdots+X_{20000}] \\ &= 1^2Var[X_1]+1^2Var[X_2]+\cdots+1^2Var[X_{20000}] \\ &= \frac{13}{4}+\frac{13}{4}+\cdots+\frac{13}{4}=20000\times\frac{13}{4}=65000. \end{aligned}$$

We therefore have 95% of the outcomes of S falling in the range

$$E[S]\pm 1.96\sqrt{Var[S]}=90000\pm 1.96\sqrt{65000}=90000\pm 500.$$

The lowest possible value of S occurs when each die is a 1 which results in $S = 20000$, while the highest possible value of S occurs when each die is a 6 which results in $S = 120000$. Thus the total range of S is $120000 - 20000 = 100000$. Despite this large range, almost all values of S will in be in a narrow band around 90000; for example 95% of the outcomes will occur in the range 90000 ± 500. This is an illustration of the law of large numbers.

Problem 4.28 *[A**] If $Z \sim N[0,1]$ it is a fact that $E\left[Z^2\right] = 1$ and $E\left[Z^4\right] = 3$. Note that if $Y = Z^2$ then $Y \sim \chi_1^2$ and so we have $E[Y] = 1$ and $E\left[Y^2\right] = 3$ or*

$$Var[Y]=E\left[Y^2\right]-E[Y]^2=3-1=2.$$

Consider generalizing this to the case where if $Y \sim \chi_r^2$ then $E[Y] = r$ and $Var[Y] = 2r$. (see the lecture notes for the relationship between the standard normal and chi-squared distributions).

Answer: If

$$Y=Z_1^2+Z_2^2+\cdots+Z_r^2$$

where $Z_i \sim N[0,1]$ are independent then $Y \sim \chi_r^2$. We therefore have

$$\begin{aligned} E[Y] &= E\left[Z_1^2+Z_2^2+\cdots+Z_r^2\right]=E\left[Z_1^2\right]+E\left[Z_2^2\right]+\cdots+E\left[Z_r^2\right] \\ &= 1+1+\cdots+1=r. \end{aligned}$$

Now

$$Var\left[Z_i^2\right] = E\left[\left(Z_i^2\right)^2\right] - E\left[Z_i^2\right]^2 = E\left[Z_i^4\right] - E\left[Z_i^2\right]^2 = 3 - 1 = 2$$

so that since the $Z_i's$ are independent

$$\begin{aligned} Var\left[Y\right] &= Var\left[Z_1^2 + Z_2^2 + \cdots + Z_r^2\right] = 1^2 Var\left[Z_1^2\right] + 1^2 Var\left[Z_2^2\right] + \cdots + 1^2 Var\left[Z_r^2\right] \\ &= 2 + 2 + \cdots + 2 = 2r. \end{aligned}$$

Problem 4.29 *[C***] Let X and W be two random variables having the following properties*

$$E\left[X\right] = 5, Var\left[X\right] = 9, E\left[W\right] = 10, Var\left[W\right] = 16, Cov\left[X,W\right] = -4.$$

What is $E\left[X^2\right], E\left[XW\right], Cov\left[W,W\right], Cov\left[W,5\right]$ and the correlation coefficient ρ between X and W? If

$$Y = 20 + 2X - 5W$$

*what is $E\left[Y\right]$ and $Var\left[Y\right]$? [B***] What is $Cov\left[Y,X\right]$?*

Answer: We have

$$\begin{aligned} E\left[X^2\right] &= E\left[X\right]^2 + Var\left[X\right] = 5^2 + 9 = 34 \\ E\left[XW\right] &= E\left[X\right]E\left[W\right] + Cov\left[X,W\right] = 5 \times 10 + -4 = 46 \\ Cov\left[W,W\right] &= Var\left[W\right] = 16, Cov\left[W,5\right] = 0 \\ \rho &= \frac{Cov\left[X,W\right]}{\sqrt{Var\left[X\right]}\sqrt{Var\left[W\right]}} = \frac{-4}{\sqrt{9}\sqrt{16}} = -\frac{1}{3}. \end{aligned}$$

For Y we have

$$\begin{aligned} E\left[Y\right] &= 20 + 2E\left[X\right] - 5E\left[W\right] = 20 + 2 \times 5 - 5 \times 10 = -20 \\ Var\left[Y\right] &= 2^2 Var\left[X\right] + (-5)^2 Var\left[W\right] + 2 \times (2) \times (-5)\, Cov\left[X,W\right] \\ &= 2^2 \times 9 + (-5)^2 \times 16 + 2 \times (2) \times (-5) \times (-4) = 516. \end{aligned}$$

For $Cov\left[Y,X\right]$ we have

$$\begin{aligned} Y - E\left[Y\right] &= 20 + 2X - 5W - E\left[20 + 2X - 5W\right] = 20 + 2X - 5W - \left(20 + 2E\left[X\right] - 5E\left[W\right]\right) \\ &= 2\left(X - E\left[X\right]\right) - 5\left(W - E\left[W\right]\right) \end{aligned}$$

so that

$$\begin{aligned} Cov\left[Y,X\right] &= E\left[\left(Y - E\left[Y\right]\right) \times \left(X - E\left[X\right]\right)\right] \\ &= E\left[\left(20 + 2X - 5W - E\left[20 + 2X - 5W\right]\right) \times \left(X - E\left[X\right]\right)\right] \\ &= E\left[\left(2\left(X - E\left[X\right]\right) - 5\left(W - E\left[W\right]\right)\right) \times \left(X - E\left[X\right]\right)\right] \\ &= 2E\left[\left(X - E\left[X\right]\right) \times \left(X - E\left[X\right]\right)\right] - 5E\left[\left(W - E\left[W\right]\right)\left(X - E\left[X\right]\right)\right] \\ &= 2E\left[\left(X - E\left[X\right]\right)^2\right] - 5E\left[\left(W - E\left[W\right]\right)\left(X - E\left[X\right]\right)\right] \\ &= 2Var\left[X\right] - 5Cov\left[W,X\right] = 2 \times 9 - 5 \times (-4) = 38. \end{aligned}$$

Alternatively we can use

$$\begin{aligned} E[YX] &= E[(20+2X-5W)X] = E\left[20X+2X^2-5XW\right] \\ &= 20E[X]+2E\left[X^2\right]-5E[XW] \\ &= 20\times 5+2\times 34-5\times 46=-62 \end{aligned}$$

so that

$$Cov[Y,X] = E[YX]-E[X]E[Y] = -62-5\times(-20) = 38.$$

Problem 4.30 *[B***] If X is a random variable with outcomes $1, 2, 3$ and with respective probabilities $\frac{1}{4}, \frac{1}{2}, \frac{1}{4}$ then calculate $E[X], E[X]^2, E\left[X^2\right], Var[X], E\left[e^X\right],$ and $e^{E[X]}$.*

Answer: We have

$$\begin{aligned} E[X] &= 1\times\frac{1}{4}+2\times\frac{1}{2}+3\times\frac{1}{4}=2,\ E[X]^2=2^2=4 \\ E\left[X^2\right] &= 1^2\times\frac{1}{4}+2^2\times\frac{1}{2}+3^2\times\frac{1}{4}=\frac{9}{2}, Var[X]=E\left[X^2\right]-E[X]^2=\frac{9}{2}-2^2=\frac{1}{2} \\ E\left[e^X\right] &= e^1\times\frac{1}{4}+e^2\times\frac{1}{2}+e^3\times\frac{1}{4}\approx 9.34,\ e^{E[X]}=e^2\approx 7.40. \end{aligned}$$

Problem 4.31 *[D***] Consider two random variables X and Y where*

$$E[X]=10, Var[X]=25,\ E[Y]=7, Var[Y]=9, Cov[X,Y]=-9.$$

Calculate $E[(3X+4)], E\left[X^2\right], E[XY], E\left[(3X+4)^2\right], Var[3X+4]$ and the correlation coefficient ρ between X and Y.

Answer: We have

$$\begin{aligned} E[(3X+4)] &= 3E[X]+4=3\times 10+4=34 \\ E\left[X^2\right] &= E[X]^2+Var[X]=10^2+25=125 \\ E[XY] &= E[X]E[Y]+Cov[X,Y]=10\times 7-9=61 \\ E\left[(3X+4)^2\right] &= E\left[(9X^2+24X+16)\right]=9E\left[X^2\right]+24E[X]+16 \\ &= 9\times 125+24\times 10+16=1381. \end{aligned}$$

From

$$Var[3X+4]=Var[3X]=9Var[X]=9\times 25=225$$

we can calculate $E\left[(3X+4)^2\right]$ as

$$E\left[(3X+4)^2\right]=E[(3X+4)]^2+Var[3X+4]=34^2+225=1381.$$

The correlation coefficient is given by

$$\rho=\frac{Cov[X,Y]}{Var[X]^{\frac{1}{2}}Var[Y]^{\frac{1}{2}}}=\frac{-9}{\sqrt{25}\sqrt{9}}=-\frac{3}{5}=-0.6.$$

Problem 4.32 *[D***] Given the information in the previous question, if*

$$R = 3X - 4Y + 10$$

calculate $E[R]$, $Var[R]$ *and* $E[R^2]$. *[A***] What is* $E[XR]$ *and* $Cov[X,R]$?

Answer: We have

$$\begin{aligned} E[R] &= E[3X - 4Y + 10] = 3E[X] - 4E[Y] + 10 = 3\times 10 - 4\times 7 + 10 = 12 \\ Var[R] &= Var[3X - 4Y + 10] = 9Var[X] + (-4)^2 Var[Y] + 2\times 3\times(-4)\,Cov[X,Y] \\ &= 9\times 25 + (-4)^2\times 9 + 2\times 3\times(-4)\times(-9) = 585 \\ E[R^2] &= E[R]^2 + Var[R] = 12^2 + 585 = 729. \end{aligned}$$

Now

$$\begin{aligned} E[XR] &= E[X(3X - 4Y + 10)] = E[3X^2 - 4XY + 10X] = 3E[X^2] - 4E[XY] + 10E[X] \\ &= 3\times 125 - 4\times 61 + 10\times 10 = 231 \\ Cov[X,R] &= E[XR] - E[X]E[R] = 231 - 10\times 12 = 111. \end{aligned}$$

Alternatively one can substitute

$$R - E[R] = 3X - 4Y + 10 - E[3X - 4Y + 10] = 3(X - E[X]) - 4(Y - E[Y])$$

into $Cov[X,R] = E[(X - E[X])(R - E[R])]$ as

$$\begin{aligned} Cov[X,R] &= E[(X - E[X])(R - E[R])] = E[(X - E[X])(3(X - E[X]) - 4(Y - E[Y]))] \\ &= 3E\left[(X - E[X])^2\right] - 4E[(X - E[X])(Y - E[Y])] \\ &= 3Var[X] - 4Cov[X,Y] = 3\times 25 - 4\times(-9) = 111. \end{aligned}$$

Problem 4.33 *[D***] Given the random variable*

$$Y = \begin{bmatrix} \textit{Outcomes} & \textit{Probabilities} \\ -2 & \frac{1}{3} \\ & \\ 4 & \frac{2}{3} \end{bmatrix}$$

calculate $E[Y]$, $E[Y^2]$, $E[Y]^2$ *and* $Var[Y]$.

Answer: We have

$$\begin{aligned} E[Y] &= -2\times\frac{1}{3} + 4\times\frac{2}{3} = 2,\ E[Y^2] = (-2)^2\times\frac{1}{3} + 4^2\times\frac{2}{3} = 12 \\ E[Y]^2 &= 2^2 = 4,\ Var[Y] = E[Y^2] - E[Y]^2 = 12 - 4 = 8. \end{aligned}$$

Problem 4.34 *[C***] Suppose that W is continuous random variable with density*

$$p(w) = e^{-w} \text{ for } w \geq 0.$$

Calculate $E[W], E[W]^2, E[W^2]$ *and* $Var[W]$. *[B***] What is* $E[W^{100}]$?

Answer: We have

$$E[W] = \int_0^\infty wp(w)\,dw = \int_0^\infty we^{-w}dw = \Gamma(2) = (2-1)! = 1$$

or using integration by parts

$$\begin{array}{ll} u = w & v' = e^{-w} \\ u' = 1 & v = -e^{-w} \end{array}$$

we have

$$E[W] = \int_0^\infty we^{-w}dw = -we^{-w}|_0^\infty + \int_0^\infty e^{-w}dw = 1$$

since

$$\int_0^\infty e^{-w}dw = -e^{-w}|_0^\infty = 0 - (-1) = 1.$$

In addition

$$E\left[W^2\right] = \int_0^\infty w^2 p(w)\,dw = \int_0^\infty w^2 e^{-w}dw = \Gamma(3) = (3-1)! = 2$$

or by using integration by parts

$$\begin{array}{ll} u = w^2 & v' = e^{-w} \\ u' = 2w & v = -e^{-w} \end{array}$$

we have

$$E\left[W^2\right] = \int_0^\infty w^2 e^{-w}dw = -w^2 e^{-w}|_0^\infty + 2\int_0^\infty we^{-w}dw = 2$$

since

$$\int_0^\infty w^1 e^{-w}dw = 1.$$

Thus

$$Var[W] = E\left[W^2\right] - E[W]^2 = 2 - 1^2 = 1.$$

In general $E[W^n] = \Gamma(n+1) = n!$ Thus $E\left[W^{100}\right] = 100!$

Problem 4.35 *[D***] Consider*

$$E[X] = 5, Var[X] = 36,\ \ E[Y] = 7, Var[Y] = 25, Cov[X,Y] = -10.$$

Calculate $E\left[X^2\right],\ E[XY],\ E[3X-4Y+7]$ *and* $Var[3X-4Y+7]$. *[B***] Suppose that* X *is the change in interest rates on Friday,* Y *is the change in the exchange rate on the following Monday, and we wish to use* X *to predict* Y *using the predictor* $Y^P = a + bX$. *We could chose say* $a = 3$ *and* $b = 4$ *but this would probably give us bad predictions. Let* $Y - Y^P$ *be the forecast error and suppose we wish* Y^P *to be on average correct or to satisfy* $E\left[Y - Y^P\right] = 0$ *and we want* $Var\left[Y - Y^P\right]$ *to be as small as possible. What values of* a *and* b *should we then chose?*

Answer: We have

$$\begin{aligned} E\left[X^2\right] &= E[X]^2 + Var[X] = 5^2 + 36 = 61 \\ E[XY] &= E[X]E[Y] + Cov[X,Y] = 5\times 7 - 10 = 25 \\ E[3X - 4Y + 7] &= 3E[X] - 4E[Y] + 7 = 3\times 5 - 4\times 7 + 7 = -6 \\ Var[3X - 4Y + 7] &= Var[3X - 4Y] = (3)^2 Var[X] + (-4)^2 Var[Y] + 2(3)(-4)Cov[X,Y] \\ &= 9\times 36 + 16\times 25 + 2\times 3\times(-4)\times(-10) = 964. \end{aligned}$$

We have

$$E\left[Y - Y^P\right] = E[Y] - E\left[Y^P\right] = 7 - E[a + bX] = 7 - a - bE[X] = 7 - a - 5b$$

so that

$$E\left[Y - Y^P\right] = 0 \Longrightarrow 7 - a - 5b = 0.$$

Furthermore

$$\begin{aligned} Var\left[Y - Y^P\right] &= Var[Y - a - bX] = Var[Y - bX] \\ &= \left(1^2\right)Var[Y] + (-b)^2 Var[X] + 2\times 1\times(-b)Cov[X,Y] = 25 + 36b^2 + 20b. \end{aligned}$$

We now minimize $Var\left[Y - Y^P\right] = f(b)$ as using the first-order condition $f'(b^*) = 0$ as

$$f(b) = 25+36b^2+20b \Longrightarrow f'(b) = 72b+20 \Longrightarrow 72b^*+20 = 0 \Longrightarrow b^* = -\frac{5}{18} \approx -0.28.$$

Now

$$7 - a^* - 5b^* = 0 \Longrightarrow a^* = 7 - 5b^* = 7 - 5\times\left(-\frac{5}{18}\right) = \frac{151}{18} = 8.39.$$

Problem 4.36 *[A**] Suppose wealth W is measured in millions of dollars, and suppose your utility from wealth is either 1) $U(W) = W^{\frac{1}{2}}$ or 2) $U(W) = W^2$. Suppose you are a millionaire so that $W = 1$. Someone offers you the following wager: toss a coin; if it's heads you get 2 million dollars $W = 2$, if it's tails you get nothing $W = 0$. Show that for $U(W) = W^{\frac{1}{2}}$ that you will not toss the coin: you are risk averse. Suppose you had to toss the coin unless you buy insurance. How much would you pay for this insurance? Show for $U(W) = W^2$ you would toss the coin: you are a risk lover.*

Answer: For $U(W) = W^{\frac{1}{2}}$ with a coin toss we have expected utility of

$$E[U(W)] = E\left[W^{\frac{1}{2}}\right] = \frac{1}{2}\times 0^{\frac{1}{2}} + \frac{1}{2}\times 2^{\frac{1}{2}} = \frac{1}{\sqrt{2}} \approx 0.71.$$

Without tossing the coin we have

$$E[U(W)] = E\left[W^{\frac{1}{2}}\right] = 1\times 1^{\frac{1}{2}} = 1.$$

Since not tossing the coin leads to higher utility ($E[U(W)] = 1$) than toss the coin ($E[U(W)] = 0.71$), you would not toss the coin. Since in both cases

$$E[W] = \frac{1}{2} \times 0 + \frac{1}{2} \times 2 = 1 \times 1 = 1.$$

The only difference is in the riskiness of the two situations, and so you are risk averse.

Let W_0 be the level of wealth that would make you indifferent to the coin toss. Since $E[U(W)] = \frac{1}{\sqrt{2}}$ with the coin toss it must be that

$$U(W_0) = W_0^{\frac{1}{2}} = \frac{1}{\sqrt{2}} \Longrightarrow W_0 = \left(\frac{1}{\sqrt{2}}\right)^2 = \frac{1}{2}$$

so that you would pay up to half your wealth, or \$500,000, for insurance not to have to toss the coin.

For $U(W) = W^2$ we have with the coin toss that

$$E[U(W)] = E\left[W^2\right] = \frac{1}{2} \times 0^2 + \frac{1}{2} \times 2^2 = 2$$

while the 1 million with certainty yields

$$E[U(W)] = E\left[W^2\right] = 1 \times 1^2 = 1$$

and so you would take the coin toss. In this case you are a risk lover.

Problem 4.37 *[A **] For a discrete random variable X with outcomes $x_1, x_2, \ldots, x_n$ and associated probabilities $p_1, p_2, \ldots, p_n$, prove Jensen's inequality: that if $f(x)$ is convex then $E[f(X)] \geq f(E[X])$. (Jensen's inequality is also valid for continuous random variables.) Apply Jensen's inequality to $E\left[X^2\right], E\left[e^X\right], E\left[X^{\frac{1}{2}}\right]$, and $E[\ln(X)]$.*

Answer: Recall (see Problem 1.30 on page 19) that if $f(x)$ is convex then for any $\lambda_i \geq 0$, $i = 1, 2, \ldots, n$ satisfying

$$\lambda_1 + \lambda_2 + \cdots + \lambda_n = 1$$

that

$$f(\lambda_1 x_1 + \lambda_2 x_2 + \cdots + \lambda_n x_n) \leq \lambda_1 f(x_1) + \lambda_2 f(x_2) + \cdots + \lambda_n f(x_n).$$

Jensen's inequality follows by letting the weights λ_i be the probabilities p_i as

$$f(p_1 x_1 + p_2 x_2 + \cdots + p_n x_n) \leq p_1 f(x_1) + p_2 f(x_2) + \cdots + p_n f(x_n)$$

where

$$\begin{aligned} E[X] &\equiv p_1 x_1 + p_2 x_2 + \cdots + p_n x_n \\ E[f(X)] &\equiv p_1 f(x_1) + p_2 f(x_2) + \cdots + p_n f(x_n) \end{aligned}$$

and so

$$f(E[X]) \leq E[f(X)] \Longrightarrow E[f(X)] \geq f(E[X]).$$

The functions $f(x) = x^2, e^x, -x^{\frac{1}{2}}, -\ln(x)$ are all convex since

$$\begin{aligned} f(x) &= x^2 \Longrightarrow f''(x) = 2 > 0 \\ f(x) &= e^x \Longrightarrow f''(x) = e^x > 0 \\ f(x) &= -x^{\frac{1}{2}} \Longrightarrow f''(x) = \frac{1}{4}x^{-\frac{3}{2}} > 0 \\ f(x) &= -\ln(x) \Longrightarrow f''(x) = \frac{1}{x^2} > 0 \end{aligned}$$

so we can apply Jensen's inequality to them. For x^2 we have $E[X^2] \geq E[X]^2$ as

$$E[X^2] = E[f(X)] \geq f(E[X]) = E[X]^2 \Longrightarrow E[X^2] \geq E[X]^2.$$

For e^x we have

$$E[e^X] = E[f(X)] \geq f(E[X]) = e^{E[X]} \Longrightarrow E[e^X] \geq e^{E[X]}.$$

The convexity of $-x^{\frac{1}{2}}$ is equivalent to the concavity of $x^{\frac{1}{2}}$ and so $E\left[X^{\frac{1}{2}}\right] \leq E[X]^{\frac{1}{2}}$ (recall for inequalities that the sign is reversed when both sides are multiplied by a negative number, e.g., $-x \geq -5 \Longrightarrow x \leq 5$) as

$$E\left[-X^{\frac{1}{2}}\right] = E[f(X)] \geq f(E[X]) = -E[X]^{\frac{1}{2}} \Longrightarrow -E\left[X^{\frac{1}{2}}\right] \geq -E[X]^{\frac{1}{2}} \Longrightarrow E\left[X^{\frac{1}{2}}\right] \leq E[X]^{\frac{1}{2}}$$

The convexity of $-\ln(x)$ is equivalent to the concavity of $\ln(x)$ and so $E[\ln(X)] \leq \ln(E[X])$ as

$$\begin{aligned} E[-\ln(X)] &= E[f(X)] \geq f(E[X]) = -\ln(E[X]) \Longrightarrow -E[\ln(X)] \geq -\ln(E[X]) \\ &\Longrightarrow E[\ln(X)] \leq \ln(E[X]). \end{aligned}$$

Problem 4.38 *[A**] A stronger version of Jensen's inequality states that if $f(x)$ is* ***strictly*** *convex then if X is not a degenerate random variable then*

$$E[f(X)] > f[E[X]].$$

Let W be wealth and let $U(W)$ be the utility of wealth. Suppose wealth is random. A risk averse household when given the choice between a random outcome for wealth W and getting the expected value of wealth $E[W]$ with probability 1 will prefer $E[W]$ and so $E[U(W)] < U(E[W])$. (Given a choice between tossing a coin and getting two million dollars if it's heads, and zero dollars if it's tails, or getting one million dollars for sure, this household would take the one million dollars for sure.) Use Jensen's inequality to show that a household with a strictly concave utility function will be risk averse. What happens if $U(W)$ is strictly convex? What can you say if $U(W) = W^{\frac{1}{2}}$ or $U(W) = e^W$?

Answer: If $U(W)$ is strictly concave then $-U(W)$ is strictly convex and so from Jensen's inequality, and assuming wealth is random and hence not a degenerate random variable, we have

$$E[-U(W)] = -E[U(W)] > -U(E[W]) \Longrightarrow E[U(W)] < U(E[W])$$

so that the household is risk averse.

If $U(W)$ is strictly convex then $E[U(W)] > U(E[W])$ and so the household prefers the random outcome W to having $E[W]$ with probability 1. In other words the household is a risk lover and would prefer tossing the coin with a chance of getting 2 million dollars.

Since $U(W) = W^{\frac{1}{2}}$ is strictly concave (since $U''(W) < 0$) this household will be risk averse. Since $U(W) = e^W$ is strictly convex (since $U''(W) > 0$) this household will be a risk lover.

Problem 4.39 *[A*] Use Stirling's approximation to prove that the Student's t distribution with r degrees of freedom approaches the standard normal as $r \to \infty$ or*

$$\lim_{r\to\infty} \frac{\Gamma\left(\frac{r+1}{2}\right)}{\sqrt{\pi \times r} \times \Gamma\left(\frac{r}{2}\right)} \left(1 + \frac{1}{r}x^2\right)^{-\frac{(r+1)}{2}} = \frac{e^{-\frac{1}{2}x^2}}{\sqrt{2\pi}}.$$

Answer: We have

$$\begin{aligned} p(x|r) &= \frac{\Gamma\left(\frac{r+1}{2}\right)}{\sqrt{\pi \times r} \times \Gamma\left(\frac{r}{2}\right)} \left(1 + \frac{1}{r}x^2\right)^{-\frac{(r+1)}{2}} \\ &= \frac{1}{\sqrt{\pi}} \frac{\Gamma\left(n + \frac{1}{2} + 1\right)}{\sqrt{2n+1} \times \Gamma(n+1)} \left(\left(1 + \frac{x^2}{r}\right)^r\right)^{-\frac{(r+1)}{2r}} \end{aligned}$$

where we define $n \equiv \frac{r-1}{2}$ (and so $r = 2n+1$) so as to use Stirling's approximation which states that

$$\lim_{n\to\infty} \frac{\Gamma(n+1)}{\sqrt{2\pi \times n} \times n^n e^{-n}} = 1.$$

Now

$$\begin{aligned} \frac{\Gamma\left(n+\frac{1}{2}+1\right)}{\sqrt{2n+1} \times \Gamma(n+1)} &= \frac{\frac{\Gamma\left(n+\frac{1}{2}+1\right)}{\sqrt{2\pi\times\left(n+\frac{1}{2}\right)}\left(n+\frac{1}{2}\right)^{n+\frac{1}{2}}e^{-\left(n+\frac{1}{2}\right)}}}{\frac{\Gamma(n+1)}{\sqrt{2\pi\times n}\times n^n e^{-n}}} \\ &\quad \times \frac{\sqrt{2\pi \times \left(n+\frac{1}{2}\right)}\left(n+\frac{1}{2}\right)^{n+\frac{1}{2}} e^{-\left(n+\frac{1}{2}\right)}}{\sqrt{2n+1}\sqrt{2\pi \times n} \times n^n e^{-n}} \\ &= \frac{\frac{\Gamma\left(n+\frac{1}{2}+1\right)}{\sqrt{2\pi\times\left(n+\frac{1}{2}\right)}\left(n+\frac{1}{2}\right)^{n+\frac{1}{2}}e^{-\left(n+\frac{1}{2}\right)}}}{\frac{\Gamma(n+1)}{\sqrt{2\pi\times n}\times n^n e^{-n}}} \frac{\left(n+\frac{1}{2}\right)^{n+1} e^{-\frac{1}{2}}}{\sqrt{2n+1} \times n^{n+\frac{1}{2}}} \\ &= \frac{\frac{\Gamma\left(n+\frac{1}{2}+1\right)}{\sqrt{2\pi\times\left(n+\frac{1}{2}\right)}\left(n+\frac{1}{2}\right)^{n+\frac{1}{2}}e^{-\left(n+\frac{1}{2}\right)}}}{\frac{\Gamma(n+1)}{\sqrt{2\pi\times n}\times n^n e^{-n}}} \frac{\left(1+\frac{1}{2n}\right)^{n+\frac{1}{2}} e^{-\frac{1}{2}}}{\left(2+\frac{1}{n}\right)^{\frac{1}{2}}} \end{aligned}$$

so that using Stirling's approximation and $\lim_{n\to\infty}\left(1+\frac{x}{n}\right)^n = e^x$ yields

$$\begin{aligned}
\lim_{n\to\infty}\frac{\Gamma\left(n+\frac{1}{2}+1\right)}{\sqrt{2n+1}\times\Gamma\left(n+1\right)} &= \frac{\lim_{n\to\infty}\left(\frac{\Gamma\left(n+\frac{1}{2}+1\right)}{\sqrt{2\pi\times\left(n+\frac{1}{2}\right)}\left(n+\frac{1}{2}\right)^{n+\frac{1}{2}}e^{-\left(n+\frac{1}{2}\right)}}\right)}{\lim_{n\to\infty}\left(\frac{\Gamma(n+1)}{\sqrt{2\pi\times n}\times n^n e^{-n}}\right)} \\
\times\frac{\left(\lim_{n\to\infty}\left(1+\frac{1}{2n}\right)^{n+\frac{1}{2}}\right)e^{-\frac{1}{2}}}{\left(\lim_{n\to\infty}\left(2+\frac{1}{n}\right)^{\frac{1}{2}}\right)} &= \frac{1}{1}\times\frac{e^{\frac{1}{2}}e^{-\frac{1}{2}}}{2^{\frac{1}{2}}} = \frac{1}{\sqrt{2}}.
\end{aligned}$$

Thus we have

$$\begin{aligned}
\lim_{r\to\infty}p\left(x|r\right) &= \frac{1}{\sqrt{\pi}}\lim_{n\to\infty}\left(\frac{\Gamma\left(n+\frac{1}{2}+1\right)}{\sqrt{2n+1}\times\Gamma\left(n+1\right)}\right)\left(\lim_{r\to\infty}\left(1+\frac{x^2}{r}\right)^r\right)^{-\left(\lim_{r\to\infty}\frac{(r+1)}{2r}\right)} \\
&= \frac{1}{\sqrt{\pi}}\times\frac{1}{\sqrt{2}}\times\left(e^{x^2}\right)^{-\frac{1}{2}} = \frac{e^{-\frac{1}{2}x^2}}{\sqrt{2\pi}}.
\end{aligned}$$

Problem 4.40 *[B***] Suppose that X is continuous random variable with density* $p\left(x\right)=3e^{-3x}$ *for* $0\le x<\infty$. *Calculate* $E\left[X\right]$ *and* $Var\left[X\right]$.

Answer: We have

$$E\left[X^2\right]=\int_0^\infty x^2\times 3e^{-3x}dx$$

so that use integration by parts with

$$\begin{aligned}
u &= x^2, v' = 3e^{-3x} \\
u' &= 2x, v = -e^{-3x}
\end{aligned}$$

we have

$$\begin{aligned}
E\left[X^2\right] &= \int_0^\infty x^2\times 3e^{-3x}dx = -x^2e^{-3x}|_0^\infty + 2\int_0^\infty x\times e^{-3x}dx \\
&= 0+\frac{2}{3}\int_0^\infty x\times 3e^{-3x}dx = \frac{2}{3}E\left[X\right].
\end{aligned}$$

To calculate

$$E\left[X\right]=\int_0^\infty x\times 3e^{-3x}dx$$

use integration by parts with

$$\begin{aligned}
u &= x, v' = 3e^{-3x} \\
u &= 1, v = -e^{-3x}.
\end{aligned}$$

We have

$$E\left[X\right]=\int_0^\infty x\times 3e^{-3x}dx = -xe^{-3x}|_0^\infty + \int_0^\infty e^{-3x}dx = \frac{1}{3} = -\frac{e^{-3x}}{3}|_0^\infty = \frac{1}{3}.$$

Thus

$$E[X] = \frac{1}{3}, E\left[X^2\right] = \frac{2}{3} \times E[X] = \frac{2}{3} \times \frac{1}{3} = \frac{2}{9}$$

and

$$Var[X] = E\left[X^2\right] - E[X]^2 = \frac{2}{9} - \left(\frac{1}{3}\right)^2 = \frac{1}{9}.$$

4.4 Econometrics

Problem 4.41 *[C***] Prove that*

$$S(\mu) = \sum_{i=1}^{n} (Y_i - \mu)^2$$

is minimized when $\mu = \hat{\mu} = \bar{Y}$ *both with calculus and without calculus.*

Answer: Using the sum rule for derivatives we have

$$\frac{dS(\mu)}{d\mu} = \sum_{i=1}^{n} \frac{d}{d\mu} (Y_i - \mu)^2 = \sum_{i=1}^{n} -2(Y_i - \mu)$$

so that the first-order conditions are

$$\begin{aligned} \frac{dS(\hat{\mu})}{d\mu} &= 0 = \sum_{i=1}^{n} -2(Y_i - \hat{\mu}) \\ &\implies -2\sum_{i=1}^{n} (Y_i - \hat{\mu}) = 0 \\ &\implies \sum_{i=1}^{n} Y_i - n \times \hat{\mu} = 0 \\ &\implies \hat{\mu} = \frac{1}{n}\sum_{i=1}^{n} Y_i = \bar{Y} \end{aligned}$$

with

$$\frac{d^2S(\mu)}{d\mu^2} = \sum_{i=1}^{n} \frac{d}{d\mu}(-2(Y_i - \mu)) = \sum_{i=1}^{n} 2 = 2n > 0$$

so that $S(\mu)$ is globally convex so that $\bar{Y}$ is a global minimizer.

Without using calculus we have (completing the square)

$$\begin{aligned} S(\mu) &= \sum_{i=1}^{n}(Y_i-\mu)^2=\sum_{i=1}^{n}\left((Y_i-\bar{Y})+(\bar{Y}-\mu)\right)^2 \\ &= \sum_{i=1}^{n}\left(Y_i-\bar{Y}\right)^2+2\sum_{i=1}^{n}\left(Y_i-\bar{Y}\right)\left(\bar{Y}-\mu\right)+\sum_{i=1}^{n}\left(\bar{Y}-\mu\right)^2 \\ &= \sum_{i=1}^{n}\left(Y_i-\bar{Y}\right)^2+2\left(\bar{Y}-\mu\right)\underbrace{\sum_{i=1}^{n}\left(Y_i-\bar{Y}\right)}_{0}+n\left(\bar{Y}-\mu\right)^2 \\ &= \sum_{i=1}^{n}\left(Y_i-\bar{Y}\right)^2+n\left(\bar{Y}-\mu\right)^2. \end{aligned}$$

Note that the term $\sum_{i=1}^{n}\left(Y_i-\bar{Y}\right)^2$ does not depend on μ while for the second term $n\left(\bar{Y}-\mu\right)^2>0$ for $\mu\neq\bar{Y}$ and $n\left(\bar{Y}-\mu\right)^2=0$ for $\mu=\bar{Y}$ so that $\mu=\bar{Y}$ is a global minimizer for $S(\mu)$.

Problem 4.42 *[B***] Consider the simple linear regression model without a constant*

$$Y_i=\beta X_i+e_i,\ i=1,2,\ldots,n.$$

The least squares estimator $\hat{\beta}$ minimizes

$$S(\beta)=\sum_{i=1}^{n}(Y_i-\beta X_i)^2.$$

Show from the first-order conditions that

$$\hat{\beta}=\frac{\sum_{i=1}^{n}X_iY_i}{\sum_{i=1}^{n}X_i^2}$$

*and that the second-order conditions for a minimum will be satisfied. Show that $\hat{\beta}$ is the least squares estimator without using calculus. [A***]Show that if $\hat{e}_i=Y_i-X_i\hat{\beta}$ is the least squares residual then*

$$\sum_{i=1}^{n}X_i\hat{e}_i=0$$

and consequently

$$\sum_{i=1}^{n}Y_i^2=\hat{\beta}^2\sum_{i=1}^{n}X_i^2+\sum_{i=1}^{n}\hat{e}_i^2.$$

Answer: Using the sum rule for derivatives we have

$$\frac{dS(\beta)}{d\beta}=\sum_{i=1}^{n}\frac{d}{d\beta}(Y_i-X_i\beta)^2=\sum_{i=1}^{n}-2X_i\left(Y_i-X_i\beta\right)$$

so that the first-order condition is

$$\begin{aligned}
\frac{dS\left(\hat{\beta}\right)}{d\beta} &= 0=\sum_{i=1}^{n}-2X_i\left(Y_i-X_i\hat{\beta}\right) \\
&\Longrightarrow -2\sum_{i=1}^{n}X_i\left(Y_i-X_i\hat{\beta}\right)=0 \\
&\Longrightarrow \sum_{i=1}^{n}X_iY_i-\hat{\beta}\left(\sum_{i=1}^{n}X_i^2\right)=0\Longrightarrow\hat{\beta}=\frac{\sum_{i=1}^{n}X_iY_i}{\sum_{i=1}^{n}X_i^2}
\end{aligned}$$

with

$$\frac{d^2S(\beta)}{d\beta^2}=-2\sum_{i=1}^{n}\frac{d}{d\beta}X_i\left(Y_i-X_i\beta\right)=2\sum_{i=1}^{n}X_i^2>0$$

so that $S(\beta)$ is globally convex so that $\hat{\beta}$ is a global minimizer.

Without using calculus we have (completing the square)

$$\begin{aligned}
S(\beta) &= \sum_{i=1}^{n}(Y_i-X_i\beta)^2=\sum_{i=1}^{n}\left(\left(Y_i-X_i\hat{\beta}\right)+X_i\left(\hat{\beta}-\beta\right)\right)^2 \\
&= \sum_{i=1}^{n}\left(Y_i-X_i\hat{\beta}\right)^2+2\sum_{i=1}^{n}\left(Y_i-X_i\hat{\beta}\right)X_i\left(\hat{\beta}-\beta\right)+\sum_{i=1}^{n}X_i^2\left(\hat{\beta}-\beta\right)^2 \\
&= \sum_{i=1}^{n}\left(Y_i-X_i\hat{\beta}\right)^2+2\left(\hat{\beta}-\beta\right)\underbrace{\sum_{i=1}^{n}X_i\left(Y_i-X_i\hat{\beta}\right)}_{0}+\left(\hat{\beta}-\beta\right)^2\sum_{i=1}^{n}X_i^2 \\
&= \sum_{i=1}^{n}\left(Y_i-X_i\hat{\beta}\right)^2+\left(\hat{\beta}-\beta\right)^2\sum_{i=1}^{n}X_i^2.
\end{aligned}$$

where the middle term is zero since

$$\begin{aligned}
\sum_{i=1}^{n}X_i\left(Y_i-X_i\hat{\beta}\right) &= \sum_{i=1}^{n}X_iY_i-\left(\sum_{i=1}^{n}X_i^2\right)\hat{\beta}=\sum_{i=1}^{n}X_iY_i-\left(\sum_{i=1}^{n}X_i^2\right)\frac{\sum_{i=1}^{n}X_iY_i}{\sum_{i=1}^{n}X_i^2} \\
&= \sum_{i=1}^{n}X_iY_i-\sum_{i=1}^{n}X_iY_i=0.
\end{aligned}$$

Now from

$$S(\beta)=\sum_{i=1}^{n}\left(Y_i-X_i\hat{\beta}\right)^2+\left(\hat{\beta}-\beta\right)^2\sum_{i=1}^{n}X_i^2$$

the term $\sum_{i=1}^{n}\left(Y_i-X_i\hat{\beta}\right)^2$ does not depend on β while the second term is positive for $\beta\neq\hat{\beta}$ and 0 for $\beta=\hat{\beta}$. It follows that $\hat{\beta}$ is a global minimizer for $S(\beta)$.

If $\hat{e}_i = Y_i - X_i\hat{\beta}$ then

$$\sum_{i=1}^{n} X_i\hat{e}_i = \sum_{i=1}^{n} X_i\left(Y_i - X_i\hat{\beta}\right) = 0$$

from above. Now since $\hat{e}_i = Y_i - X_i\hat{\beta} \Longrightarrow Y_i = X_i\hat{\beta} + \hat{e}_i$ we have

$$\begin{aligned}\sum_{i=1}^{n} Y_i^2 &= \sum_{i=1}^{n}\left(X_i\hat{\beta} + \hat{e}_i\right)^2 = \sum_{i=1}^{n}\left(X_i^2\hat{\beta}^2 + 2X_i\hat{\beta}\hat{e}_i + \hat{e}_i^2\right) \\ &= \hat{\beta}^2\sum_{i=1}^{n} X_i^2 + 2\hat{\beta}\underbrace{\sum_{i=1}^{n} X_i\hat{e}_i}_{0} + \sum_{i=1}^{n}\hat{e}_i^2 = \hat{\beta}^2\sum_{i=1}^{n} X_i^2 + \sum_{i=1}^{n}\hat{e}_i^2.\end{aligned}$$

Problem 4.43 *[B***] For the simple linear regression model* $Y_i = \alpha + \beta X_i + e_i$, $i = 1, 2, \ldots, n$ *the least squares estimators* $\hat{\alpha}$ *and* $\hat{\beta}$ *are the values of* α *and* β *which minimize the sum of squares function*

$$S(\alpha, \beta) = \sum_{i=1}^{n}(Y_i - \alpha - \beta X_i)^2.$$

Show that

$$\begin{aligned}\hat{\alpha} &= \bar{Y} - \hat{\beta}\bar{X} \\ \hat{\beta} &= \frac{n\sum_{i=1}^{n} X_iY_i - \left(\sum_{i=1}^{n} X_i\right)\left(\sum_{i=1}^{n} Y_i\right)}{n\sum_{i=1}^{n} X_i^2 - \left(\sum_{i=1}^{n} X_i\right)^2} \\ &= \frac{\sum_{i=1}^{n}\left(X_i - \bar{X}\right)\left(Y_i - \bar{Y}\right)}{\sum_{i=1}^{n}\left(X_i - \bar{X}\right)^2}.\end{aligned}$$

Show that $S(\alpha, \beta)$ *is globally convex as long as the* X_i *'s are not all identical.*

Answer: We have using the sum and chain rules that

$$\begin{aligned}\frac{\partial S(\alpha, \beta)}{\partial \alpha} &= -2\sum_{i=1}^{n}(Y_i - \alpha - \beta X_i) \\ \frac{\partial S(\alpha, \beta)}{\partial \beta} &= -2\sum_{i=1}^{n} X_i(Y_i - \alpha - \beta X_i)\end{aligned}$$

so that the first-order conditions for a minimum are

$$\begin{aligned}
\frac{\partial S\left(\hat{\alpha},\hat{\beta}\right)}{\partial\alpha} &= -2\sum_{i=1}^{n}\left(Y_i-\hat{\alpha}-\hat{\beta}X_i\right)=0 \\
&\Longrightarrow n\hat{\alpha}+\left(\sum_{i=1}^{n}X_i\right)\hat{\beta}=\sum_{i=1}^{n}Y_i \\
\frac{\partial S\left(\hat{\alpha},\hat{\beta}\right)}{\partial\beta} &= -2\sum_{i=1}^{n}X_i\left(Y_i-\hat{\alpha}-\hat{\beta}X_i\right)=0 \\
&\Longrightarrow \left(\sum_{i=1}^{n}X_i\right)\hat{\alpha}+\left(\sum_{i=1}^{n}X_i^2\right)\hat{\beta}=\sum_{i=1}^{n}X_iY_i.
\end{aligned}$$

From the first first-order condition we have

$$n\hat{\alpha}+\left(\sum_{i=1}^{n}X_i\right)\hat{\beta}=\sum_{i=1}^{n}Y_i\Longrightarrow\hat{\alpha}+\bar{X}\hat{\beta}=\bar{Y}\Longrightarrow\hat{\alpha}=\bar{Y}-\bar{X}\hat{\beta}.$$

Writing both first-order conditions in matrix notation we have

$$\begin{bmatrix} n & \sum_{i=1}^{n}X_i \\ \sum_{i=1}^{n}X_i & \sum_{i=1}^{n}X_i^2 \end{bmatrix}\begin{bmatrix}\hat{\alpha}\\ \hat{\beta}\end{bmatrix}=\begin{bmatrix}\sum_{i=1}^{n}Y_i \\ \sum_{i=1}^{n}X_iY_i\end{bmatrix}.$$

Solving for $\hat{\beta}$ using Cramer's rule we find that

$$\hat{\beta}=\frac{n\sum_{i=1}^{n}X_iY_i-\left(\sum_{i=1}^{n}X_i\right)\left(\sum_{i=1}^{n}Y_i\right)}{n\sum_{i=1}^{n}X_i^2-\left(\sum_{i=1}^{n}X_i\right)^2}.$$

Now

$$\begin{aligned}
n\sum_{i=1}^{n}\left(X_i-\bar{X}\right)\left(Y_i-\bar{Y}\right) &= n\sum_{i=1}^{n}X_iY_i-\left(\sum_{i=1}^{n}X_i\right)\left(\sum_{i=1}^{n}Y_i\right) \\
n\sum_{i=1}^{n}\left(X_i-\bar{X}\right)^2 &= n\sum_{i=1}^{n}X_i^2-\left(\sum_{i=1}^{n}X_i\right)^2
\end{aligned}$$

and so

$$\hat{\beta}=\frac{n\sum_{i=1}^{n}\left(X_i-\bar{X}\right)\left(Y_i-\bar{Y}\right)}{n\sum_{i=1}^{n}\left(X_i-\bar{X}\right)^2}=\frac{\sum_{i=1}^{n}\left(X_i-\bar{X}\right)\left(Y_i-\bar{Y}\right)}{\sum_{i=1}^{n}\left(X_i-\bar{X}\right)^2}.$$

The Hessian of $S\left(\alpha,\beta\right)$ is given by

$$H=\begin{bmatrix} 2n & 2\sum_{i=1}^{n}X_i \\ 2\sum_{i=1}^{n}X_i & 2\sum_{i=1}^{n}X_i^2 \end{bmatrix}.$$

Using leading principal minors we have

$$\begin{aligned} M_1 &= 2n > 0 \\ M_2 &= \det[H] = 4\left(n\sum_{i=1}^{n} X_i^2 - \left(\sum_{i=1}^{n} X_i\right)^2\right) = 4n\sum_{i=1}^{n}\left(X_i - \bar{X}\right)^2 > 0 \end{aligned}$$

as long as not all X_i 's are not all identical since in that case there will be at least one $X_i \neq \bar{X}$ and so $\left(X_i - \bar{X}\right)^2 > 0$. Thus $\hat{\alpha}, \hat{\beta}$ is a unique global minimizer.

Problem 4.44 *[A ***] Consider the general linear regression model $Y = X\beta + e$ where rank $[X] = p$ so that $X^T X$ is a positive definite matrix. Prove with and without calculus that the value of β that minimizes the sum of squares*

$$S(\beta) = (Y - X\beta)^T (Y - X\beta)$$

is

$$\hat{\beta} = \left(X^T X\right)^{-1} X^T Y.$$

Answer: We have

$$Y - X\beta = Y - X\hat{\beta} + X\hat{\beta} - X\beta = Y - X\hat{\beta} + X\left(\hat{\beta} - \beta\right) = \hat{e} + X\left(\hat{\beta} - \beta\right)$$

where

$$\hat{e} = Y - X\hat{\beta} = Y - X\left(X^T X\right)^{-1} X^T Y = \left(I - X\left(X^T X\right)^{-1} X^T\right) Y = (I - P)Y$$

and where $I - P$ is symmetric, idempotent with

$$\begin{aligned} X^T\hat{e} &= X^T(I - P)Y = \left(X^T - X^T X\left(X^T X\right)^{-1} X^T\right) Y = \left(X^T - X^T\right) Y = 0 \\ \hat{e}^T X &= \left(X^T\hat{e}\right)^T = 0^T = 0. \end{aligned}$$

Now

$$\begin{aligned} S(\beta) &= (Y - X\beta)^T (Y - X\beta) = \left(\hat{e} + X\left(\hat{\beta} - \beta\right)\right)^T \left(\hat{e} + X\left(\hat{\beta} - \beta\right)\right) \\ &= \hat{e}^T\hat{e} + \hat{e}^T X\left(\hat{\beta} - \beta\right) + \left(\hat{\beta} - \beta\right)^T X^T\hat{e} + \left(\hat{\beta} - \beta\right)^T X^T X\left(\hat{\beta} - \beta\right) \\ &= \hat{e}^T\hat{e} + \left(\hat{\beta} - \beta\right)^T X^T X\left(\hat{\beta} - \beta\right). \end{aligned}$$

Now $\hat{e}^T\hat{e}$ does not depend on β, and since $X^T X$ is positive definite we have for $\hat{\beta} \neq \beta$ that

$$\begin{aligned} S(\beta) &= \hat{e}^T\hat{e} + \left(\hat{\beta} - \beta\right)^T X^T X\left(\hat{\beta} - \beta\right) > \hat{e}^T\hat{e} \\ S\left(\hat{\beta}\right) &= \hat{e}^T\hat{e} + \left(\hat{\beta} - \hat{\beta}\right)^T X^T X\left(\hat{\beta} - \hat{\beta}\right) = \hat{e}^T\hat{e} \end{aligned}$$

so that $S(\beta) > S\left(\hat{\beta}\right)$ for $\beta \neq \hat{\beta}$ so that $\hat{\beta}$ is a global minimizer for $S(\beta)$.

Using calculus we have the gradient

$$\nabla S(\beta) = -2X^T (Y - X\beta)$$

so that the first-order conditions $\nabla S\left(\hat{\beta}\right) = 0$ yield

$$\begin{aligned} -2X^T\left(Y - X\hat{\beta}\right) &= 0 \Longrightarrow X^TY - X^TX\hat{\beta} = 0 \Longrightarrow \left(X^TX\right)\hat{\beta} = X^TY \\ &\Longrightarrow \hat{\beta} = \left(X^TX\right)^{-1} X^TY. \end{aligned}$$

This is a global minimizer since the Hessian of $S(\beta)$ is

$$H(\beta) = 2X^TX$$

which is positive definite since X^TX is positive definite.

Problem 4.45 *[B***] Let $\hat{Y}_i = \hat{\alpha} + \hat{\beta}X_i$ be the fitted value and let $\hat{e}_i = Y_i - \hat{\alpha} - \hat{\beta}X_i = Y_i - \hat{Y}_i$ be the least squares residual. Show that*

$$\sum_{i=1}^{n} \hat{e}_i = 0,\ \bar{Y} = \hat{\alpha} + \hat{\beta}\bar{X},\ \sum_{i=1}^{n} X_i\hat{e}_i = 0, \sum_{i=1}^{n} \hat{Y}_i\hat{e}_i = 0.$$

Use this to show that

$$\sum_{i=1}^{n} \left(Y_i - \bar{Y}\right)^2 = \sum_{i=1}^{n} \left(\hat{Y}_i - \bar{Y}\right)^2 + \sum_{i=1}^{n} \hat{e}_i^2 = \hat{\beta}^2 \sum_{i=1}^{n} \left(X_i - \bar{X}\right)^2 + \sum_{i=1}^{n} \hat{e}_i^2$$

and from this that $0 \leq R^2 \leq 1$ where

$$R^2 = \frac{\sum_{i=1}^{n}\left(\hat{Y}_i - \bar{Y}\right)^2}{\sum_{i=1}^{n}\left(Y_i - \bar{Y}\right)^2}.$$

Answer: From the first-order conditions we have

$$\begin{aligned} \frac{\partial S\left(\hat{\alpha}, \hat{\beta}\right)}{\partial \alpha} &= -2\sum_{i=1}^{n} \overbrace{\left(Y_i - \hat{\alpha} - \hat{\beta}X_i\right)}^{\hat{e}_i} = 0 \\ &\Longrightarrow \sum_{i=1}^{n} \hat{e}_i = 0 \\ &\Longrightarrow n\hat{\alpha} + \left(\sum_{i=1}^{n} X_i\right)\hat{\beta} = \sum_{i=1}^{n} Y_i \\ &\Longrightarrow \bar{Y} = \hat{\alpha} + \hat{\beta}\bar{X} \\ \frac{\partial S\left(\hat{\alpha}, \hat{\beta}\right)}{\partial \beta} &= -2\sum_{i=1}^{n} X_i \overbrace{\left(Y_i - \hat{\alpha} - \hat{\beta}X_i\right)}^{\hat{e}_i} = 0 \\ &\Longrightarrow \sum_{i=1}^{n} X_i\hat{e}_i = 0. \end{aligned}$$

Therefore

$$\sum_{i=1}^{n} \hat{Y}_i \hat{e}_i = \sum_{i=1}^{n} \left(\hat{\alpha} + \hat{\beta} X_i\right) \hat{e}_i = \hat{\alpha} \sum_{i=1}^{n} \hat{e}_i + \hat{\beta} \sum_{i=1}^{n} X_i \hat{e}_i = 0.$$

Now

$$\begin{aligned} Y_i &= \hat{Y}_i + \hat{e}_i \Longrightarrow Y_i - \bar{Y} = \hat{Y}_i - \bar{Y} + \hat{e}_i \Longrightarrow \left(Y_i - \bar{Y}\right)^2 = \left(\hat{Y}_i - \bar{Y}\right)^2 + \hat{e}_i^2 + 2\left(\hat{Y}_i - \bar{Y}\right)\hat{e}_i \\ &\Longrightarrow \sum_{i=1}^{n} \left(Y_i - \bar{Y}\right)^2 = \sum_{i=1}^{n} \left(\hat{Y}_i - \bar{Y}\right)^2 + \sum_{i=1}^{n} \hat{e}_i^2 + 2\sum_{i=1}^{n} \left(\hat{Y}_i - \bar{Y}\right)\hat{e}_i. \end{aligned}$$

But

$$\sum_{i=1}^{n} \left(\hat{Y}_i - \bar{Y}\right)\hat{e}_i = \sum_{i=1}^{n} \hat{Y}_i \hat{e}_i - \bar{Y} \sum_{i=1}^{n} \hat{e}_i = 0$$

so that

$$\sum_{i=1}^{n} \left(Y_i - \bar{Y}\right)^2 = \sum_{i=1}^{n} \left(\hat{Y}_i - \bar{Y}\right)^2 + \sum_{i=1}^{n} \hat{e}_i^2.$$

Also note that since $\hat{Y}_i = \hat{\alpha} + \hat{\beta} X_i$ and $\bar{Y} = \hat{\alpha} + \hat{\beta} \bar{X}$ it follows that

$$\hat{Y}_i - \bar{Y} = \hat{\beta}\left(X_i - \bar{X}\right)$$

and so

$$\sum_{i=1}^{n} \left(\hat{Y}_i - \bar{Y}\right)^2 = \hat{\beta}^2 \sum_{i=1}^{n} \left(X_i - \bar{X}\right)^2.$$

Dividing both sides by $\sum_{i=1}^{n} \left(Y_i - \bar{Y}\right)^2$ we have

$$1 = \frac{\sum_{i=1}^{n} \left(\hat{Y}_i - \bar{Y}\right)^2 + \sum_{i=1}^{n} \hat{e}_i^2}{\sum_{i=1}^{n} \left(Y_i - \bar{Y}\right)^2} = R^2 + \frac{\sum_{i=1}^{n} \hat{e}_i^2}{\sum_{i=1}^{n} \left(Y_i - \bar{Y}\right)^2}$$

so that since $\sum_{i=1}^{n} \left(\hat{Y}_i - \bar{Y}\right)^2 \geq 0$ and $\sum_{i=1}^{n} \left(Y_i - \bar{Y}\right)^2 > 0$ we have

$$R^2 = \frac{\sum_{i=1}^{n} \left(\hat{Y}_i - \bar{Y}\right)^2}{\sum_{i=1}^{n} \left(Y_i - \bar{Y}\right)^2} \geq 0$$

and since $\sum_{i=1}^{n} \hat{e}_i^2 \geq 0$ we have

$$R^2 = 1 - \frac{\sum_{i=1}^{n} \hat{e}_i^2}{\sum_{i=1}^{n} \left(Y_i - \bar{Y}\right)^2} \leq 1.$$

Problem 4.46 *[A***] Now consider the* **weighted** *sum of squares function given by*

$$S_W(\alpha,\beta)=\sum_{i=1}^{n} w_i\left(Y_i-\alpha-\beta X_i\right)^2$$

where the weights w_i satisfy $\sum_{i=1}^{n} w_i = n$. Show that if $\hat{\beta}$ minimizes $S_W(\alpha,\beta)$, then

$$\hat{\beta}=\frac{n\sum_{i=1}^{n} w_iX_iY_i-\left(\sum_{i=1}^{n} w_iX_i\right)\left(\sum_{i=1}^{n} w_iY_i\right)}{n\sum_{i=1}^{n} w_iX_i^2-\left(\sum_{i=1}^{n} w_iX_i\right)^2}.$$

Answer: We have using the sum and chain rules that

$$\begin{aligned}\frac{\partial S(\alpha,\beta)}{\partial\alpha} &= -2\sum_{i=1}^{n} w_i\left(Y_i-\alpha-\beta X_i\right)\\ \frac{\partial S(\alpha,\beta)}{\partial\beta} &= -2\sum_{i=1}^{n} w_iX_i\left(Y_i-\alpha-\beta X_i\right)\end{aligned}$$

so that the first-order conditions for a minimum are

$$\begin{aligned}-2\sum_{i=1}^{n} w_i\left(Y_i-\hat{\alpha}-\hat{\beta}X_i\right) &= 0 \Longrightarrow \overbrace{\left(\sum_{i=1}^{n} w_i\right)}^{n}\hat{\alpha}+\left(\sum_{i=1}^{n} w_iX_i\right)\hat{\beta}=\sum_{i=1}^{n} w_iY_i\\ -2\sum_{i=1}^{n} w_iX_i\left(Y_i-\hat{\alpha}-\hat{\beta}X_i\right) &= 0 \Longrightarrow \left(\sum_{i=1}^{n} w_iX_i\right)\hat{\alpha}+\left(\sum_{i=1}^{n} w_iX_i^2\right)\hat{\beta}=\sum_{i=1}^{n} w_iX_iY_i\end{aligned}$$

or in matrix notation

$$\begin{bmatrix} n & \sum_{i=1}^{n} w_iX_i \\ \sum_{i=1}^{n} w_iX_i & \sum_{i=1}^{n} w_iX_i^2\end{bmatrix}\begin{bmatrix}\hat{\alpha}\\ \hat{\beta}\end{bmatrix}=\begin{bmatrix}\sum_{i=1}^{n} w_iY_i\\ \sum_{i=1}^{n} w_iX_iY_i\end{bmatrix}.$$

Solving for $\hat{\beta}$ using Cramer's rule we find that

$$\hat{\beta}=\frac{n\sum_{i=1}^{n} w_iX_iY_i-\left(\sum_{i=1}^{n} w_iX_i\right)\left(\sum_{i=1}^{n} w_iY_i\right)}{n\sum_{i=1}^{n} w_iX_i^2-\left(\sum_{i=1}^{n} w_iX_i\right)^2}.$$

Problem 4.47 *[A*] Let p_i $i=1,2,3,\ldots,m$ be the probability that an individual randomly chosen from the general population has an income in category i (say \$20,000/year to \$25,000/year) and where there are m income categories. Alternatively p_i is the proportion of the population in income class i. These probabilities describe the income distribution of society. You as an econometrician do not know these probabilities $p_1,p_2,\ldots,p_m$. However, you have surveyed n people taken randomly from the population. Let n_i be the number of people in your sample in income category i. Clearly $n_1+n_2+\cdots+n_m=n$. You would like to use these numbers to estimate the p_i 's and you decide to do this by choosing $\hat{p}_i$ for $i=1,2,\ldots,m$ to maximize the log likelihood given by*

$$l\left(p_1,p_2,\ldots,p_m\right)=\sum_{i=1}^{m} n_i\ln\left(p_i\right).$$

Of course you realize that probabilities sum to 1 so that the following constraint holds

$$p_1 + p_2 + \cdots + p_m = 1.$$

What is the Lagrangian for this constrained maximization problem? From the first-order conditions show that the maximum likelihood estimates are given by $\hat{p}_i = \frac{n_i}{n}$. *Why is this sensible? If* $l^*(n_1, n_2, \ldots, n_m) = l(\hat{p}_1, \hat{p}_2, \ldots, \hat{p}_m)$, *what is* $\frac{\partial l^*}{\partial n_i}$ *equal to?*

Answer: We have

$$\mathcal{L}(\lambda, p_1, p_2, \ldots, p_m) = \sum_{i=1}^{m} n_i \ln(p_i) + \lambda\left(1 - \sum_{i=1}^{m} p_i\right)$$

so that the first-order conditions yield

$$\begin{aligned} 1 - \sum_{i=1}^{m} \hat{p}_i &= 0 \Longrightarrow 1 = \sum_{i=1}^{m} \hat{p}_i \\ \frac{n_i}{\hat{p}_i} - \hat{\lambda} &= 0 \Longrightarrow n_i = \hat{\lambda}\hat{p}_i \text{ for } i = 1, 2, \ldots, n. \end{aligned}$$

It follows then that

$$n_i = \hat{\lambda}\hat{p}_i \Longrightarrow n = \sum_{i=1}^{m} n_i = \sum_{i=1}^{m} \hat{\lambda}\hat{p}_i = \hat{\lambda}\sum_{i=1}^{m} \hat{p}_i = \hat{\lambda}$$

so that $\hat{\lambda} = n$ and $\hat{p}_i = \frac{n_i}{\hat{\lambda}} = \frac{n_i}{n}$.

From the envelope theorem we have

$$\begin{aligned} \frac{\partial l^*}{\partial n_i} &= \frac{\partial}{\partial n_i}\mathcal{L}(\lambda, p_1, p_2, \ldots, p_m)\,|_{p_i=\hat{p}_i,\lambda=\hat{\lambda}} = \frac{\partial}{\partial n_i}\left(\sum_{i=1}^{m} n_i \ln(p_i) + \lambda\left(1 - \sum_{i=1}^{m} p_i\right)\right)|_{p_i=\hat{p}_i,\lambda=\hat{\lambda}} \\ &= \ln(p_i)\,|_{p_i=\hat{p}_i,\lambda=\hat{\lambda}} = \ln(\hat{p}_i). \end{aligned}$$

Problem 4.48 *[A *]An entropy measure of the inequality of income is given by*

$$E(p_1, p_2, \ldots, p_m) = -\sum_{i=1}^{m} p_i \ln(p_i)$$

where the greater is E the greater is the income inequality. Show that the minimum of this measure of income inequality occurs when $p_i = \frac{1}{m}$ *(remember the constraint that probabilities sum to one!) Show that the estimated income inequality satisfies*

$$E(\hat{p}_1, \hat{p}_2, \ldots, \hat{p}_m) = -\frac{1}{n} \times l(\hat{p}_1, \hat{p}_2, \ldots, \hat{p}_m).$$

Answer: The Lagrangian is

$$\mathcal{L}(\lambda, p_1, p_2, \ldots, p_m) = -\sum_{i=1}^{m} p_i \ln(p_i) + \lambda\left(1 - \sum_{i=1}^{m} p_i\right).$$

The first-order conditions yield

$$\begin{aligned} 1 - \sum_{i=1}^{m} \hat{p}_i &= 0 \Longrightarrow 1 = \sum_{i=1}^{m} \hat{p}_i \\ 1 + \ln(\hat{p}_i) - \hat{\lambda} &= 0 \Longrightarrow \hat{p}_i = e^{\hat{\lambda}-1} \text{ for } i = 1, 2, \ldots, m. \end{aligned}$$

We then have

$$1 = \sum_{i=1}^{m} \hat{p}_i \Longrightarrow \sum_{i=1}^{m} e^{\hat{\lambda}-1} = 1 \Longrightarrow m e^{\hat{\lambda}-1} = 1 \Longrightarrow \hat{\lambda} = 1 - \ln(m)$$

and so

$$\hat{p}_i = e^{\hat{\lambda}-1} = e^{-\ln(m)} = \frac{1}{m} \text{for } i = 1, 2, \ldots, m.$$

The objective function is convex $-\sum_{i=1}^{m} p_i \ln(p_i)$ since the Hessian is

$$H(p_1, p_2, \ldots, p_m) = \begin{bmatrix} \frac{1}{p_1} & 0 & \cdots & 0 \\ 0 & \frac{1}{p_2} & \ddots & \vdots \\ \vdots & \ddots & \ddots & 0 \\ 0 & \cdots & 0 & \frac{1}{p_m} \end{bmatrix}$$

and so is diagonal with positive elements along the diagonal and hence is positive definite. Since the constraint is linear it then follows that the solution to the first-order conditions is a global minimizer.

Problem 4.49 *[D***] If $n = 10$ and*

$$\sum_{i=1}^{10} X_i = 60\ , \sum_{i=1}^{10} X_i^2 = 520, \sum_{i=1}^{10} X_i Y_i = 360, \sum_{i=1}^{10} Y_i = 120, \sum_{i=1}^{10} Y_i^2 = 1690$$

then what is $\bar{X}, \bar{Y}, \sum_{i=1}^{10}(4X_i + 3Y_i)$ and $\sum_{i=1}^{10}(4X_i + 3Y_i)^2$. If $\hat{\sigma}_X^2 = \frac{1}{10}\sum_{i=1}^{10}(X_i - \bar{X})^2$ then what is $\hat{\sigma}_X^2$? If

$$\hat{\beta} = \frac{\sum_{i=1}^{10}(X_i - \bar{X})(Y_i - \bar{Y})}{\sum_{i=1}^{10}(X_i - \bar{X})^2}$$

then what is $\hat{\beta}$? Find the value of b which minimizes $S(b) = \sum_{i=1}^{10}(Y_i - bX_i)^2$.

Answer: We have

$$\begin{aligned}
\bar{X} &= \frac{1}{10}\sum_{i=1}^{10} X_i = 6,\ \bar{Y} = \frac{1}{10}\sum_{i=1}^{10} Y_i = 12 \\
\sum_{i=1}^{10}(4X_i + 3Y_i) &= 4\sum_{i=1}^{10} X_i + 3\sum_{i=1}^{10} Y_i = 4\times 60 + 3\times 120 = 600 \\
\sum_{i=1}^{10}(4X_i + 3Y_i)^2 &= \sum_{i=1}^{10}\left(16X_i^2 + 24X_iY_i + 9Y_i^2\right) = 16\sum_{i=1}^{10} X_i^2 + 24\sum_{i=1}^{10} X_iY_i + 9\sum_{i=1}^{10} Y_i^2 \\
&= 16\times 520 + 24\times 360 + 9\times 1690 = 32170
\end{aligned}$$

and

$$\hat{\sigma}_X^2 = \frac{1}{10}\sum_{i=1}^{10}\left(X_i - \bar{X}\right)^2 = \frac{1}{10}\left(\sum_{i=1}^{10} X_i^2 - 10\bar{X}^2\right) = \frac{1}{10}\sum_{i=1}^{10} X_i^2 - \bar{X}^2 = \frac{1}{10}520 - 6^2 = 16.$$

For $\hat{\beta}$ we have

$$\begin{aligned}
\hat{\beta} &= \frac{\sum_{i=1}^{10}\left(X_i - \bar{X}\right)\left(Y_i - \bar{Y}\right)}{\sum_{i=1}^{10}\left(X_i - \bar{X}\right)^2} = \frac{\sum_{i=1}^{10} X_iY_i - 10\times\bar{X}\bar{Y}}{\sum_{i=1}^{10} X_i^2 - 10\times\bar{X}^2} \\
&= \frac{360 - 10\times 6\times 12}{520 - 10\times 6^2} = -\frac{9}{4}.
\end{aligned}$$

For $S(b)$ we have

$$\begin{aligned}
S(b) &= \sum_{i=1}^{10}\left(Y_i - bX_i\right)^2 = \sum_{i=1}^{10}\left(Y_i^2 - 2bX_iY_i + b^2X_i^2\right) \\
&= \sum_{i=1}^{10} Y_i^2 - 2b\sum_{i=1}^{10} X_iY_i + b^2\sum_{i=1}^{10} X_i^2 \\
&= 1690 - 720b + 520b^2
\end{aligned}$$

so that

$$S'(b^*) = 0 \Longrightarrow -720 + 1040b^* = 0 \Longrightarrow b^* = \frac{720}{1040} = \frac{9}{13}$$

with $S''(b) = 1040$ so that $b^* = \frac{9}{13}$ is a global minimizer.

Problem 4.50 *If* $n = 10$,

$$\sum_{i=1}^{10} X_i = 500,\ \sum_{i=1}^{10} X_i^2 = 27250,\ \sum_{i=1}^{10} X_iY_i = 30500,\ \sum_{i=1}^{10} Y_i = 700 \text{ and } \sum_{i=1}^{10} Y_i^2 = 55000$$

then calculate $\bar{X}$, $\hat{\sigma}^2 = \frac{1}{10}\sum_{i=1}^{10}\left(X_i - \bar{X}\right)^2, \sum_{i=1}^{10}\left(2X_i - 3Y_i\right)^2$ *and*

$$\hat{\beta} = \frac{\sum_{i=1}^{10}\left(X_i - \bar{X}\right)\left(Y_i - \bar{Y}\right)}{\sum_{i=1}^{10}\left(X_i - \bar{X}\right)^2}.$$

Answer: We have

$$\begin{aligned}
\bar{X} &= \frac{1}{10} \times 500 = 50,\ \hat{\sigma}^2 = \frac{1}{10} 27250 - 50^2 = 225 \\
\sum_{i=1}^{10} (2X_i - 3Y_i)^2 &= \sum_{i=1}^{10} (4X_i^2 - 12X_iY_i + 9Y_i^2) = 4\sum_{i=1}^{10} X_i^2 - 12\sum_{i=1}^{10} X_iY_i + 9\sum_{i=1}^{10} Y_i^2 \\
&= 4 \times 27250 - 12 \times 30500 + 9 \times 55000 = 238000 \\
\hat{\beta} &= \frac{\sum_{i=1}^n X_iY_i - n\bar{X}\bar{Y}}{\sum_{i=1}^n X_i^2 - n\bar{X}^2} = \frac{30500 - 10 \times 50 \times 70}{27250 - 10 \times 50^2} = -2.
\end{aligned}$$

Chapter 5

Difference Equations

Problem 5.1 *[D***] Consider the difference equation*

$$Y_t = 100 + 1.3Y_{t-1} - 0.4Y_{t-2} \text{ with } Y_0 = Y_1 = 100.$$

*Calculate Y_t for $t = 2, 3, 4, 5, 6$. What is the equilibrium value Y^*? If the solution is written as $Y_t = Y^* + A_1 r_1^t + A_2 r_2^t$, calculate r_1 and r_2. Use r_1 and r_2 to determine if this difference equation is stable or unstable. [C**] Calculate A_1 and A_2.*

Answer: The equilibrium value Y^* is determined by

$$Y^* = 100 + 1.3Y^* - 0.4Y^* \Longrightarrow Y^* = 1000.$$

The characteristic polynomial is

$$r^2 - 1.3r + 0.4 = 0$$

with roots

$$r_1 = 0.5, r_2 = 0.8.$$

The difference equation is stable since

$$|r_1| = 0.5 < 1, \ |r_2| = 0.8 < 1.$$

As well

$$\begin{aligned} Y_2 &= 100 + 1.3 \times 100 - 0.4 \times 100 = 190 \\ Y_3 &= 100 + 1.3 \times 190 - 0.4 \times 100 = 307 \\ Y_4 &= 100 + 1.3 \times 307 - 0.4 \times 190 = 423.1 \\ Y_5 &= 100 + 1.3 \times 423.1 - 0.4 \times 307 = 527.23 \\ Y_6 &= 100 + 1.3 \times 527.23 - 0.4 \times 423.1 = 616.16. \end{aligned}$$

To calculate A_1 and A_2 we have

$$\begin{aligned} Y_0 &= 100 = 1000 + A_1 + A_2 \Longrightarrow A_1 + A_2 = -900 \\ Y_1 &= 100 = 1000 + A_1 r_1 + A_2 r_2 \Longrightarrow r_1 A_1 + r_2 A_2 = -900 \end{aligned}$$

or in matrix notation

$$\begin{bmatrix} 1 & 1 \\ 0.5 & 0.8 \end{bmatrix} \begin{bmatrix} A_1 \\ A_2 \end{bmatrix} = \begin{bmatrix} -900 \\ -900 \end{bmatrix}$$

so that

$$\begin{bmatrix} A_1 \\ A_2 \end{bmatrix} = \begin{bmatrix} 1 & 1 \\ 0.5 & 0.8 \end{bmatrix}^{-1} \begin{bmatrix} -900 \\ -900 \end{bmatrix} = \begin{bmatrix} 600 \\ -1500 \end{bmatrix}$$

so the solution is

$$Y_t = 1000 + 600 \times 0.5^t - 1500 \times 0.8^t$$

which is shown below.

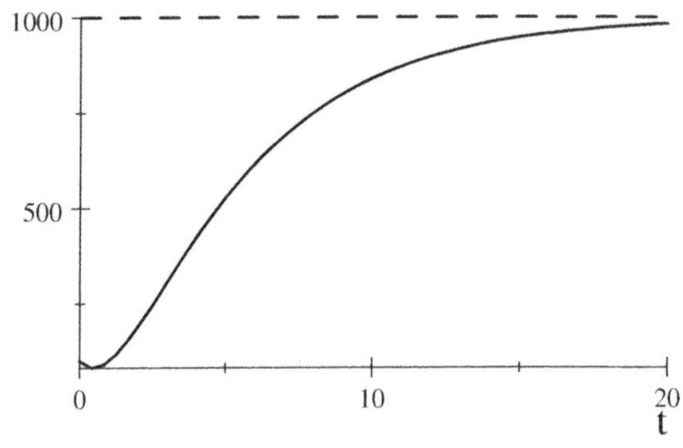

Problem 5.2 *[A**] A second-order difference equation*

$$Y_t = a_0 + a_1 Y_{t-1} + a_2 Y_{t-2}$$

can always be rewritten as a first-order vector difference equation as

$$\begin{bmatrix} Y_t \\ Y_{t-1} \end{bmatrix} = \begin{bmatrix} a_1 & a_2 \\ 1 & 0 \end{bmatrix} \begin{bmatrix} Y_{t-1} \\ Y_{t-2} \end{bmatrix} + \begin{bmatrix} a_0 \\ 0 \end{bmatrix}$$

or as $X_t = a + AX_{t-1}$ where

$$X_t = \begin{bmatrix} Y_t \\ Y_{t-1} \end{bmatrix}, A = \begin{bmatrix} a_1 & a_2 \\ 1 & 0 \end{bmatrix}, a = \begin{bmatrix} a_0 \\ 0 \end{bmatrix}.$$

(This turns out to be a very useful idea. For example the Kalman filter, used in econometrics and also to put a man on the moon, is based on this idea!) Show that the equilibrium value of X_t is

$$X^* = (I - A)^{-1} a$$

and that

$$X_t = X^* + A^t (X_0 - X^*) .$$

Show that if the eigenvalues of A are less than 1 in absolute value then X_t is stable. Show that the eigenvalues of A are the roots of the characteristic polynomial of Y_t.

Answer: If $X_t = X_{t-1} = X^*$ we have

$$X_t = a+AX_{t-1} \Longrightarrow X^* = a+AX^* \Longrightarrow X^*-AX^* = a \Longrightarrow (I - A) X^* = a \Longrightarrow X^* = (I - A)^{-1} a.$$

Thus if $\tilde{X}_t = X_t - X^*$ we have

$$X_t = a + AX_{t-1},\ X^* = a + AX^* \Longrightarrow \tilde{X}_t = A\tilde{X}_{t-1}.$$

Thus

$$\tilde{X}_t = A\tilde{X}_{t-1} \Longrightarrow \tilde{X}_t = AA\tilde{X}_{t-2} = A^2\tilde{X}_{t-2} \Longrightarrow \cdots \Longrightarrow \tilde{X}_t = A^t\tilde{X}_0$$

or using $\tilde{X}_t = X_t - X^*$ we have

$$X_t = X^* + A^t (X_0 - X^*) .$$

Since $A = C\Lambda C^{-1}$ where Λ is a diagonal matrix with the eigenvalues along the diagonal so that $A^t = C\Lambda^t C^{-1}$ and hence $A^t \to 0$ *iff* $\Lambda^t \to 0$ which requires all eigenvalues be less than 1 in absolute value. The eigenvalues of A are given by the roots of

$$\det\left[\begin{bmatrix} a_1 & a_2 \\ 1 & 0 \end{bmatrix} - \lambda\begin{bmatrix} 1 & 0 \\ 0 & 1 \end{bmatrix}\right] = \det\begin{bmatrix} a_1 - \lambda & a_2 \\ 1 & -\lambda \end{bmatrix} = \lambda^2 - a_1\lambda - a_2$$

which is the characteristic polynomial of Y_t.

Problem 5.3 *[B**] Consider the first-order linear difference equation $Y_t = \alpha + Y_{t-1}$. Show that the equilibrium value Y^* does not exist for $\alpha \neq 0$. Show that Y_t is unstable and follows a linear trend for $\alpha \neq 0$. How does Y_t behave when $\alpha = 0$? Now consider the second-order difference equation*

$$Y_t = \alpha + \phi Y_{t-1} + (1 - \phi) Y_{t-2}.$$

Show that the equilibrium value Y^* does not exist for $\alpha \neq 0$. Show that Y_t is unstable and follows a linear trend for $\alpha \neq 2$.

Answer: Suppose an equilibrium value Y^* exists for $\alpha \neq 0$. Then we have a contradiction to $\alpha \neq 0$ as

$$Y^* = \alpha + Y^* \Longrightarrow \alpha = 0.$$

Define $\Delta Y_t \equiv Y_t - Y_{t-1}$. Then we have

$$Y_t = \alpha + Y_{t-1} \Longrightarrow Y_t - Y_{t-1} = \alpha \Longrightarrow \Delta Y_t = \alpha$$

and so

$$\Delta Y_t + \Delta Y_{t-1} + \cdots + \Delta Y_1 = \alpha + \alpha + \cdots + \alpha = \alpha t.$$

But

$$\begin{aligned} \Delta Y_t + \Delta Y_{t-1} + \cdots + \Delta Y_1 &= (Y_t - Y_{t-1}) + (Y_{t-1} - Y_{t-2}) + \cdots + (Y_2 - Y_1) + (Y_1 - Y_0) \\ &= Y_t - Y_0 \end{aligned}$$

(this is like the fundamental theorem of integral calculus!) and so

$$Y_t - Y_0 = \alpha t \Longrightarrow Y_t = Y_0 + \alpha t.$$

If $\alpha = 0$ the $Y_t = Y_0$ for all t so that effectively there are an infinite number of equilibrium values for Y_t.

For

$$Y_t = \alpha + \phi Y_{t-1} + (1 - \phi) Y_{t-2}$$

suppose an equilibrium value Y^* exists for $\alpha \neq 0$. Then we have

$$Y^* = \alpha + \phi Y^* + (1 - \phi) Y^* \Longrightarrow \alpha = 0$$

which contradicts $\alpha \neq 0$.

Subtracting Y_{t-1} from both sides we have

$$\begin{aligned} Y_t &= \alpha + \phi Y_{t-1} + (1 - \phi) Y_{t-2} \\ &\Longrightarrow Y_t - Y_{t-1} = \alpha + (\phi - 1) Y_{t-1} + (1 - \phi) Y_{t-2} \\ &\Longrightarrow \Delta Y_t = \alpha + (\phi - 1) Y_{t-1} - (\phi - 1) Y_{t-2} \\ &\Longrightarrow \Delta Y_t = \alpha + (\phi - 1)(Y_{t-1} - Y_{t-2}) \\ &\Longrightarrow \Delta Y_t = \alpha + (\phi - 1) \Delta Y_{t-1} \end{aligned}$$

or

$$\Delta Y_t = \alpha + (\phi - 1) \Delta Y_{t-1}.$$

Thus ΔY_t follows a first-order linear difference equation. As long as the coefficient on ΔY_{t-1} is not 1, or

$$\phi - 1 \neq 1 \Longrightarrow \phi \neq 2$$

the equilibrium value for ΔY_t exists as

$$\Delta Y^* = \alpha + (\phi - 1) \Delta Y^* \Longrightarrow \Delta Y^* = \frac{\alpha}{1 - (\phi - 1)} = \frac{\alpha}{2 - \phi}$$

with solution

$$\Delta Y_t = \Delta Y^* + (\phi - 1)^t (\Delta Y_0 - \Delta Y^*).$$

Now to solve for Y_t use

$$\sum_{\tau=1}^{t} \Delta Y_\tau = \Delta Y_t + \Delta Y_{t-1} + \cdots + \Delta Y_1 = Y_t - Y_0$$

as

$$Y_t - Y_0 = \sum_{\tau=1}^{t} \Delta Y_\tau = \sum_{\tau=1}^{t} (\Delta Y^* + (\phi - 1)^\tau (\Delta Y_0 - \Delta Y^*)) = \Delta Y^* \times t + (\Delta Y_0 - \Delta Y^*) \sum_{\tau=1}^{t} (\phi - 1)^\tau .$$

Now from the geometric series

$$\sum_{\tau=1}^{t} \lambda^\tau = \lambda \sum_{\tau=0}^{t-1} \lambda^\tau = \lambda \left(1 + \lambda + \lambda^2 + \cdots \lambda^{t-1}\right) = \lambda \frac{1 - \lambda^t}{1 - \lambda}$$

with $\lambda = \phi - 1$ we have

$$\sum_{\tau=1}^{t} (\phi - 1)^\tau = (\phi - 1) \frac{1 - (\phi - 1)^t}{1 - (\phi - 1)} = (\phi - 1) \frac{1 - (\phi - 1)^t}{2 - \phi}$$

so that

$$Y_t = Y_0 + \Delta Y^* \times t + (\Delta Y_0 - \Delta Y^*) \sum_{\tau=1}^{t} (\phi - 1)^\tau = Y_0 + \frac{\alpha}{2 - \phi} \times t + \frac{\left(\Delta Y_0 - \frac{\alpha}{2-\phi}\right)(\phi - 1)}{2 - \phi} \left(1 - (\phi - 1)^t\right).$$

Problem 5.4 *[A**] Consider the difference equation*

$$Y_t = 100 + 1.3 Y_{t-1} - 0.4 Y_{t-2}$$

where $Y_0 = Y_1 = 100$. Write this in the form $X_t = a + A X_{t-1}$ considered in the previous question and show that Y_t is stable by calculating the eigenvalues of A. Write down the solution $X_t = X^* + A^t (X_0 - X^*)$ for this model.

Answer: Here

$$X_t = \begin{bmatrix} Y_t \\ Y_{t-1} \end{bmatrix}, \ A = \begin{bmatrix} 1.3 & -0.4 \\ 1 & 0 \end{bmatrix}, \ a = \begin{bmatrix} 100 \\ 0 \end{bmatrix}.$$

The eigenvalues of A are the roots of

$$\det \left[\begin{bmatrix} 1.3 & -0.4 \\ 1 & 0 \end{bmatrix} - \lambda \begin{bmatrix} 1 & 0 \\ 0 & 1 \end{bmatrix} \right] = \det \begin{bmatrix} a_1 - \lambda & a_2 \\ 1 & -\lambda \end{bmatrix} = \lambda^2 - 1.3\lambda + 0.4 = 0$$

and so

$$\lambda_1 = 0.8, \lambda_2 = 0.5.$$

Note that these are identical to the roots of the characteristic polynomial (which is also the same) of

$$r^2 - 1.3r + 0.4 = 0.$$

Since both eigenvalues are less than 1 in absolute value Y_t is stable. Alternatively $A^t = C\Lambda^t C^{-1}$ as (the computer calculates this)

$$A^t = \begin{bmatrix} 2.6667 & -1.6667 \\ 3.3333 & -3.3333 \end{bmatrix} \begin{bmatrix} 0.8^t & 0 \\ 0 & 0.5^t \end{bmatrix} \begin{bmatrix} 1.0 & -0.5 \\ 1.0 & -0.8 \end{bmatrix} \to \begin{bmatrix} 0 & 0 \\ 0 & 0 \end{bmatrix}$$

since

$$\Lambda^t = \begin{bmatrix} 0.8^t & 0 \\ 0 & 0.5^t \end{bmatrix} \to \begin{bmatrix} 0 & 0 \\ 0 & 0 \end{bmatrix}.$$

The equilibrium value $X^* = (I - A)^{-1} a$ is

$$X^* = \begin{bmatrix} Y^* \\ Y^* \end{bmatrix} = \left(\begin{bmatrix} 1 & 0 \\ 0 & 1 \end{bmatrix} - \begin{bmatrix} 1.3 & -0.4 \\ 1 & 0 \end{bmatrix} \right)^{-1} \begin{bmatrix} 100 \\ 0 \end{bmatrix} = \begin{bmatrix} 1000 \\ 1000 \end{bmatrix}$$

and so the equilibrium value is $Y^* = 1000$, as we calculated earlier.

The starting values are $Y_0 = Y_1 = 100$ or

$$X_0 = \begin{bmatrix} 100 \\ 100 \end{bmatrix}$$

so that $X_t = X^* + A^t (X_0 - X^*)$ becomes

$$\begin{bmatrix} Y_t \\ Y_{t-1} \end{bmatrix} = \begin{bmatrix} 1000 \\ 1000 \end{bmatrix} + \begin{bmatrix} 1.3 & -0.4 \\ 1 & 0 \end{bmatrix}^t \left(\begin{bmatrix} 1000 \\ 1000 \end{bmatrix} - \begin{bmatrix} 100 \\ 100 \end{bmatrix} \right)$$

or

$$\begin{bmatrix} Y_t \\ Y_{t-1} \end{bmatrix} = \begin{bmatrix} 1000 \\ 1000 \end{bmatrix} + \begin{bmatrix} 1.3 & -0.4 \\ 1 & 0 \end{bmatrix}^t \begin{bmatrix} 900 \\ 900 \end{bmatrix}$$

or replacing A^t by $C\Lambda^t C^{-1}$ given above

Problem 5.5 *[D***] Consider the difference equation*

$$Y_t = 100 + 0.3Y_{t-1} + 0.4Y_{t-2} \text{ with } Y_0 = Y_1 = 100.$$

Calculate Y_t for $t = 2, 3, 4, 5, 6$. What is the equilibrium value Y^*? If the solution is written as $Y_t = Y^* + A_1 r_1^t + A_2 r_2^t$, calculate r_1 and r_2. Use r_1 and r_2 to determine if this difference equation is stable or unstable.

Answer: The equilibrium value Y^* is determined by

$$Y^* = 100 + 0.3Y^* + 0.4Y^* \Longrightarrow Y^* = \frac{100}{0.3} \approx 333.33.$$

The characteristic polynomial is

$$r^2 - 0.3r - 0.4 = 0 \Longrightarrow r_1 = -0.5, r_2 = 0.8.$$

The difference equation is stable since

$$|r_1| = 0.5 < 1, \ |r_2| = 0.8 < 1.$$

As well

$$\begin{aligned}
Y_2 &= 100 + 0.3 \times 100 + 0.4 \times 100 = 170, Y_3 = 100 + 0.3 \times 170 + 0.4 \times 100 = 191 \\
Y_4 &= 100 + 0.3 \times 191 + 0.4 \times 170 = 225.3, Y_5 = 100 + 0.3 \times 225.3 + 0.4 \times 191 = 243.99 \\
Y_6 &= 100 + 0.3 \times 243.99 + 0.4 \times 225.3 = 263.32.
\end{aligned}$$

To calculate A_1 and A_2 we have

$$\begin{aligned}
Y_0 &= 100 = 333.33 + A_1 + A_2 \Longrightarrow A_1 + A_2 = -233.33 \\
Y_1 &= 100 = 333.33 + A_1 r_1 + A_2 r_2 \Longrightarrow r_1 A_1 + r_2 A_2 = -233.33
\end{aligned}$$

or in matrix notation

$$\begin{bmatrix} 1 & 1 \\ -0.5 & 0.8 \end{bmatrix} \begin{bmatrix} A_1 \\ A_2 \end{bmatrix} = \begin{bmatrix} -233.33 \\ -233.33 \end{bmatrix}$$

so that

$$\begin{bmatrix} A_1 \\ A_2 \end{bmatrix} = \begin{bmatrix} 1 & 1 \\ -0.5 & 0.8 \end{bmatrix}^{-1} \begin{bmatrix} -233.33 \\ -233.33 \end{bmatrix} = \begin{bmatrix} 35.897 \\ -269.23 \end{bmatrix}$$

so the solution is

$$Y_t = 333.33 + 35.897 \times (-0.5)^t - 269.23 \times (0.8)^t$$

which is shown below.

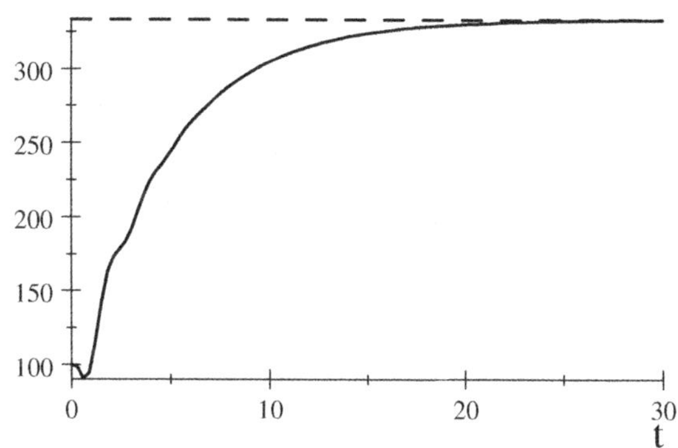

Problem 5.6 *[D***] Consider the second-order difference equation*

$$Y_t = 50 + 1.2 Y_{t-1} - 0.3 Y_{t-2}, \text{ with } Y_1 = Y_o = 400.$$

Calculate Y_t for $t = 2, 3, 4$. What is the equilibrium value Y^*? If the solution is written as $Y_t = Y^* + A_1 r_1^t + A_2 r_2^t$, calculate r_1 and r_2. Use r_1 and r_2 to determine if this difference equation is stable or unstable.

Answer: The equilibrium value Y^* is then

$$Y^* = \frac{50}{1-(1.2)-(-0.3)} = 500.$$

The characteristic polynomial has roots as

$$r^2 - 1.2r + 0.3 = 0 \Longrightarrow r_1, r_2 = \frac{-(-1.2) \pm \sqrt{(-1.2)^2 - 4 \times 0.3}}{2} \Longrightarrow r_1 = 0.355, r_2 = 0.845.$$

Since $\mid r_1 \mid < 1$ and $\mid r_2 \mid < 1$ this difference equation is stable; that is no matter what the starting values Y_t will converge to the equilibrium value $Y^* = 500$.

Problem 5.7 *[D***] Consider the second-order difference equation*

$$Y_t = -50 + 0.6Y_{t-1} + 0.5Y_{t-2} \text{ with } Y_1 = 450, Y_o = 400.$$

What is the equilibrium value Y^*? If the solution is written as $Y_t = Y^* + A_1 r_1^t + A_2 r_2^t$, calculate r_1 and r_2. Use r_1 and r_2 to determine if this difference equation is stable or unstable.

Answer: The equilibrium value Y^* is then

$$Y^* = \frac{-50}{1-(0.6)-(0.5)} = 500.$$

The characteristic polynomial has roots as

$$r^2 - 0.6r - 0.5 = 0 \Longrightarrow r_1 = 1.07, r_2 = -0.47.$$

Since $r_1 = 1.07 > 1$ it follows that the difference equation is unstable.

Problem 5.8 *[D***] Consider the difference equation*

$$Y_t = 10 + 1.2Y_{t-1} - 0.3Y_{t-2} \text{ with } Y_0 = Y_1 = 50.$$

Calculate Y_t for $t = 2, 3, 4$. By calculating the appropriate roots, determine if Y_t is stable or unstable. If Y_t is written as

$$Y_t = Y^* + A_1 r_1^t + A_2 r_2^t$$

what are the values of Y^*, A_1, A_2, r_1, r_2?

Answer: We have

$$\begin{aligned}
Y_2 &= 10 + 1.2 \times 50 - 0.3 \times 50 = 55 \\
Y_3 &= 10 + 1.2 \times 55 - 0.3 \times 50 = 61 \\
Y_4 &= 10 + 1.2 \times 61 - 0.3 \times 55 = 66.7.
\end{aligned}$$

The characteristic polynomial has roots as

$$r^2 - 1.2r + 0.3 = 0 \Longrightarrow r_1 = 0.35505,\ r_2 = 0.84495.$$

Since $|r_1| < 1$ and $|r_1| < 1$ it follows that Y_t is stable. The equilibrium value Y^* is found as

$$Y^* = 10 + 1.2Y^* - 0.3Y^* \Longrightarrow Y^* = 100.$$

To find A_1 and A_2 we have

$$\begin{aligned} 50 &= 100 + A_1 (0.35505)^0 + A_2 (0.84495)^0 \\ 50 &= 100 + A_1 (0.35505)^1 + A_2 (0.84495)^1 \end{aligned}$$

or in matrix notation

$$\begin{bmatrix} 1 & 1 \\ 0.35505 & 0.84495 \end{bmatrix} \begin{bmatrix} A_1 \\ A_2 \end{bmatrix} = \begin{bmatrix} -50 \\ -50 \end{bmatrix}$$

or $A_1 = 15.825$, $A_2 = -65.825$ so that

$$Y_t = 100 + 15.825 (0.35505)^t - 65.825 (0.84495)^t .$$

This is shown in the figure below.

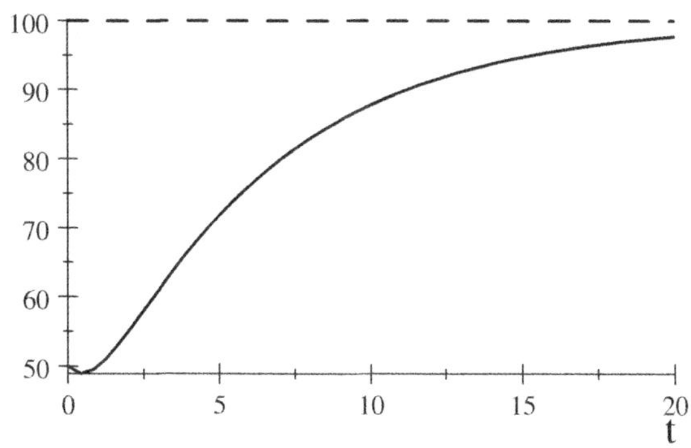

Problem 5.9 *[D***] Consider the difference equation*

$$Y_t = 10 + 1.2Y_{t-1} - 0.3Y_{t-2} \textit{ with } Y_0 = Y_1 = 50.$$

Calculate Y_t for $t = 2, 3, 4$. By calculating the appropriate roots, determine if Y_t is stable or unstable. If Y_t is written as

$$Y_t = Y^* + A_1 r_1^t + A_2 r_2^t$$

what are the values of Y^*, A_1, A_2, r_1, r_2?

Answer: We have

$$\begin{aligned} Y_2 &= 10 + 1.2 \times 50 - 0.3 \times 50 = 55 \\ Y_3 &= 10 + 1.2 \times 55 - 0.3 \times 50 = 61 \\ Y_4 &= 10 + 1.2 \times 61 - 0.3 \times 55 = 66.7. \end{aligned}$$

The characteristic polynomial has roots as

$$r^2 - 1.2r + 0.3 = 0 \Longrightarrow r_1 = 0.35505, \ r_2 = 0.84495.$$

Since $|r_1| < 1$ and $|r_1| < 1$ it follows that Y_t is stable. The equilibrium value Y^* satisfies

$$Y^* = 10 + 1.2Y^* - 0.3Y^* \Longrightarrow Y^* = 100.$$

To find A_1 and A_2 we have

$$\begin{aligned} 50 &= 100 + A_1 (0.35505)^0 + A_2 (0.84495)^0 \\ 50 &= 100 + A_1 (0.35505)^1 + A_2 (0.84495)^1 \end{aligned}$$

or in matrix notation

$$\begin{bmatrix} 1 & 1 \\ 0.35505 & 0.84495 \end{bmatrix} \begin{bmatrix} A_1 \\ A_2 \end{bmatrix} = \begin{bmatrix} -50 \\ -50 \end{bmatrix}$$

or $A_1 = 15.825$, $A_2 = -65.825$ so that

$$Y_t = 100 + 15.825 (0.35505)^t - 65.825 (0.84495)^t .$$

Chapter 6

Differential Equations

Problem 6.1 *[D***] Consider the differential equation*

$$Y''(t) + 5Y'(t) + 4Y(t) = 36.$$

Calculate the equilibrium value Y^*. *If* $Y(t) = Y^* + A_1 e^{r_1 t} + A_2 e^{r_2 t}$, *calculate* r_1 *and* r_2 *and from these determine whether or not this differential equation is stable. [B**] If* $A_1 = A_2 = 1$ *calculate*

$$\int_0^\infty (Y(t) - Y^*)^2 \, dt.$$

*[B***] If* $Y(0) = 20$ *and* $Y'(0) = -1$ *find* A_1 *and* A_2 *and write down the solution* $Y(t)$.

Answer: The equilibrium value Y^* is given by $Y^* = \frac{36}{4} = 9$. The characteristic polynomial has roots as

$$r^2 + 5r + 4 = 0 \Longrightarrow r_1 = -4, r_2 = -1.$$

Since $r_1 < 0$ and $r_2 < 0$ the differential equation is stable.

We have

$$\begin{aligned}\int_0^\infty (Y(t) - Y^*)^2 \, dt &= \int_0^\infty \left(e^{-t} + e^{-4t}\right)^2 dt = \int_0^\infty \left(e^{-2t} + 2e^{-5t} + e^{-8t}\right) dt \\ &= \int_0^\infty e^{-2t} dt + 2\int_0^\infty e^{-5t} dt + \int_0^\infty e^{-8t} dt = \frac{1}{2} + 2 \times \frac{1}{5} + \frac{1}{8} = \frac{41}{40}.\end{aligned}$$

We have

$$20 = 9 + A_1 + A_2 \text{ and } -1 = -1 \times A_1 + -4 \times A_2$$

so that in matrix notation

$$\begin{bmatrix} 1 & 1 \\ -1 & -4 \end{bmatrix} \begin{bmatrix} A_1 \\ A_2 \end{bmatrix} = \begin{bmatrix} 11 \\ -1 \end{bmatrix}$$

and

$$\begin{bmatrix} A_1 \\ A_2 \end{bmatrix} = \begin{bmatrix} 1 & 1 \\ -1 & -4 \end{bmatrix}^{-1} \begin{bmatrix} 11 \\ -1 \end{bmatrix} = \begin{bmatrix} \frac{43}{3} \\ -\frac{10}{3} \end{bmatrix}.$$

The solution is

$$Y(t) = 9 + \frac{43}{3}e^{-t} - \frac{10}{3}e^{-4t}$$

which is shown below.

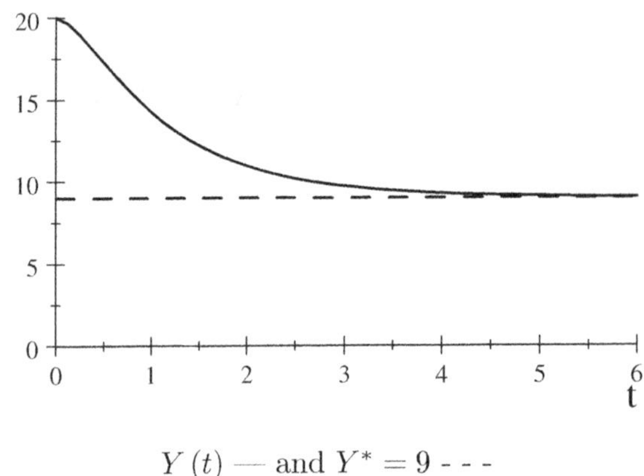

$Y(t)$ — and $Y^* = 9$ - - -

Problem 6.2 *[D***] Consider the differential equation*

$$Y''(t) + 3Y'(t) + Y(t) = 4 \text{ with } Y(0) = 2, Y'(0) = 1$$

*If $Y(t) = Y^* + A_1 e^{r_1 t} + A_2 e^{r_2 t}$, calculate Y^* and explain what it is. Calculate r_1 and r_2 and determine whether or not this differential equation is stable. [B***] Calculate A_1 and A_2.*

Answer: The equilibrium value Y^* is given by $Y^* = 4$. The characteristic polynomial has roots as

$$r^2 + 3r + 1 = 0 \Longrightarrow r_1 = -\frac{3}{2} + \frac{1}{2}\sqrt{5} = -0.38197 < 0, r_2 = -\frac{3}{2} - \frac{1}{2}\sqrt{5} \approx -2.618 < 0.$$

Since $r_1 < 0$ and $r_2 < 0$ the differential equation is stable.

We have

$$\begin{aligned} Y(0) &= 2 = 4 + A_1 e^{r_1 0} + A_2 e^{r_2 0} \Longrightarrow A_1 + A_2 = -2 \\ Y'(0) &= 1 = r_1 A_1 e^{r_1 0} + r_2 A_2 e^{r_2 0} \Longrightarrow r_1 A_1 + r_2 A_2 = 1 \end{aligned}$$

so that in matrix notation

$$\begin{bmatrix} 1 & 1 \\ -\frac{3}{2} + \frac{1}{2}\sqrt{5} & -\frac{3}{2} - \frac{1}{2}\sqrt{5} \end{bmatrix} \begin{bmatrix} A_1 \\ A_2 \end{bmatrix} = \begin{bmatrix} -2 \\ 1 \end{bmatrix}$$

so that

$$\begin{bmatrix} A_1 \\ A_2 \end{bmatrix} = \begin{bmatrix} 1 & 1 \\ -\frac{3}{2}+\frac{1}{2}\sqrt{5} & -\frac{3}{2}-\frac{1}{2}\sqrt{5} \end{bmatrix}^{-1} \begin{bmatrix} -2 \\ 1 \end{bmatrix} = \begin{bmatrix} -1.8944 \\ -0.10557 \end{bmatrix}.$$

Problem 6.3 *[D***] Consider*

$$Y''(t) + 8Y'(t) + 2Y(t) = 6$$

with $Y(0) = 2, Y'(0) = 3$. If $Y(t) = Y^* + A_1 e^{r_1 t} + A_2 e^{r_2 t}$, calculate Y^* and explain what it is. Calculate r_1 and r_2 and determine whether or not this differential equation is stable. Calculate A_1 and A_2.

Answer: We have $Y^* = \frac{6}{2} = 3$. The roots of the characteristic polynomial are found as

$$r^2+8r+2 = 0 \Longrightarrow r_1 = -4+\sqrt{14} \approx -0.25834 < 0, r_2 = -4-\sqrt{14} \approx -7.7417 < 0.$$

Since $r_1 < 0$ and $r_2 < 0$ we conclude that $Y(t)$ stable.

To solve for A_1 and A_2 note that from

$$Y(t) = 3 + A_1 e^{r_1 t} + A_2 e^{r_2 t}$$

it follows that

$$\begin{bmatrix} 1 & 1 \\ -4+\sqrt{14} & -4-\sqrt{14} \end{bmatrix} \begin{bmatrix} A_1 \\ A_2 \end{bmatrix} = \begin{bmatrix} -1 \\ 3 \end{bmatrix}$$

so

$$\begin{bmatrix} A_1 \\ A_2 \end{bmatrix} = \begin{bmatrix} 1 & 1 \\ -4+\sqrt{14} & -4-\sqrt{14} \end{bmatrix}^{-1} \begin{bmatrix} -1 \\ 3 \end{bmatrix} = \begin{bmatrix} -\frac{1}{28}\sqrt{14}-\frac{1}{2} \\ \frac{1}{28}\sqrt{14}-\frac{1}{2} \end{bmatrix}.$$

The solution is therefore

$$Y(t) = 3 + \left(-\frac{1}{2} - \frac{1}{28}\sqrt{14}\right) e^{\left(-4+\sqrt{14}\right)t} - \frac{1}{28}\left(-1+\sqrt{14}\right)\sqrt{14}e^{-\left(4+\sqrt{14}\right)t}$$

as shown below.

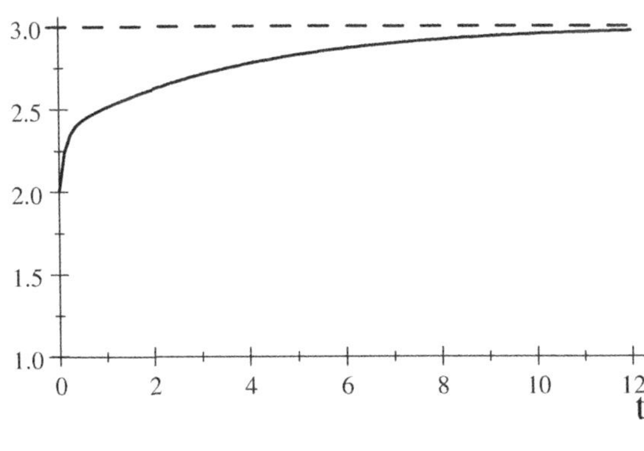

$Y(t)$ — and $Y^* = 3$ - - -

Problem 6.4 *[D***] Consider*

$$Y''(t) - 3Y'(t) - 2Y(t) = -4 \text{ with } Y(0) = 1,\ Y'(0) = 2.$$

*If $Y(t) = Y^* + A_1 e^{r_1 t} + A_2 e^{r_2 t}$, calculate Y^* and explain what it is. Calculate r_1 and r_2 and determine whether or not this differential equation is stable. [B***]Calculate A_1 and A_2.*

Answer: We have $Y^* = \frac{-4}{-2} = 2$. Now

$$r^2 - 3r - 2 = 0 \Longrightarrow r_1 = \frac{3}{2} + \frac{1}{2}\sqrt{17} \approx 3.56 > 0, r_2 = \frac{3}{2} - \frac{1}{2}\sqrt{17} \approx -0.56 < 0.$$

Since $r_1 \approx 3.56 > 0$ it follows that (even though $r_2 < 0$) that the differential equation is unstable.

To solve for A_1 and A_2 note that from

$$Y(t) = 2 + A_1 e^{r_1 t} + A_2 e^{r_2 t}$$

it follows that

$$\begin{bmatrix} 1 & 1 \\ \frac{3}{2} + \frac{1}{2}\sqrt{17} & \frac{3}{2} - \frac{1}{2}\sqrt{17} \end{bmatrix} \begin{bmatrix} A_1 \\ A_2 \end{bmatrix} = \begin{bmatrix} -1 \\ 2 \end{bmatrix}$$

so

$$\begin{aligned} \begin{bmatrix} A_1 \\ A_2 \end{bmatrix} &= \begin{bmatrix} 1 & 1 \\ \frac{3}{2} + \frac{1}{2}\sqrt{17} & \frac{3}{2} - \frac{1}{2}\sqrt{17} \end{bmatrix}^{-1} \begin{bmatrix} -1 \\ 2 \end{bmatrix} \\ &= \begin{bmatrix} \frac{7}{34}\sqrt{17} - \frac{1}{2} \\ -\frac{7}{34}\sqrt{17} - \frac{1}{2} \end{bmatrix}. \end{aligned}$$

Thus

$$Y(t) = 2 - 1.3489 \exp\left(-\frac{1}{2}\left(-3 + \sqrt{17}\right)t\right) + 0.34887 e^{\frac{1}{2}(3+\sqrt{17})t}$$

which is shown below.

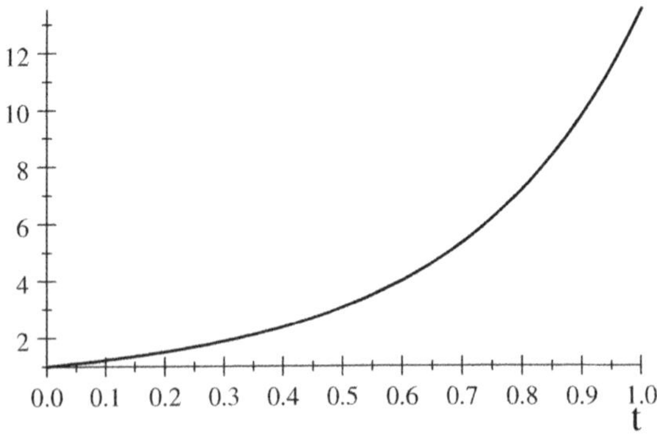

$Y(t)$ — with $Y^* = 0$

Problem 6.5 *[C***] Consider the differential equation*

$$Y''(t) + 6Y'(t) + 4Y(t) = 16, \text{ with } Y(0) = 1, Y'(0) = 1$$

Determine if this differential equation is stable or unstable. If

$$Y(t) = Y^* + A_1 e^{r_1 t} + A_2 e^{r_2 t}$$

what are the values of Y^*, A_1, A_2, r_1, r_2?

Answer: Setting $Y''(t) = Y'(t) = 0$ we have

$$4Y^* = 16 \Longrightarrow Y^* = 4.$$

The characteristic polynomial has roots as

$$r^2 + 6r + 4 = 0 \Longrightarrow r_1 = -3 + \sqrt{5} \approx -0.764 < 0, \ r_2 = -3 - \sqrt{5} \approx -5.236 < 0.$$

Since $r_1 < 0$ and $r_2 < 0$ the system is stable. To find A_1 and A_2 we have

$$Y(t) = 4 + A_1 e^{\left(-3+\sqrt{5}\right)t} + A_2 e^{\left(-3-\sqrt{5}\right)t}$$

so that

$$\begin{aligned} Y(0) &= 1 = 4 + A_1 + A_2 \\ Y'(0) &= 1 = \left(-3+\sqrt{5}\right) A_1 + \left(-3-\sqrt{5}\right) A_2 \end{aligned}$$

or in matrix notation

$$\begin{bmatrix} 1 & 1 \\ (-3+\sqrt{5}) & (-3-\sqrt{5}) \end{bmatrix} \begin{bmatrix} A_1 \\ A_2 \end{bmatrix} = \begin{bmatrix} -3 \\ 1 \end{bmatrix}$$

with solution $A_1 = -\frac{3}{2} - \frac{4}{5}\sqrt{5} \approx -3.2889$ and $A_2 = \frac{4}{5}\sqrt{5} - \frac{3}{2} \approx 0.28885$ so that the solution takes the form

$$Y(t) = 4 + \left(-\frac{3}{2} - \frac{4}{5}\sqrt{5}\right) e^{\left(-3+\sqrt{5}\right)t} + \left(\frac{4}{5}\sqrt{5} - \frac{3}{2}\right) e^{\left(-3-\sqrt{5}\right)t}.$$

Problem 6.6 *[A**] Solve the Solow growth model where*

$$Q(t) = L(t)^{\frac{1}{2}} K(t)^{\frac{1}{2}}$$

and where $\delta = 0.05, n = 0.01, s = 0.1, k(0) = 1.$

Answer: For this model $q = f(k) = k^{\frac{1}{2}}$ so that

$$k'(t) = sf(k(t)) - (n+\delta) k(t) \Longrightarrow k'(t) = 0.1k(t)^{\frac{1}{2}} - 0.06k(t).$$

Setting $k'(t) = 0$ we solve for k^* as

$$0.1\,(k^*)^{\frac{1}{2}} - 0.06k^* = 0 \Longrightarrow k^* = \left(\frac{0.1}{0.06}\right)^{\frac{1}{1-\frac{1}{2}}} = \frac{25}{9} \approx 2.78.$$

If $k(t) = Y(t)^2$ then $k'(t) = 2Y(t)\,Y'(t)$ and

$$\begin{aligned} k'(t) &= 0.1k(t)^{\frac{1}{2}} - 0.06k(t) \Longrightarrow 2Y(t)\,Y'(t) = 0.1\left(Y(t)^2\right)^{\frac{1}{2}} - 0.06Y(t)^2 \\ &\Longrightarrow Y'(t) = 0.05 - 0.03Y(t) \Longrightarrow Y'(t) + 0.03Y(t) = 0.05. \end{aligned}$$

Since $k(0) = 1$ it follows that $Y(0) = k(0)^{\frac{1}{2}} = 1$ and hence $Y(t)$ follows the first-order linear differential equation

$$Y'(t) + 0.03Y(t) = 0.05,\ Y(0) = 1$$

which has a solution

$$Y(t) = \frac{0.1}{0.06} + \left(1 - \frac{0.1}{0.06}\right)e^{-0.5\times 0.06t} = \frac{5}{3} - \frac{2}{3}e^{-0.03t}$$

so that $k(t)$ evolves as

$$k(t) = Y(t)^2 = \left(\frac{5}{3} - \frac{2}{3}e^{-0.03t}\right)^2$$

which is shown below.

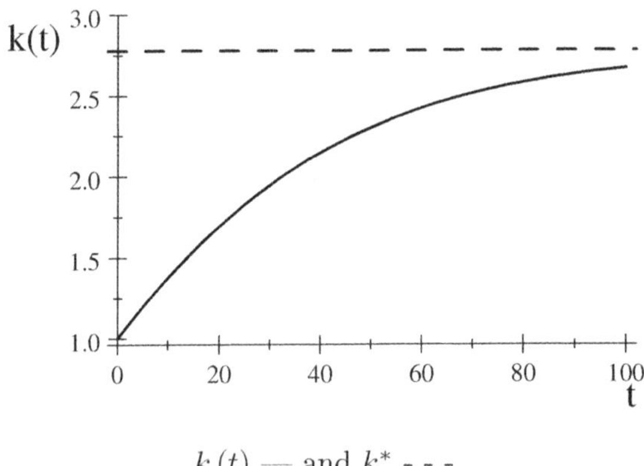

$k(t)$ — and k^* - - -

Chapter 7

Complex Variables and Trigonometry

Problem 7.1 *[C***] Show that* $\frac{1}{i} = -i$.

Answer: Multiplying the numerator and denominator by i we have

$$\frac{1}{i} \times \frac{i}{i} = \frac{i}{i \times i} = \frac{i}{-1} = -i$$

since $i = \sqrt{-1}$ so that $i \times i = -1$.

Problem 7.2 *If* $\frac{1}{a+bi} = c + di$ *then what are* c *and* d*? Show that if* $r = |a + bi|$ *then*

$$|c + di| = \left| \frac{1}{a + bi} \right| = \frac{1}{r}.$$

Answer: Multiplying the numerator and denominator by $a - bi = \overline{a + bi}$ we have using $i \times i = -1$

$$\begin{aligned} c + di &= \frac{1}{a + bi} = \frac{1}{a + bi} \times \frac{a - bi}{a - bi} = \frac{a - bi}{a^2 - b^2 i^2} = \frac{a - bi}{a^2 + b^2} \\ &= \frac{a}{a^2 + b^2} - \frac{b}{a^2 + b^2} i \Longrightarrow c = \frac{a}{a^2 + b^2}, d = -\frac{b}{a^2 + b^2}. \end{aligned}$$

Now $r = \sqrt{a^2 + b^2}$ and so we have

$$c + di = \frac{a}{a^2 + b^2} - \frac{b}{a^2 + b^2} i = \frac{a}{r^2} - \frac{b}{r^2} i$$

$$|c + di| = \left| \frac{a}{r^2} - \frac{b}{r^2} i \right| = \sqrt{\left(\frac{a}{r^2} \right)^2 + \left(\frac{b}{r^2} \right)^2} = \frac{1}{r^2} \sqrt{a^2 + b^2} = \frac{1}{r^2} \times r = \frac{1}{r}.$$

Problem 7.3 *[C***] What is* $e^{1.7i}$*?*

Answer: Using Euler's theorem we have

$$e^{1.7i} = \cos(1.7) + i\sin(1.7) = -0.12884 + 0.99166i.$$

Problem 7.4 *[C***] Given the complex number* $\frac{2}{3} - \frac{1}{2}i$ *calculate the absolute value* $\left|\frac{2}{3} - \frac{1}{2}i\right|$ *and the conjugate* $\overline{\frac{2}{3} - \frac{1}{2}i}$. *Will* $\left(\frac{2}{3} - \frac{1}{2}i\right)^t \to 0$ *as* $t \to \infty$*?*

Answer: We have

$$\left|\frac{2}{3} - \frac{1}{2}i\right| = \sqrt{\left(\frac{2}{3}\right)^2 + \left(-\frac{1}{2}\right)^2} = \frac{5}{6}, \ \overline{\frac{2}{3} - \frac{1}{2}i} = \frac{2}{3} + \frac{1}{2}i.$$

Since $\left|\frac{2}{3} - \frac{1}{2}i\right| = \frac{5}{6} < 1$ it follows that $\left(\frac{2}{3} - \frac{1}{2}i\right)^t \to 0$ as $t \to \infty$.

Problem 7.5 *[B***] What is* $\left|e^{a+bi}\right|$ *and when will* $e^{(a+bi)t} \to 0$ *as* $t \to \infty$*? Explain. If* $a + bi = 7e^{3i}$*, then what are* a *and* b *equal to? If* $a + bi = e^{3+4i}$ *then what are* a *and* b *equal to?*

Answer: We have

$$e^{a+bi} = e^a e^{bi} = e^a(\cos(b) + i\sin(b)) = e^a\cos(b) + ie^a\sin(b)$$

so that

$$\begin{aligned} \left|e^{a+bi}\right| &= \sqrt{(e^a\cos(b))^2 + (e^a\sin(b))^2} = \sqrt{e^{2a}\cos(b)^2 + e^{2a}\sin(b)^2} \\ &= e^a\sqrt{\cos(b)^2 + \sin(b)^2} = e^a. \end{aligned}$$

Thus

$$\left|e^{(a+bi)t}\right| = e^{at} \to 0$$

if and only if $a < 0$.

If $a + bi = 7e^{3i}$ then

$$a + bi = 7e^{3i} = 7(\cos(3) + i\sin(3)) \Longrightarrow a = 7\cos(3), \ b = 7\sin(3).$$

If $a + bi = e^{3+4i}$ then

$$a + bi = e^{3+4i} = e^3\left(e^{4i}\right) = e^3(\cos(4) + i\sin(4))$$

and so $a = e^3\cos(4)$ and $b = e^3\sin(4)$.

Problem 7.6 *[C***] Given the complex number* $0.8 - 0.7i$ *calculate the absolute value* $|0.8 - 0.7i|$ *and the conjugate* $\overline{0.8 - 0.7i}$. *Will* $(0.8 - 0.7i)^t \to 0$ *as* $t \to \infty$*?*

Answer: We have

$$|0.8 - 0.7i| = \sqrt{0.8^2 + (-0.7)^2} \approx 1.06.$$

The conjugate is

$$\overline{0.8 - 0.7i} = 0.8 + 0.7i.$$

Since $|0.8 - 0.7i| > 1$ it follows that $(0.8 - 0.7i)^t$ does not go to 0 as $t \to \infty$.

Problem 7.7 *[B***] Prove that* $\overline{e^{\theta i}} = e^{-\theta i}$ *and* $\left|e^{\theta i}\right| = 1$.

Answer: We have

$$\overline{e^{\theta i}} = \overline{\cos(\theta) + i\sin(\theta)} = \cos(\theta) - i\sin(\theta) = \cos(-\theta) + i\sin(-\theta) = e^{-\theta i}$$

so that

$$\left|e^{\theta i}\right| = \sqrt{e^{\theta i} \times \overline{e^{\theta i}}} = \sqrt{e^{\theta i} \times e^{-\theta i}} = \sqrt{e^{\theta i - \theta i}} = \sqrt{e^0} = \sqrt{1} = 1.$$

Problem 7.8 *[C***] Calculate* $\left(\frac{1}{2} + \frac{1}{2}i\right)^6$ *directly and using the polar form.*

Answer: From the polar form

$$\frac{1}{2} + \frac{1}{2}i = \frac{1}{\sqrt{2}}e^{i\frac{\pi}{4}} \Longrightarrow \left(\frac{1}{2} + \frac{1}{2}i\right)^t = \left(\frac{1}{\sqrt{2}}\right)^t e^{it\frac{\pi}{4}}$$

we have

$$\left(\frac{1}{2} + \frac{1}{2}i\right)^6 = \left(\frac{1}{\sqrt{2}}\right)^6 e^{i6\frac{\pi}{4}} = \frac{1}{8}\left(\cos\left(\frac{3}{2}\pi\right) + i\sin\left(\frac{3}{2}\pi\right)\right) = -\frac{1}{8}i.$$

To do the direct calculation use

$$\left(\frac{1}{2} + \frac{1}{2}i\right) \times \left(\frac{1}{2} + \frac{1}{2}i\right) = \frac{i}{2}$$

so that

$$\left(\frac{1}{2} + \frac{1}{2}i\right)^6 = \left(\left(\frac{1}{2} + \frac{1}{2}i\right)^2\right)^3 = \left(\frac{i}{2}\right)^3 = \frac{i \times i}{8} \times i = -\frac{1}{8}i.$$

Problem 7.9 *[C**] What is* $\ln\left(\frac{1}{2} + \frac{1}{2}i\right)$*?*

Answer: From the polar form

$$\frac{1}{2} + \frac{1}{2}i = \frac{1}{\sqrt{2}}e^{i\frac{\pi}{4}}$$

we have

$$\ln\left(\frac{1}{2} + \frac{1}{2}i\right) = \ln\left(\frac{1}{\sqrt{2}}e^{i\frac{\pi}{4}}\right) = \ln\left(\frac{1}{\sqrt{2}}\right) + \ln\left(e^{i\frac{\pi}{4}}\right) = -\frac{1}{2}\ln(2) + i\frac{\pi}{4}.$$

Actually things are a bit more complicated. The log of a complex number takes on an infinite number of values. To see this let k be any integer. Then

$$e^{2\pi ik} = \cos(2\pi k) + i\sin(2\pi k) = 1.$$

It follows then that

$$\frac{1}{2} + \frac{1}{2}i = \frac{1}{\sqrt{2}}e^{i\frac{\pi}{4}} \times e^{2\pi ik} = \frac{1}{\sqrt{2}}e^{i\frac{\pi}{4} + 2\pi ki}$$

and

$$\ln\left(\frac{1}{2} + \frac{1}{2}i\right) = \ln\left(\frac{1}{\sqrt{2}}e^{i\frac{\pi}{4} + 2\pi ki}\right) = -\frac{1}{2}\ln(2) + i\frac{\pi}{4} + 2\pi ik.$$

Problem 7.10 *[B*] Prove that*

$$\tan(x)^2 + 1 = \frac{1}{\cos(x)^2}.$$

Answer: Since

$$\tan(x) = \frac{\sin(x)}{\cos(x)}$$

we have

$$\begin{aligned}\tan(x)^2 &= \frac{\sin(x)^2}{\cos(x)^2} = \frac{1-\cos(x)^2}{\cos(x)^2} = \frac{1}{\cos(x)^2} - 1 \\ &\implies \tan(x)^2 + 1 = \frac{1}{\cos(x)^2}.\end{aligned}$$

Problem 7.11 *[B*] Prove that*

$$\frac{d\tan(x)}{dx} = \frac{1}{\cos(x)^2} \text{ and } \frac{d\arctan(x)}{dx} = \frac{1}{1+x^2}.$$

Answer: Using the quotient rule

$$\begin{aligned}\frac{d\tan(x)}{dx} &= \frac{\cos(x)\frac{d\sin(x)}{dx} - \sin(x)\frac{d\cos(x)}{dx}}{\cos(x)^2} \\ &= \frac{\cos(x)\times\cos(x) - \sin(x)\times(-\sin(x))}{\cos(x)^2} \\ &= \frac{\overbrace{\cos(x)^2 + \sin(x)^2}^{1}}{\cos(x)^2} = \frac{1}{\cos(x)^2}.\end{aligned}$$

Now since $\tan(x)$ and $\arctan(x)$ are inverse functions we have

$$\tan(\arctan(x)) = x$$

so that differentiating both sides and using the chain rule yields

$$\tan'(\arctan(x)) \times \frac{d\arctan(x)}{dx} = 1.$$

Now since

$$\tan'(x) = \frac{1}{\cos(x)^2} = \tan(x)^2 + 1$$

we have

$$\begin{aligned}\tan'(\arctan(x)) &= \frac{1}{\cos(\arctan(x))^2} = \tan(\arctan(x))^2 + 1 \\ &= x^2 + 1 = 1 + x^2.\end{aligned}$$

Thus

$$\left(1+x^2\right)\frac{d\arctan(x)}{dx} = 1$$
$$\implies \frac{d\arctan(x)}{dx} = \frac{1}{1+x^2}.$$

Problem 7.12 *[A*] The Cauchy distribution (which is a t distribution with* 1 *degree of freedom) has a density of the form*

$$p(x) = c\frac{1}{1+x^2}, \text{ for } -\infty < x < \infty.$$

Show that $c = \frac{1}{\pi}$, *that*

$$\Pr[a < X < b] = \frac{\arctan(b) - \arctan(a)}{\pi}$$

and that in particular $\Pr[-1 < X < 1] = \frac{1}{2}$.

Answer: Since the area under a density must be 1 we have

$$1 = \int_{-\infty}^{\infty} p(x)\,dx = \int_{-\infty}^{\infty} c\frac{1}{1+x^2}dx = c\int_{-\infty}^{\infty}\frac{1}{1+x^2}dx.$$

Now since $\arctan'(x) = \frac{1}{1+x^2}$ the anti-derivative of $\frac{1}{1+x^2}$ is $\arctan(x)$ so that

$$\int_{-\infty}^{\infty}\frac{1}{1+x^2}dx = \arctan(x)\,|_{-\infty}^{\infty} = \arctan(\infty) - \arctan(-\infty) = \frac{\pi}{2} - \left(-\frac{\pi}{2}\right) = \pi$$

so that

$$1 = c\int_{-\infty}^{\infty}\frac{1}{1+x^2}dx = c\times\pi \Longrightarrow c = \frac{1}{\pi}.$$

Thus the Cauchy density takes the form

$$p(x) = \frac{1}{\pi}\frac{1}{1+x^2}$$

and is shown below.

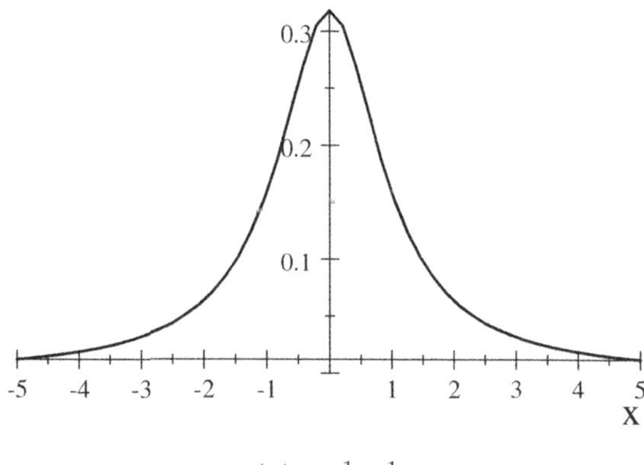

$p(x) = \frac{1}{\pi}\frac{1}{1+x^2}$

The Cauchy distribution looks like the normal distribution but has much thicker tails than the normal.

We now have

$$\Pr\left[a < X < b\right] = \int_a^b \frac{1}{\pi}\frac{1}{1+x^2}dx = \frac{1}{\pi}\int_a^b \frac{1}{1+x^2}dx = \frac{1}{\pi}\left(\arctan\left(b\right) - \arctan\left(a\right)\right).$$

Thus

$$\Pr\left[-1 < X < 1\right] = \frac{1}{\pi}\left(\arctan\left(1\right) - \arctan\left(-1\right)\right) = \frac{1}{\pi}\left(\frac{\pi}{4} - \left(-\frac{\pi}{4}\right)\right) = \frac{1}{2}.$$

Problem 7.13 *[B***] Solve the following system of equations for A_1 and A_2*

$$\begin{aligned} A_1 + A_2 &= \alpha \\ \left(\frac{1}{2} + \frac{1}{2}i\right) \times A_1 + \left(\frac{1}{2} - \frac{1}{2}i\right) A_2 &= \beta \end{aligned}$$

where α and β are real. Show that if $A_1 = c + di$ then $A_2 = c - di$. (Here A_1 and A_2 are conjugates of each other.)

Answer: In matrix notation we have

$$\begin{bmatrix} 1 & 1 \\ \frac{1}{2} + \frac{1}{2}i & \frac{1}{2} - \frac{1}{2}i \end{bmatrix}\begin{bmatrix} A_1 \\ A_2 \end{bmatrix} = \begin{bmatrix} \alpha \\ \beta \end{bmatrix}.$$

Here

$$\det\begin{bmatrix} 1 & 1 \\ \frac{1}{2} + \frac{1}{2}i & \frac{1}{2} - \frac{1}{2}i \end{bmatrix} = 1 \times \left(\frac{1}{2} - \frac{1}{2}i\right) - 1 \times \left(\frac{1}{2} + \frac{1}{2}i\right) = -i.$$

Using $i \times i = -1$ we have

$$\frac{1}{-i} = \frac{1}{-i}\frac{i}{i} = \frac{1}{-1 \times -1}i = i$$

so that using Cramer's rule

$$\begin{aligned} A_1 &= \frac{\det\begin{bmatrix} \alpha & 1 \\ \beta & \frac{1}{2} - \frac{1}{2}i \end{bmatrix}}{-i} = \frac{1}{-i} \times \left(\alpha\frac{1}{2} - \beta - i\alpha\frac{1}{2}\right) \\ &= i \times \left(\alpha\frac{1}{2} - \beta - i\alpha\frac{1}{2}\right) = \frac{\alpha}{2} + \left(\frac{\alpha}{2} - \beta\right)i \end{aligned}$$

and

$$\begin{aligned} A_2 &= \frac{\det\begin{bmatrix} 1 & \alpha \\ \frac{1}{2} + \frac{1}{2}i & \beta \end{bmatrix}}{-i} = \frac{1}{-i} \times \left(\beta - \alpha\frac{1}{2} - i\alpha\frac{1}{2}\right) \\ &= i \times \left(\beta - \alpha\frac{1}{2} - i\alpha\frac{1}{2}\right) = \frac{\alpha}{2} + \left(\beta - \frac{\alpha}{2}\right)i = \frac{\alpha}{2} - \left(\frac{\alpha}{2} - \beta\right)i. \end{aligned}$$

Thus if $A_1 = c + di$ we have

$$c = \frac{\alpha}{2} \text{ and } d = \frac{\alpha}{2} - \beta$$

and $A_2 = c - di$.

Problem 7.14 *[B***] Solve the following system of equations for A_1 and A_2*

$$\begin{aligned} A_1 + A_2 &= \alpha \\ (a + bi) \times A_1 + (a - bi) A_2 &= \beta \end{aligned}$$

where α and β are real. Show that if $A_1 = c + di$ then $A_2 = c - di$ (A_1 and A_2 are conjugates of each other). (This is a common problem in the solution of both second-order difference and differential equations.)

Answer: In matrix notation we have

$$\begin{bmatrix} 1 & 1 \\ a + bi & a - bi \end{bmatrix} \begin{bmatrix} A_1 \\ A_2 \end{bmatrix} = \begin{bmatrix} \alpha \\ \beta \end{bmatrix}.$$

Here

$$\det \begin{bmatrix} 1 & 1 \\ a + bi & a - bi \end{bmatrix} = 1 \times (a - bi) - 1 \times (a + bi) = -2bi.$$

Using $i \times i = -1$ we have

$$\frac{1}{-2bi} = \frac{1}{-2bi} \frac{i}{i} = \frac{1}{2b} i$$

so that using Cramer's rule

$$\begin{aligned} A_1 &= \frac{\det \begin{bmatrix} \alpha & 1 \\ \beta & a - bi \end{bmatrix}}{-2bi} = \frac{1}{-2bi} \times (\alpha a - \beta - i\alpha b) \\ &= \frac{1}{2b} i \times (\alpha a - \beta - i\alpha b) = \frac{\alpha}{2} + \left(\frac{\alpha a - \beta}{2b} \right) i \end{aligned}$$

and

$$\begin{aligned} A_2 &= \frac{\det \begin{bmatrix} 1 & \alpha \\ a + bi & \beta \end{bmatrix}}{-2bi} = \frac{1}{-2bi} \times (\beta - \alpha a - i\alpha b) \\ &= \frac{1}{2b} i \times (\beta - \alpha a - i\alpha b) = \frac{\alpha}{2} + \left(\frac{\beta - \alpha a}{2b} \right) i = \frac{\alpha}{2} - \left(\frac{\alpha a - \beta}{2b} \right) i. \end{aligned}$$

Thus if $A_1 = c + di$ we have

$$c = \frac{\alpha}{2} \text{ and } d = \left(\frac{\alpha a - \beta}{2b} \right)$$

and $A_2 = c - di$.

7.1 Difference Equations with Complex Roots

Problem 7.15 *[D***] Consider the difference equation*

$$Y_t = 100 + \frac{3}{2}Y_{t-1} - \frac{3}{4}Y_{t-2}.$$

Find the equilibrium value Y^* and determine whether this difference equation is stable or not. If $Y_0 = Y_1 = 100$, calculate Y_t for $t = 2, 3, 4, 5, 6$.

Answer: The equilibrium value Y^* is determined by

$$\begin{aligned} Y^* &= 100 + \frac{3}{2}Y^* - \frac{3}{4}Y^* \\ &\implies Y^* = 400. \end{aligned}$$

The characteristic polynomial is

$$r^2 - \frac{3}{2}r + \frac{3}{4} = 0$$

with complex roots

$$r_1 = \frac{3}{4} + \frac{\sqrt{3}}{4}i,\ r_2 = \frac{3}{4} - \frac{\sqrt{3}}{4}i.$$

The difference equation is stable since

$$|r_1| = |r_2| = \sqrt{\left(\frac{3}{4}\right)^2 + \left(\frac{\sqrt{3}}{4}\right)^2} = \frac{\sqrt{3}}{2} \approx 0.866 < 1.$$

As well

$$\begin{aligned} Y_2 &= 100 + \frac{3}{2} \times 100 - \frac{3}{4} \times 100 = 175 \\ Y_3 &= 100 + \frac{3}{2} \times 175 - \frac{3}{4} \times 100 = 287.5 \\ Y_4 &= 100 + \frac{3}{2} \times 287.5 - \frac{3}{4} \times 175 = 400 \\ Y_5 &= 100 + \frac{3}{2} \times 400 - \frac{3}{4} \times 287.5 = 484.38 \\ Y_6 &= 100 + \frac{3}{2} \times 484.38 - \frac{3}{4} \times 400 = 526.57. \end{aligned}$$

Problem 7.16 *[C**] Consider the following version of the accelerator/multiplier model*

$$\begin{aligned} C_t &= 50 + \frac{3}{5}Y_{t-1}, I_t = 50 + \frac{2}{5}(Y_{t-1} - Y_{t-2}) \\ G_t &= 50, Y_t = C_t + I_t + G_t. \end{aligned}$$

Show that Y_t follows a second-order difference equation

$$Y_t = a_0 + a_1 Y_{t-1} + a_2 Y_{t-2}$$

and calculate a_o, a_1, a_2. Find the equilibrium value of Y_t. Calculate Y_t for $t = 2, 3, 4, 5, 6$ when $Y_0 = 200$ and $Y_1 = 200$. Find the roots of the quadratic polynomial for this difference equation. Show that the economy is stable.

Answer: We have

$$Y_t = C_t + I_t + G_t = 50 + \frac{3}{5} Y_{t-1} + 50 + \frac{2}{5}\left(Y_{t-1} - Y_{t-2}\right) + 50 = 150 + Y_{t-1} - \frac{2}{5} Y_{t-2}.$$

The equilibrium value comes from

$$Y^* = 150 + Y^* - 0.4Y^* \Longrightarrow Y^* = \frac{150}{\frac{2}{5}} = 375.$$

As well

$$\begin{aligned}
Y_2 &= 150 + 200 - \frac{2}{5} \times 200 = 270, Y_3 = 150 + 270 - \frac{2}{5} \times 200 = 340 \\
Y_4 &= 150 + 340 - \frac{2}{5} \times 270 = 382, Y_5 = 150 + 382 - \frac{2}{5} \times 340 = 396 \\
Y_6 &= 150 + 396 - \frac{2}{5} \times 382 = 393.2.
\end{aligned}$$

We have

$$r^2 - r + \frac{2}{5} = 0 \Longrightarrow r_1 = \frac{1}{2} + \frac{\sqrt{15}}{10} i, r_2 = \frac{1}{2} - \frac{\sqrt{15}}{10} i$$

so that

$$|r_1| = |r_2| = \left| \frac{1}{2} + \frac{\sqrt{15}}{10} i \right| = \sqrt{\left(\frac{1}{2}\right)^2 + \left(\frac{\sqrt{15}}{10}\right)^2} \approx 0.63 < 1$$

and the economy is stable.

Problem 7.17 *[C**] Suppose that Y_t is GNP, C_t is consumption, I_t is investment and G_t is government expenditure and that*

$$\begin{aligned}
C_t &= 100 + 0.6 Y_{t-1}, \; I_t = 50 + 0.5\left(Y_{t-1} - Y_{t-2}\right) \\
G_t &= 50, \; Y_t = C_t + I_t + G_t.
\end{aligned}$$

Find the difference equation that Y_t follows, its equilibrium value Y^* and the solution to the difference equation. Is the economy stable?

Answer: It now follows that

$$Y_t = C_t + I_t + G_t = 100 + 0.6Y_{t-1} + 50 + 0.5\left(Y_{t-1} - Y_{t-2}\right) + 50$$

or

$$Y_t = 200 + 1.1Y_{t-1} - 0.5Y_{t-2}$$

which is a second-order difference equation.

The equilibrium value of GNP is thus

$$Y^* = \frac{200}{1 - 1.1 - (-0.5)} = 500.$$

We then have

$$r^2 - 1.1r + 0.5 = 0 \Longrightarrow r_1 = 0.55 + 0.44i,\ r_2 = 0.55 - 0.44i.$$

Since

$$|r_1| = |r_2| = \sqrt{0.55^2 + 0.44^2} \approx 0.70$$

the difference equation is stable.

$$\begin{aligned}\begin{bmatrix} Y_t \\ Y_{t-1} \end{bmatrix} &= \begin{bmatrix} 1000 \\ 1000 \end{bmatrix} + \begin{bmatrix} 2.6667 & -1.6667 \\ 3.3333 & -3.3333 \end{bmatrix}\begin{bmatrix} 0.8^t & 0 \\ 0 & 0.5^t \end{bmatrix}\begin{bmatrix} 1.0 & -0.5 \\ 1.0 & -0.8 \end{bmatrix}\begin{bmatrix} 900 \\ 900 \end{bmatrix} \\ &= \begin{bmatrix} 1000 + 1200.0 \times 0.8^t - 300.01 \times 0.5^t \\ 1000 + 1500.0 \times 0.8^t - 599.99 \times 0.5^t \end{bmatrix}.\end{aligned}$$

Problem 7.18 *[D***] Consider the difference equation*

$$Y_t = 100 + Y_{t-1} - \frac{1}{2}Y_{t-2} \text{ with } Y_0 = Y_1 = 100.$$

*Calculate Y_t for $t = 2, 3, 4, 5, 6$. If the solution is written as $Y_t = Y^* + A_1 r_1^t + A_2 r_2^t$, calculate Y^* and explain what it is. Calculate r_1 and r_2. Determine if this difference equation is stable by calculating $|r_1|$ and $|r_2|$. What does the fact that r_1 and r_2 are complex tell you, or if they are not complex, what does that tell you? Calculate A_1 and A_2. [A***] Show that the solution can be written as*

$$Y_t = 200 - 200\left(\frac{1}{\sqrt{2}}\right)^{t+1} \cos\left(\frac{\pi}{4}(t-1)\right).$$

Answer: The equilibrium value Y^* is determined by

$$Y^* = 100 + Y^* - \frac{1}{2}Y^* \Longrightarrow Y^* = 200.$$

The characteristic polynomial has complex roots as

$$r^2 - r + \frac{1}{2} = 0 \Longrightarrow r_1 = \frac{1}{2} + \frac{1}{2}i, r_2 = \frac{1}{2} - \frac{1}{2}i.$$

The difference equation is stable since

$$|r_1| = |r_2| = \sqrt{\left(\frac{1}{2}\right)^2 + \left(\frac{1}{2}\right)^2} = \frac{\sqrt{2}}{2} \approx 0.71 < 1.$$

As well

$$\begin{aligned} Y_2 &= 100 + 100 - \frac{1}{2} \times 100 = 150, Y_3 = 100 + 150 - \frac{1}{2} \times 100 = 200 \\ Y_4 &= 100 + 200 - \frac{1}{2} \times 150 = 225, Y_5 = 100 + 225 - \frac{1}{2} \times 200 = 225 \\ Y_6 &= 100 + 225 - \frac{1}{2} \times 225 = 212.5. \end{aligned}$$

To calculate A_1 and A_2 we have

$$\begin{aligned} Y_0 &= 100 = 200 + A_1 + A_2 \Longrightarrow A_1 + A_2 = -100 \\ Y_1 &= 100 = 200 + A_1 r_1 + A_2 r_2 \Longrightarrow r_1 A_1 + r_2 A_2 = -100 \end{aligned}$$

or in matrix notation

$$\begin{bmatrix} 1 & 1 \\ \frac{1}{2} + \frac{1}{2}i & \frac{1}{2} - \frac{1}{2}i \end{bmatrix} \begin{bmatrix} A_1 \\ A_2 \end{bmatrix} = \begin{bmatrix} -100 \\ -100 \end{bmatrix}$$

so that

$$\begin{bmatrix} A_1 \\ A_2 \end{bmatrix} = \begin{bmatrix} 1 & 1 \\ \frac{1}{2} + \frac{1}{2}i & \frac{1}{2} - \frac{1}{2}i \end{bmatrix}^{-1} \begin{bmatrix} -100 \\ -100 \end{bmatrix} = \begin{bmatrix} -50 + 50i \\ -50 - 50i \end{bmatrix}.$$

Thus

$$\begin{aligned} Y_t &= 200 + (-50 + 50i)\left(\frac{1}{2} + \frac{1}{2}i\right)^t + (-50 - 50i)\left(\frac{1}{2} - \frac{1}{2}i\right)^t \\ &= 200 - 100\left(\left(\frac{1}{2} - \frac{1}{2}i\right)\left(\frac{1}{2} + \frac{1}{2}i\right)^t + \left(\frac{1}{2} + \frac{1}{2}i\right)\left(\frac{1}{2} - \frac{1}{2}i\right)^t\right) \\ &= 200 - 100\left(\frac{1}{\sqrt{2}}e^{-i\frac{\pi}{4}} \times \left(\frac{1}{\sqrt{2}}\right)^t e^{i\frac{\pi}{4}t} + \frac{1}{\sqrt{2}}e^{i\frac{\pi}{4}} \times \left(\frac{1}{\sqrt{2}}\right)^t e^{-i\frac{\pi}{4}t}\right) \\ &= 200 - 200\left(\frac{1}{\sqrt{2}}\right)^{t+1}\left(\frac{e^{i\frac{\pi}{4}(t-1)} + e^{-i\frac{\pi}{4}(t-1)}}{2}\right) = 200 - 200\left(\frac{1}{\sqrt{2}}\right)^{t+1} \cos\left(\frac{\pi}{4}(t-1)\right). \end{aligned}$$

Problem 7.19 *[D***] Consider the second-order difference equation*

$$Y_t = 50 + 1.8Y_{t-1} - 1.3Y_{t-2} \text{ with } Y_1 = 100, Y_o = 120.$$

If $Y_0 = Y_1 = 100$, calculate Y_t for $t = 2, 3, 4$. What is the equilibrium value Y^*? If the solution is written as $Y_t = Y^* + A_1 r_1^t + A_2 r_2^t$, calculate r_1 and r_2. Use r_1 and r_2 to determine if this difference equation is stable or unstable.

Answer: The equilibrium value Y^* is

$$Y^* = \frac{50}{1-(1.8)-(-1.3)} = 100.$$

The characteristic polynomial has roots as

$$r^2 - 1.8r + 1.3 = 0 \Longrightarrow r_1 = 0.9 + 0.7i,\ r_2 = 0.9 - 0.7i.$$

Since

$$| r_1 |=| r_2 |= \sqrt{0.9^2 + 0.7^2} \approx 1.14 > 1$$

the difference equation is unstable.

Problem 7.20 *Consider the second-order difference equation*

$$Y_t = 50 + 1.1Y_{t-1} - 0.6Y_{t-2} \text{ with } Y_1 = 50, Y_o = 60.$$

Calculate Y_t *for* $t = 2, 3, 4$. *What is the equilibrium value* Y^*? *If the solution is written as* $Y_t = Y^* + A_1 r_1^t + A_2 r_2^t$, *calculate* r_1 *and* r_2. *Use* r_1 *and* r_2 *to determine if this difference equation is stable or unstable.*

Answer: The equilibrium value Y^* is then

$$Y^* = \frac{50}{1-(1.1)-(-0.6)} = 100.$$

The characteristic polynomial has roots as

$$\begin{aligned} r^2 - 1.1r + 0.6 &= 0 \Longrightarrow r_1, r_2 = \frac{--(1.1) \pm \sqrt{(-1.1)^2 - 4 \times 0.6}}{2} \\ &\Longrightarrow r_1 = 0.55 - 0.55i,\ r_2 = 0.55 + 0.55i. \end{aligned}$$

Since

$$| r_1 |=| r_2 |= \sqrt{(0.55)^2 + (0.55)^2} \approx 0.78 < 1$$

the difference equation is stable.

Problem 7.21 *[D***] Consider the second-order difference equation*

$$Y_t = 5 + \frac{1}{2}Y_{t-1} - \frac{1}{8}Y_{t-2} \text{ with } Y_0 = 4, Y_1 = 6$$

Find the equilibrium value Y^*. *Show that* Y_t *is stable. [A***] Write* Y_t *in the form*

$$Y_t = Y^* + A_1 r_1^t + A_2 r_2^t$$

and simplify this to remove all complex terms. Show that the solution will exhibit a wave pattern that repeats itself every 8 periods.

Answer: The equilibrium value Y^* is

$$Y^* = \frac{5}{1 - \frac{1}{2} + \frac{1}{8}} = 8.$$

The characteristic polynomial has roots as

$$\begin{aligned} r^2 - \frac{1}{2}r + \frac{1}{8} &= 0 \\ \implies r_1, r_2 &= \frac{\frac{1}{2} \pm \sqrt{\left(\frac{1}{2}\right)^2 - 4 \times \frac{1}{8}}}{2} \\ \implies r_1 &= \frac{1}{4} + \frac{1}{4}i, \; r_2 = \frac{1}{4} - \frac{1}{4}i. \end{aligned}$$

If the polar form for the roots is $\rho e^{\pm\theta i}$ then $\rho = \frac{1}{2\sqrt{2}} = 0.354$ since

$$|r_1| = |r_2| = \sqrt{\left(\frac{1}{4}\right)^2 + \left(\frac{1}{4}\right)^2} = \frac{1}{2\sqrt{2}} = 0.354$$

and

$$\theta = \arctan\left(\frac{\frac{1}{4}}{\frac{1}{4}}\right) = \arctan(1) = \frac{\pi}{4}$$

so that

$$r_1 = \frac{1}{4} + \frac{1}{4}i = \frac{1}{2\sqrt{2}}e^{\frac{\pi}{4}i}, r_2 = \frac{1}{4} - \frac{1}{4}i = \frac{1}{2\sqrt{2}}e^{-\frac{\pi}{4}i}.$$

Stability requires $|r_1| = |r_2| < 1$ and

$$|r_1| = |r_2| = \sqrt{\left(\frac{1}{4}\right)^2 + \left(\frac{1}{4}\right)^2} = \frac{1}{2\sqrt{2}} = 0.354 < 1$$

so it follows that Y_t is stable.

The general solution is then

$$Y_t = Y^* + A_1 r_1^t + A_2 r_2^t = 8 + A_1\left(\frac{1}{4} + \frac{1}{4}i\right)^t + A_2\left(\frac{1}{4} - \frac{1}{4}i\right)^t$$

where A_1 and A_2 depend on the starting values.

Given $Y_0 = 4, Y_1 = 6$ we have

$$\begin{aligned} 4 &= 8 + A_1\left(\frac{1}{4} + \frac{1}{4}i\right)^0 + A_2\left(\frac{1}{4} - \frac{1}{4}i\right)^0 \\ 6 &= 8 + A_1\left(\frac{1}{4} + \frac{1}{4}i\right)^1 + A_2\left(\frac{1}{4} - \frac{1}{4}i\right)^1 \end{aligned}$$

so that in matrix notation

$$\begin{bmatrix} 1 & 1 \\ \frac{1}{4} + \frac{1}{4}i & \frac{1}{4} - \frac{1}{4}i \end{bmatrix}\begin{bmatrix} A_1 \\ A_2 \end{bmatrix} = \begin{bmatrix} -4 \\ -2 \end{bmatrix}$$

or

$$\begin{bmatrix} A_1 \\ A_2 \end{bmatrix} = \begin{bmatrix} 1 & 1 \\ \frac{1}{4}+\frac{1}{4}i & \frac{1}{4}-\frac{1}{4}i \end{bmatrix}^{-1} \begin{bmatrix} -4 \\ -2 \end{bmatrix} = \begin{bmatrix} -2+2i \\ -2-2i \end{bmatrix}$$

where

$$A_1 = -2+2i,\ A_2 = -2-2i.$$

We thus have the solution

$$\begin{aligned}
Y_t &= Y^* + A_1\left(\frac{1}{4}+\frac{1}{4}i\right)^t + A_2\left(\frac{1}{4}-\frac{1}{4}i\right)^t \\
&= 8 + A_1\left(\frac{1}{2\sqrt{2}}e^{\frac{\pi}{4}i}\right)^t + A_2\left(\frac{1}{2\sqrt{2}}e^{-\frac{\pi}{4}i}\right)^t \\
&= 8 + A_1\left(\frac{1}{2\sqrt{2}}\right)^t e^{\frac{\pi}{4}ti} + A_2\left(\frac{1}{2\sqrt{2}}\right)^t e^{-\frac{\pi}{4}ti} \\
&= 8 + (-2+2i)\left(\frac{1}{2\sqrt{2}}\right)^t e^{\frac{\pi}{4}it} + (-2-2i)\left(\frac{1}{2\sqrt{2}}\right)^t e^{-\frac{\pi}{4}it} \\
&= 8 + \left(\frac{1}{2\sqrt{2}}\right)^t (-2+2i)\left(\cos\left(\frac{\pi}{4}t\right) + i\sin\left(\frac{\pi}{4}t\right)\right) \\
&\quad + \left(\frac{1}{2\sqrt{2}}\right)^t (-2-2i)\left(\cos\left(-\frac{\pi}{4}t\right) + i\sin\left(-\frac{\pi}{4}t\right)\right) \\
&= 8 + \left(\frac{1}{2\sqrt{2}}\right)^t (-2+2i)\left(\cos\left(\frac{\pi}{4}t\right) + i\sin\left(\frac{\pi}{4}t\right)\right) \\
&\quad + \left(\frac{1}{2\sqrt{2}}\right)^t (-2-2i)\left(\cos\left(\frac{\pi}{4}t\right) - i\sin\left(\frac{\pi}{4}t\right)\right) \\
&= 8 + \left(\frac{1}{2\sqrt{2}}\right)^t \left(-4\times\cos\left(\frac{\pi}{4}t\right) - 4\times\sin\left(\frac{\pi}{4}t\right)\right)
\end{aligned}$$

or

$$Y_t = 8 - 4\left(\frac{1}{2\sqrt{2}}\right)^t \left(\cos\left(\frac{\pi}{4}t\right) + \sin\left(\frac{\pi}{4}t\right)\right).$$

Since $\cos(\theta+2\pi) = \cos(\theta)$ and $\sin(\theta+2\pi) = \sin(\theta)$ the term $\cos\left(\frac{\pi}{4}t\right) + \sin\left(\frac{\pi}{4}t\right)$ has a period P satisfying

$$\frac{\pi}{4}(t+P) = \frac{\pi}{4}t + 2\pi \Longrightarrow P = \frac{2\pi}{\frac{\pi}{4}} = 8.$$

Problem 7.22 *[D***] Consider a second-order difference equation of the form*

$$Y_t = 3 + \frac{1}{2}Y_{t-1} - \frac{1}{4}Y_{t-2} \text{ with } Y_0 = 2, Y_1 = 3.$$

*Find the equilibrium value Y^*. Show that Y_t is stable. [A***] Write Y_t in the form*

$$Y_t = Y^* + A_1 r_1^t + A_2 r_2^t$$

and simplify this to remove all complex terms. Show that the solution will exhibit a wave pattern that repeats itself every 6 periods.

Answer: The equilibrium value Y^* is

$$Y^* = \frac{3}{1 - \frac{1}{2} + \frac{1}{4}} = 4.$$

The characteristic polynomial has roots as

$$\begin{aligned} r^2 - \frac{1}{2}r + \frac{1}{4} &= 0 \\ \implies r_1, r_2 &= \frac{\frac{1}{2} \pm \sqrt{\left(\frac{1}{2}\right)^2 - 4 \times \frac{1}{4}}}{2} \\ \implies r_1 &= \frac{1}{4} + \frac{\sqrt{3}}{4}i,\ r_2 = \frac{1}{4} - \frac{\sqrt{3}}{4}. \end{aligned}$$

If the polar form for the roots is $\rho e^{\pm\theta i}$ then $\rho = \frac{1}{2}$ since

$$|r_1| = |r_2| = \sqrt{\left(\frac{1}{4}\right)^2 + \left(\frac{\sqrt{3}}{4}\right)^2} = \frac{1}{2}$$

and $\theta = \frac{\pi}{3}$ since

$$\theta = \arctan\left(\frac{\frac{\sqrt{3}}{4}}{\frac{1}{4}}\right) = \arctan\left(\sqrt{3}\right) = \frac{\pi}{3}$$

so that

$$r_1 = \frac{1}{4} + \frac{\sqrt{3}}{4}i = \frac{1}{2}e^{\frac{\pi}{3}i}, r_2 = \frac{1}{4} - \frac{\sqrt{3}}{4}i = \frac{1}{2}e^{-\frac{\pi}{3}i}.$$

Stability requires $|r_1| = |r_2| < 1$ and

$$|r_1| = |r_2| = \frac{1}{2} < 1$$

so it follows that Y_t is stable.

The general solution is then

$$Y_t = Y^* + A_1 r_1^t + A_2 r_2^t = 4 + A_1\left(\frac{1}{4} + \frac{\sqrt{3}}{4}i\right)^t + A_2\left(\frac{1}{4} - \frac{\sqrt{3}}{4}i\right)^t$$

where A_1 and A_2 depend on the starting values Y_o and Y_1 as

$$\begin{aligned} 2 &= 4 + A_1\left(\frac{1}{4} + \frac{\sqrt{3}}{4}i\right)^0 + A_2\left(\frac{1}{4} - \frac{\sqrt{3}}{4}i\right)^0 \\ 3 &= 4 + A_1\left(\frac{1}{4} + \frac{\sqrt{3}}{4}i\right)^1 + A_2\left(\frac{1}{4} - \frac{\sqrt{3}}{4}i\right)^1 \end{aligned}$$

so that in matrix notation

$$\begin{bmatrix} 1 & 1 \\ \frac{1}{4}+\frac{\sqrt{3}}{4}i & \frac{1}{4}-\frac{\sqrt{3}}{4}i \end{bmatrix}\begin{bmatrix} A_1 \\ A_2 \end{bmatrix}=\begin{bmatrix} -2 \\ -1 \end{bmatrix}$$

or

$$\begin{bmatrix} A_1 \\ A_2 \end{bmatrix}=\begin{bmatrix} 1 & 1 \\ \frac{1}{4}+\frac{\sqrt{3}}{4}i & \frac{1}{4}-\frac{\sqrt{3}}{4}i \end{bmatrix}^{-1}\begin{bmatrix} -2 \\ -1 \end{bmatrix}=\begin{bmatrix} -1+\frac{1}{\sqrt{3}}i \\ -1-\frac{1}{\sqrt{3}}i \end{bmatrix}$$

and so

$$A_1=-1+\frac{1}{\sqrt{3}}i,\ A_2=-1-\frac{1}{\sqrt{3}}i.$$

We thus have the solution

$$\begin{aligned}
Y_t &= Y^*+A_1\left(\frac{1}{4}+\frac{\sqrt{3}}{4}i\right)^t+A_2\left(\frac{1}{4}-\frac{\sqrt{3}}{4}i\right)^t \\
&= 4+A_1\left(\frac{1}{2}e^{\frac{\pi}{3}i}\right)^t+A_2\left(\frac{1}{2}e^{\frac{\pi}{3}i}\right)^t \\
&= 4+\left(-1+\frac{1}{\sqrt{3}}i\right)\left(\frac{1}{2}\right)^t e^{\frac{\pi}{3}it}+\left(-1-\frac{1}{\sqrt{3}}i\right)\left(\frac{1}{2}\right)^t e^{-\frac{\pi}{3}it} \\
&= 4+\left(\frac{1}{2}\right)^t\left(-1+\frac{1}{\sqrt{3}}i\right)\left(\cos\left(\frac{\pi}{3}t\right)+i\sin\left(\frac{\pi}{3}t\right)\right) \\
&\quad +\left(\frac{1}{2}\right)^t\left(-1-\frac{1}{\sqrt{3}}i\right)\left(\cos\left(-\frac{\pi}{3}t\right)+i\sin\left(-\frac{\pi}{3}t\right)\right) \\
&= 4+\left(\frac{1}{2}\right)^t\left(-2\times\cos\left(\frac{\pi}{3}t\right)-2\times\frac{1}{\sqrt{3}}\sin\left(\frac{\pi}{3}t\right)\right)
\end{aligned}$$

or

$$Y_t=4-2\left(\frac{1}{2}\right)^t\left(\cos\left(\frac{\pi}{3}t\right)+\frac{1}{\sqrt{3}}\sin\left(\frac{\pi}{3}t\right)\right)$$

as shown in the figure below.

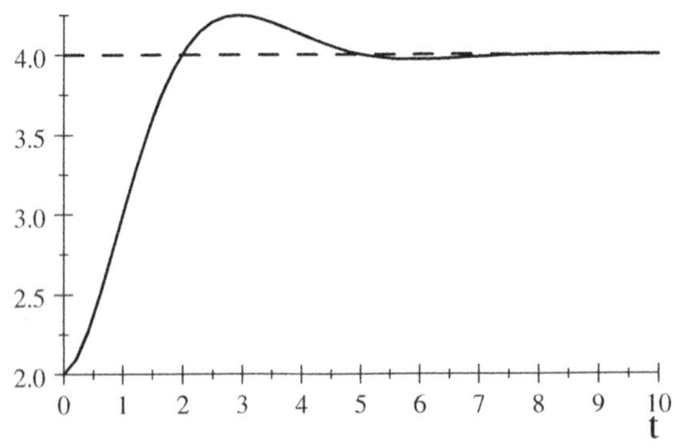

Since $\cos(\theta + 2\pi) = \cos(\theta)$ and $\sin(\theta + 2\pi) = \sin(\theta)$ the term

$$\cos\left(\frac{\pi}{3}t\right) + \frac{1}{\sqrt{3}}\sin\left(\frac{\pi}{3}t\right)$$

has a period P satisfying

$$\begin{aligned}\frac{\pi}{3}(t+P) &= \frac{\pi}{3}t + 2\pi \\ &\Longrightarrow P = \frac{2\pi}{\frac{\pi}{3}} = 6.\end{aligned}$$

Problem 7.23 *Determine if*

$$Y_t = 100 + 1.5Y_{t-1} - 0.625Y_{t-2}$$

is stable or unstable.

Answer: We have

$$\begin{aligned} r^2 - 1.5r + 0.625 &= 0 \Longrightarrow r = \frac{1.5 \pm \sqrt{(-1.5)^2 - 4 \times 0.625}}{2} \\ &\Longrightarrow r_1 = \frac{3}{4} + \frac{1}{4}i,\ r_2 = \frac{3}{4} - \frac{1}{4}i. \end{aligned}$$

This is stable since

$$|r_1| = |r_2| = \sqrt{\left(\frac{3}{4}\right)^2 + \left(\frac{1}{4}\right)^2} \approx 0.79 < 1.$$

Problem 7.24 *[D***] Consider a second-order difference equation of the form*

$$Y_t = 7 + \frac{1}{2}Y_{t-1} - \frac{1}{12}Y_{t-2} \text{ with } Y_0 = 10, Y_1 = 11.$$

*Find the equilibrium value Y^*. Show that Y_t is stable. [A***] Write Y_t in the form*

$$Y_t = Y^* + A_1 r_1^t + A_2 r_2^t$$

and simplify this to remove all complex terms. Show that the solution will exhibit a wave pattern that repeats itself every 12 periods.

Answer: The equilibrium value Y^* is

$$Y^* = \frac{7}{1 - \frac{1}{2} + \frac{1}{12}} = 12.$$

The characteristic polynomial has roots as

$$\begin{aligned} r^2 - \frac{1}{2}r + \frac{1}{12} &= 0 \\ &\Longrightarrow r_1, r_2 = \frac{\frac{1}{2} \pm \sqrt{\left(\frac{1}{2}\right)^2 - 4 \times \frac{1}{12}}}{2} = \frac{1}{4} \pm \frac{1}{4\sqrt{3}}i \\ &\Longrightarrow r_1 = \frac{1}{4} + \frac{1}{4\sqrt{3}}i, r_2 = \frac{1}{4} - \frac{1}{4\sqrt{3}}i. \end{aligned}$$

If the polar form for the roots is $\rho e^{\pm\theta i}$ then $\rho = \frac{1}{2\sqrt{3}}$ since

$$|r_1| = |r_2| = \sqrt{\left(\frac{1}{4}\right)^2 + \left(\frac{1}{4\sqrt{3}}\right)^2} = \frac{1}{2\sqrt{3}}$$

and

$$\theta = \arctan\left(\frac{\frac{1}{4\sqrt{3}}}{\frac{1}{4}}\right) = \arctan\left(\frac{1}{\sqrt{3}}\right) = \frac{\pi}{6}$$

so that

$$r_1 = \frac{1}{4} + \frac{1}{4\sqrt{3}}i = \frac{1}{2\sqrt{3}}e^{\frac{\pi}{6}i}, r_2 = \frac{1}{4} - \frac{1}{4\sqrt{3}}i = \frac{1}{2\sqrt{3}}e^{-\frac{\pi}{6}i}.$$

Stability requires $|r_1| = |r_2| < 1$ and

$$|r_1| = |r_2| = \sqrt{\left(\frac{1}{4}\right)^2 + \left(\frac{1}{4\sqrt{3}}\right)^2} = \frac{1}{2\sqrt{3}} \approx 0.29 < 1$$

it so it follows that Y_t is stable.

The general solution is then

$$Y_t = Y^* + A_1 r_1^t + A_2 r_2^t = 12 + A_1\left(\frac{1}{4} + \frac{1}{4\sqrt{3}}i\right)^t + A_2\left(\frac{1}{4} - \frac{1}{4\sqrt{3}}i\right)^t$$

where A_1 and A_2 depend on the starting values $Y_o = 10$ and $Y_1 = 11$ as

$$\begin{aligned} 10 &= 12 + A_1\left(\frac{1}{4} + \frac{1}{4\sqrt{3}}i\right)^0 + A_2\left(\frac{1}{4} - \frac{1}{4\sqrt{3}}i\right)^0 \\ 11 &= 12 + A_1\left(\frac{1}{4} + \frac{1}{4\sqrt{3}}i\right)^1 + A_2\left(\frac{1}{4} - \frac{1}{4\sqrt{3}}i\right)^1 \end{aligned}$$

so that in matrix notation

$$\begin{bmatrix} 1 & 1 \\ \frac{1}{4} + \frac{1}{4\sqrt{3}}i & \frac{1}{4} - \frac{1}{4\sqrt{3}}i \end{bmatrix}\begin{bmatrix} A_1 \\ A_2 \end{bmatrix} = \begin{bmatrix} -2 \\ -1 \end{bmatrix}$$

or

$$\begin{bmatrix} A_1 \\ A_2 \end{bmatrix} = \begin{bmatrix} 1 & 1 \\ \frac{1}{4} + \frac{1}{4\sqrt{3}}i & \frac{1}{4} - \frac{1}{4\sqrt{3}}i \end{bmatrix}^{-1}\begin{bmatrix} -2 \\ -1 \end{bmatrix} = \begin{bmatrix} -1 + i\sqrt{3} \\ -1 - i\sqrt{3} \end{bmatrix}$$

where

$$\begin{aligned} A_1 &= -1 + i\sqrt{3} \\ A_2 &= -1 - i\sqrt{3}. \end{aligned}$$

We thus have the solution

$$\begin{aligned}
Y_t &= Y^* + A_1\left(\frac{1}{4}+\frac{1}{4\sqrt{3}}i\right)^t + A_2\left(\frac{1}{4}-\frac{1}{4\sqrt{3}}i\right)^t = 12 + A_1\left(\frac{1}{2\sqrt{3}}e^{\frac{\pi}{6}it}\right)^t + A_2\left(\frac{1}{2\sqrt{3}}e^{-\frac{\pi}{6}it}\right)^t \\
&= 12 + \left(-1+i\sqrt{3}\right)\left(\frac{1}{2\sqrt{3}}\right)^t e^{\frac{\pi}{6}it} + \left(-1-i\sqrt{3}\right)\left(\frac{\frac{1}{2}}{\sqrt{3}}\right)^t e^{-\frac{\pi}{6}it} \\
&= 12 + \left(\frac{1}{2\sqrt{3}}\right)^t \left(-1+i\sqrt{3}\right)\left(\cos\left(\frac{\pi}{6}t\right)+i\sin\left(\frac{\pi}{6}t\right)\right) \\
&\quad + \left(\frac{1}{2\sqrt{3}}\right)^t \left(-1-i\sqrt{3}\right)\left(\cos\left(-\frac{\pi}{6}t\right)+i\sin\left(-\frac{\pi}{6}t\right)\right) \\
&= 12 + \left(\frac{1}{2\sqrt{3}}\right)^t \left(-2\times\cos\left(\frac{\pi}{6}t\right) + -2\times\sqrt{3}\sin\left(\frac{\pi}{6}t\right)\right)
\end{aligned}$$

or

$$Y_t = 12 - 2\left(\frac{1}{2\sqrt{3}}\right)^t\left(\cos\left(\frac{\pi}{6}t\right)+\sqrt{3}\sin\left(\frac{\pi}{6}t\right)\right).$$

Since $\cos(\theta+2\pi) = \cos(\theta)$ and $\sin(\theta+2\pi) = \sin(\theta)$ the term

$$\cos\left(\frac{\pi}{6}t\right)+\sqrt{3}\sin\left(\frac{\pi}{6}t\right)$$

has a period P satisfying

$$\frac{\pi}{6}(t+P) = \frac{\pi}{6}t + 2\pi \Longrightarrow P = \frac{2\pi}{\frac{\pi}{6}} = 12.$$

Problem 7.25 *[B***] Consider a second-order difference equation of the form*

$$Y_t = \alpha + \phi Y_{t-1} - \frac{\phi^2}{2}Y_{t-2}$$

where $\phi > 0$ and for given starting values Y_0, Y_1. Find the equilibrium value Y^*. Show that the roots of the characteristic polynomial are given by

$$r_1 = \frac{\phi}{2}+\frac{\phi}{2}i = \frac{\phi}{\sqrt{2}}e^{\frac{\pi}{4}i}, r_2 = \frac{\phi}{2}-\frac{\phi}{2}i = \frac{\phi}{\sqrt{2}}e^{-\frac{\pi}{4}i}.$$

*Show that Y_t is stable if $\phi < \sqrt{2}$ and unstable otherwise. [A***] Show that*

$$Y_t = Y^* + 2\times\left(\frac{\phi}{\sqrt{2}}\right)^t\left(\frac{Y_0^*}{2}\times\cos\left(\frac{\pi}{4}t\right) - \left(\frac{Y_0^*}{2}-\frac{Y_1^*}{\phi}\right)\times\sin\left(\frac{\pi}{4}t\right)\right)$$

where $Y_0^* = Y_0 - Y^*$ and $Y_1^* = Y_1 - Y^*$. Show the term involving $\cos(\ )$ *and* $\sin(\ )$ *repeats itself every 8 periods.*

Answer: The equilibrium value Y^* is

$$Y^* = \frac{\alpha}{1-\phi+\frac{\phi^2}{2}}.$$

The characteristic polynomial has roots as

$$\begin{aligned} r^2 - \phi r + \frac{\phi^2}{2} &= 0 \\ \implies r_1, r_2 &= \frac{\phi \pm \sqrt{\phi^2 - 4 \times \frac{\phi^2}{2}}}{2} = \frac{\phi}{2} \pm \frac{\phi}{2} i \\ \implies r_1 &= \frac{\phi}{2} + \frac{\phi}{2} i, r_2 = \frac{\phi}{2} - \frac{\phi}{2} i. \end{aligned}$$

If the polar form for the roots is $\rho e^{\pm \theta i}$ then $\rho = \frac{\phi}{\sqrt{2}}$ since

$$\rho = |r_1| = |r_2| = \sqrt{\left(\frac{\phi}{2}\right)^2 + \left(\frac{\phi}{2}\right)^2} = \frac{\phi}{\sqrt{2}}$$

and

$$\theta = \arctan\left(\frac{\frac{\phi}{2}}{\frac{\phi}{2}}\right) = \arctan(1) = \frac{\pi}{4}$$

so that

$$\begin{aligned} r_1 &= \frac{\phi}{2} + \frac{\phi}{2} i = \frac{\phi}{\sqrt{2}} e^{\frac{\pi}{4} i} \\ r_2 &= \frac{\phi}{2} - \frac{\phi}{2} i = \frac{\phi}{\sqrt{2}} e^{-\frac{\pi}{4} i}. \end{aligned}$$

Stability requires $|r_1| = |r_2| < 1$ and

$$|r_1| = |r_2| = \sqrt{\left(\frac{\phi}{2}\right)^2 + \left(\frac{\phi}{2}\right)^2} = \frac{\phi}{\sqrt{2}}.$$

Thus Y_t is stable if and only if $\phi < \sqrt{2}$.

The general solution is then

$$Y_t = Y^* + A_1 r_1^t + A_2 r_2^t = Y^* + A_1 \left(\frac{\phi}{2} + \frac{\phi}{2} i\right)^t + A_2 \left(\frac{\phi}{2} - \frac{\phi}{2} i\right)^t$$

where A_1 and A_2 depend on the starting values.

Given Y_o and Y_1 we have

$$\begin{aligned} Y_o &= Y^* + A_1 \left(\frac{\phi}{2} + \frac{\phi}{2} i\right)^0 + A_2 \left(\frac{\phi}{2} - \frac{\phi}{2} i\right)^0 \\ Y_1 &= Y^* + A_1 \left(\frac{\phi}{2} + \frac{\phi}{2} i\right)^1 + A_2 \left(\frac{\phi}{2} - \frac{\phi}{2} i\right)^1 \end{aligned}$$

so that in matrix notation

$$\begin{bmatrix} 1 & 1 \\ \frac{\phi}{2} + \frac{\phi}{2} i & \frac{\phi}{2} - \frac{\phi}{2} i \end{bmatrix} \begin{bmatrix} A_1 \\ A_2 \end{bmatrix} = \begin{bmatrix} Y_0 - Y^* \\ Y_1 - Y^* \end{bmatrix} = \begin{bmatrix} Y_0^* \\ Y_1^* \end{bmatrix}$$

or

$$\begin{bmatrix} A_1 \\ A_2 \end{bmatrix} = \begin{bmatrix} 1 & 1 \\ \frac{\phi}{2} + \frac{\phi}{2}i & \frac{\phi}{2} - \frac{\phi}{2}i \end{bmatrix}^{-1} \begin{bmatrix} Y_0^* \\ Y_1^* \end{bmatrix} = \begin{bmatrix} c + di \\ c - di \end{bmatrix}$$

where

$$\begin{aligned} A_1 &= c + di = \frac{Y_0^*}{2} + \left(\frac{Y_0^*}{2} - \frac{Y_1^*}{\phi} \right) i \\ A_2 &= c - di = \frac{Y_0^*}{2} - \left(\frac{Y_0^*}{2} - \frac{Y_1^*}{\phi} \right) i. \end{aligned}$$

We thus have the solution

$$\begin{aligned} Y_t &= Y^* + A_1 \left(\frac{\phi}{2} + \frac{\phi}{2} i \right)^t + A_2 \left(\frac{\phi}{2} - \frac{\phi}{2} i \right)^t \\ &= Y^* + A_1 \left(\frac{\phi}{\sqrt{2}} e^{\frac{\pi}{4} i} \right)^t + A_2 \left(\frac{\phi}{\sqrt{2}} e^{-\frac{\pi}{4} i} \right)^t \\ &= Y^* + A_1 \left(\frac{\phi}{\sqrt{2}} \right)^t e^{\frac{\pi}{4} ti} + A_2 \left(\frac{\phi}{\sqrt{2}} \right)^t e^{-\frac{\pi}{4} ti} \\ &= Y^* + (c + di) \left(\frac{\phi}{\sqrt{2}} \right)^t e^{\frac{\pi}{4} it} + (c - di) \left(\frac{\phi}{\sqrt{2}} \right)^t e^{-\frac{\pi}{4} it} \\ &= Y^* + \left(\frac{\phi}{\sqrt{2}} \right)^t (c + di) \left(\cos\left(\frac{\pi}{4} t \right) + i \sin\left(\frac{\pi}{4} t \right) \right) \\ &\quad + \left(\frac{\phi}{\sqrt{2}} \right)^t (c - di) \left(\cos\left(-\frac{\pi}{4} t \right) + i \sin\left(-\frac{\pi}{4} t \right) \right) \\ &= Y^* + \left(\frac{\phi}{\sqrt{2}} \right)^t (c + di) \left(\cos\left(\frac{\pi}{4} t \right) + i \sin\left(\frac{\pi}{4} t \right) \right) \\ &\quad + \left(\frac{\phi}{\sqrt{2}} \right)^t (c - di) \left(\cos\left(\frac{\pi}{4} t \right) - i \sin\left(\frac{\pi}{4} t \right) \right) \\ &= Y^* + 2 \times \left(\frac{\phi}{\sqrt{2}} \right)^t \left(c \times \cos\left(\frac{\pi}{4} t \right) - d \times \sin\left(\frac{\pi}{4} t \right) \right) \end{aligned}$$

or

$$Y_t = Y^* + 2 \times \left(\frac{\phi}{\sqrt{2}} \right)^t \left(\frac{Y_0^*}{2} \times \cos\left(\frac{\pi}{4} t \right) - \left(\frac{Y_0^*}{2} - \frac{Y_1^*}{\phi} \right) \times \sin\left(\frac{\pi}{4} t \right) \right).$$

Since $\cos(\theta + 2\pi) = \cos(\theta)$ and $\sin(\theta + 2\pi) = \sin(\theta)$ the term

$$c \times \cos\left(\frac{\pi}{4} t \right) - d \times \sin\left(\frac{\pi}{4} t \right)$$

has a period P satisfying

$$\frac{\pi}{4} (t + P) = \frac{\pi}{4} t + 2\pi \Longrightarrow P = \frac{2\pi}{\frac{\pi}{4}} = 8.$$

Problem 7.26 *[B***] Consider a second-order difference equation of the form*

$$Y_t = \alpha + \phi Y_{t-1} - \phi^2 Y_{t-2}$$

where $\phi > 0$ *and for given starting values* Y_0, Y_1*.Find the equilibrium value* Y^**. Show that the roots of the characteristic polynomial are given by*

$$r_1 = \frac{\phi}{2} + \frac{\sqrt{3}\phi}{2}i = \phi e^{\frac{\pi}{3}i}, r_2 = \frac{\phi}{2} - \frac{\sqrt{3}\phi}{2}i = \phi e^{-\frac{\pi}{3}i}.$$

Show that Y_t *is stable if* $\phi < 1$ *and unstable otherwise. [A***] Show that*

$$Y_t = Y^* + 2\phi^t \left(\frac{Y_0^*}{2} \times \cos\left(\frac{\pi}{3}t\right) - \frac{1}{\sqrt{3}} \left(\frac{Y_0^*}{2} - \frac{Y_1^*}{\phi} \right) \times \sin\left(\frac{\pi}{3}t\right) \right)$$

where $Y_0^* = Y_0 - Y^*$ *and* $Y_1^* = Y_1 - Y^*$*. Show the term involving* $\cos(\,)$ *and* $\sin(\,)$ *repeats itself every* 6 *periods.*

Answer: The equilibrium value Y^* is

$$Y^* = \frac{\alpha}{1 - \phi + \phi^2}.$$

The characteristic polynomial has roots as

$$r^2 - \phi r + \phi^2 = 0 \Longrightarrow r_1, r_2 = \frac{\phi \pm \sqrt{\phi^2 - 4\phi^2}}{2} \Longrightarrow r_1 = \frac{\phi}{2} + \frac{\sqrt{3}\phi}{2}i,\ r_2 = \frac{\phi}{2} - \frac{\sqrt{3}\phi}{2}i.$$

If the polar form for the roots is $\rho e^{\pm\theta i}$ then $\rho = \phi$ since

$$\rho = |r_1| = |r_2| = \sqrt{\left(\frac{\phi}{2}\right)^2 + \left(\frac{\sqrt{3}\phi}{2}\right)^2} = \phi$$

and $\theta = \frac{\pi}{3}$ since

$$\theta = \arctan\left(\frac{\frac{\sqrt{3}\phi}{2}}{\frac{\phi}{2}} \right) = \arctan\left(\sqrt{3}\right) = \frac{\pi}{3}$$

so that

$$r_1 = \frac{\phi}{2} + \frac{\sqrt{3}\phi}{2}i = \phi e^{\frac{\pi}{3}i}, r_2 = \frac{\phi}{2} - \frac{\sqrt{3}\phi}{2}i = \phi e^{-\frac{\pi}{3}i}.$$

Stability requires $|r_1| = |r_2| < 1$ and

$$|r_1| = |r_2| = \sqrt{\left(\frac{\phi}{2}\right)^2 + \left(\frac{\sqrt{3}\phi}{2}\right)^2} = \phi$$

Thus Y_t is stable if and only if $\phi < 1$.

The general solution is then

$$\begin{aligned} Y_t &= Y^* + A_1 r_1^t + A_2 r_2^t \\ &= Y^* + A_1 \left(\frac{\phi}{2} + \frac{\sqrt{3}\phi}{2} i \right)^t + A_2 \left(\frac{\phi}{2} - \frac{\sqrt{3}\phi}{2} i \right)^t \end{aligned}$$

where A_1 and A_2 depend on the starting values Y_o and Y_1 as

$$\begin{aligned} Y_o &= Y^* + A_1 \left(\frac{\phi}{2} + \frac{\sqrt{3}\phi}{2} i \right)^0 + A_2 \left(\frac{\phi}{2} - \frac{\sqrt{3}\phi}{2} i \right)^0 \\ Y_1 &= Y^* + A_1 \left(\frac{\phi}{2} + \frac{\sqrt{3}\phi}{2} i \right)^1 + A_2 \left(\frac{\phi}{2} - \frac{\sqrt{3}\phi}{2} i \right)^1 \end{aligned}$$

so that in matrix notation

$$\begin{bmatrix} 1 & 1 \\ \frac{\phi}{2} + \frac{\sqrt{3}\phi}{2} i & \frac{\phi}{2} - \frac{\sqrt{3}\phi}{2} i \end{bmatrix} \begin{bmatrix} A_1 \\ A_2 \end{bmatrix} = \begin{bmatrix} Y_o - Y^* \\ Y_1 - Y^* \end{bmatrix} = \begin{bmatrix} Y_0^* \\ Y_1^* \end{bmatrix}$$

or

$$\begin{bmatrix} A_1 \\ A_2 \end{bmatrix} = \begin{bmatrix} 1 & 1 \\ \frac{\phi}{2} + \frac{\sqrt{3}\phi}{2} i & \frac{\phi}{2} - \frac{\sqrt{3}\phi}{2} i \end{bmatrix}^{-1} \begin{bmatrix} Y_0^* \\ Y_1^* \end{bmatrix} = \begin{bmatrix} c + di \\ c - di \end{bmatrix}$$

where

$$\begin{aligned} A_1 &= c + di = \frac{Y_0^*}{2} + \frac{1}{\sqrt{3}} \left(\frac{Y_0^*}{2} - \frac{Y_1^*}{\phi} \right) i \\ A_2 &= c - di = \frac{Y_0^*}{2} - \frac{1}{\sqrt{3}} \left(\frac{Y_0^*}{2} - \frac{Y_1^*}{\phi} \right) i. \end{aligned}$$

We thus have the solution

$$\begin{aligned} Y_t &= Y^* + A_1 \left(\frac{\phi}{2} + \frac{\sqrt{3}\phi}{2} i \right)^t + A_2 \left(\frac{\phi}{2} + \frac{\sqrt{3}\phi}{2} i \right)^t \\ &= Y^* + A_1 \left(\phi e^{\frac{\pi}{3} i} \right)^t + A_2 \left(\phi e^{\frac{\pi}{3} i} \right)^t = Y^* + (c + di) \phi^t e^{\frac{\pi}{3} it} + (c - di) \phi^t e^{\frac{\pi}{3} it} \\ &= Y^* + \phi^t (c + di) \left(\cos\left(\frac{\pi}{3} t \right) + i \sin\left(\frac{\pi}{3} t \right) \right) \\ &\quad + \phi^t (c - di) \left(\cos\left(-\frac{\pi}{3} t \right) + i \sin\left(-\frac{\pi}{3} t \right) \right) \\ &= Y^* + \phi^t (c + di) \left(\cos\left(\frac{\pi}{3} t \right) + i \sin\left(\frac{\pi}{3} t \right) \right) \\ &\quad + \phi^t (c - di) \left(\cos\left(\frac{\pi}{3} t \right) - i \sin\left(\frac{\pi}{3} t \right) \right) \\ &= Y^* + 2\phi^t \left(c \times \cos\left(\frac{\pi}{3} t \right) - d \times \sin\left(\frac{\pi}{3} t \right) \right). \end{aligned}$$

or

$$\begin{aligned} Y_t &= Y^* + 2\phi^t \left(c \times \cos\left(\frac{\pi}{3}t\right) - d \times \sin\left(\frac{\pi}{3}t\right)\right) \\ &= Y^* + 2\phi^t \left(\frac{Y_0^*}{2} \times \cos\left(\frac{\pi}{3}t\right) - \frac{1}{\sqrt{3}}\left(\frac{Y_0^*}{2} - \frac{Y_1^*}{\phi}\right) \times \sin\left(\frac{\pi}{3}t\right)\right). \end{aligned}$$

Since $\cos(\theta + 2\pi) = \cos(\theta)$ and $\sin(\theta + 2\pi) = \sin(\theta)$ the term

$$c \times \cos\left(\frac{\pi}{3}t\right) - d \times \sin\left(\frac{\pi}{3}t\right)$$

has a period P satisfying

$$\frac{\pi}{3}(t + P) = \frac{\pi}{3}t + 2\pi \Longrightarrow P = \frac{2\pi}{\frac{\pi}{3}} = 6.$$

Problem 7.27 *[B***] Consider a second-order difference equation of the form*

$$Y_t = \alpha + \phi Y_{t-1} - \frac{\phi^2}{3} Y_{t-2}$$

*where $\phi > 0$ and for given starting values Y_0, Y_1. Find the equilibrium value Y^*. [A***] Show that the roots of the characteristic polynomial are given by*

$$r_1 = \frac{\phi}{2} + \frac{\phi}{2\sqrt{3}}i = \frac{\phi}{\sqrt{3}}e^{\frac{\pi}{6}i}, r_2 = \frac{\phi}{2} - \frac{\phi}{2\sqrt{3}}i = \frac{\phi}{\sqrt{3}}e^{-\frac{\pi}{6}i}.$$

Show that Y_t is stable if $\phi < \sqrt{3}$ and unstable otherwise. Show that

$$Y_t = Y^* + 2\left(\frac{\phi}{\sqrt{3}}\right)^t \left(\frac{Y^*}{2} \times \cos\left(\frac{\pi}{6}t\right) - \sqrt{3}\left(\frac{Y_0^*}{2} - \frac{Y_1^*}{\phi}\right) \times \sin\left(\frac{\pi}{6}t\right)\right)$$

where $Y_0^* = Y_0 - Y^*$ and $Y_1^* = Y_1 - Y^*$. Show the term involving $\cos(\,)$ *and* $\sin(\,)$ *repeats itself every* 12 *periods.*

Answer: The equilibrium value Y^* is

$$Y^* = \frac{\alpha}{1 - \phi + \frac{\phi^2}{3}}.$$

The characteristic polynomial has roots as

$$\begin{aligned} r^2 - \phi r + \frac{\phi^2}{3} &= 0 \\ \Longrightarrow r_1, r_2 &= \frac{\phi \pm \sqrt{\phi^2 - 4 \times \frac{\phi^2}{3}}}{2} = \frac{\phi}{2} \pm \frac{\phi}{2\sqrt{3}}i \\ \Longrightarrow r_1 &= \frac{\phi}{2} + \frac{\phi}{2\sqrt{3}}i, r_2 = \frac{\phi}{2} - \frac{\phi}{2\sqrt{3}}i. \end{aligned}$$

If the polar form for the roots is $\rho e^{\pm\theta i}$ then $\rho = \frac{\phi}{\sqrt{3}}$ since

$$|r_1| = |r_2| = \sqrt{\left(\frac{\phi}{2}\right)^2 + \left(\frac{\phi}{2\sqrt{3}}\right)^2} = \frac{\phi}{\sqrt{3}}$$

and $\theta = \frac{\pi}{6}$ since

$$\theta = \arctan\left(\frac{\frac{\phi}{2}}{\frac{\sqrt{3}\phi}{2}}\right) = \arctan\left(\frac{1}{\sqrt{3}}\right) = \frac{\pi}{6}$$

so that

$$\begin{aligned} r_1 &= \frac{\phi}{2} + \frac{\phi}{2\sqrt{3}}i = \frac{\phi}{\sqrt{3}}e^{\frac{\pi}{6}i} \\ r_2 &= \frac{\phi}{2} - \frac{\phi}{2\sqrt{3}}i = \frac{\phi}{\sqrt{3}}e^{-\frac{\pi}{6}i}. \end{aligned}$$

Stability requires $|r_1| = |r_2| < 1$ and

$$|r_1| = |r_2| = \sqrt{\left(\frac{\phi}{2}\right)^2 + \left(\frac{\phi}{2\sqrt{3}}\right)^2} = \frac{\phi}{\sqrt{3}}.$$

It follows that Y_t is stable if and only if $\phi < \sqrt{3}$.

The general solution is then

$$Y_t = Y^* + A_1 r_1^t + A_2 r_2^t = Y^* + A_1\left(\frac{\phi}{2} + \frac{\phi}{2\sqrt{3}}i\right)^t + A_2\left(\frac{\phi}{2} - \frac{\phi}{2\sqrt{3}}i\right)^t$$

where A_1 and A_2 depend on the starting values Y_o, Y_1 as

$$\begin{aligned} Y_o &= Y^* + A_1\left(\frac{\phi}{2} + \frac{\phi}{2\sqrt{3}}i\right)^0 + A_2\left(\frac{\phi}{2} - \frac{\phi}{2\sqrt{3}}i\right)^0 \\ Y_1 &= Y^* + A_1\left(\frac{\phi}{2} + \frac{\phi}{2\sqrt{3}}i\right)^1 + A_2\left(\frac{\phi}{2} - \frac{\phi}{2\sqrt{3}}i\right)^1 \end{aligned}$$

so that in matrix notation

$$\begin{bmatrix} 1 & 1 \\ \frac{\phi}{2} + \frac{\phi}{2\sqrt{3}}i & \frac{\phi}{2} - \frac{\phi}{2\sqrt{3}}i \end{bmatrix} \begin{bmatrix} A_1 \\ A_2 \end{bmatrix} = \begin{bmatrix} Y_o - Y^* \\ Y_1 - Y^* \end{bmatrix} = \begin{bmatrix} Y_0^* \\ Y_1^* \end{bmatrix}$$

or

$$\begin{bmatrix} A_1 \\ A_2 \end{bmatrix} = \begin{bmatrix} 1 & 1 \\ \frac{\phi}{2} + \frac{\phi}{2\sqrt{3}}i & \frac{\phi}{2} - \frac{\phi}{2\sqrt{3}}i \end{bmatrix}^{-1} \begin{bmatrix} Y_0^* \\ Y_1^* \end{bmatrix} = \begin{bmatrix} c + di \\ c - di \end{bmatrix}$$

where

$$\begin{aligned} A_1 &= c + di = \frac{Y_0^*}{2} + \sqrt{3}\left(\frac{Y_0^*}{2} - \frac{Y_1^*}{\phi}\right)i \\ A_2 &= c - di = \frac{Y_0^*}{2} - \sqrt{3}\left(\frac{Y_0^*}{2} - \frac{Y_1^*}{\phi}\right)i. \end{aligned}$$

We thus have the solution

$$\begin{aligned}
Y_t &= Y^* + A_1\left(\frac{\phi}{2} + \frac{\phi}{2\sqrt{3}}i\right)^t + A_2\left(\frac{\phi}{2} - \frac{\phi}{2\sqrt{3}}i\right)^t \\
&= Y^* + A_1\left(\frac{\phi}{\sqrt{3}}e^{\frac{\pi}{6}it}\right)^t + A_2\left(\frac{\phi}{\sqrt{3}}e^{-\frac{\pi}{6}it}\right)^t \\
&= Y^* + (c+di)\left(\frac{\phi}{\sqrt{3}}\right)^t e^{\frac{\pi}{6}it} + (c-di)\left(\frac{\phi}{\sqrt{3}}\right)^t e^{-\frac{\pi}{6}it} \\
&= Y^* + \left(\frac{\phi}{\sqrt{3}}\right)^t (c+di)\left(\cos\left(\frac{\pi}{6}t\right) + i\sin\left(\frac{\pi}{6}t\right)\right) \\
&\quad + \left(\frac{\phi}{\sqrt{3}}\right)^t (c-di)\left(\cos\left(-\frac{\pi}{6}t\right) + i\sin\left(-\frac{\pi}{6}t\right)\right) \\
&= Y^* + \left(\frac{\phi}{\sqrt{3}}\right)^t (c+di)\left(\cos\left(\frac{\pi}{6}t\right) + i\sin\left(\frac{\pi}{6}t\right)\right) \\
&\quad + \left(\frac{\phi}{\sqrt{3}}\right)^t (c-di)\left(\cos\left(\frac{\pi}{6}t\right) - i\sin\left(\frac{\pi}{6}t\right)\right) \\
&= Y^* + 2\left(\frac{\phi}{\sqrt{3}}\right)^t \left(c\times\cos\left(\frac{\pi}{6}t\right) - d\times\sin\left(\frac{\pi}{6}t\right)\right)
\end{aligned}$$

or

$$\begin{aligned}
Y_t &= Y^* + 2\left(\frac{\phi}{\sqrt{3}}\right)^t \left(c\times\cos\left(\frac{\pi}{6}t\right) - d\times\sin\left(\frac{\pi}{6}t\right)\right) \\
&= Y^* + 2\left(\frac{\phi}{\sqrt{3}}\right)^t \left(\frac{Y^*}{2}\times\cos\left(\frac{\pi}{6}t\right) - \sqrt{3}\left(\frac{Y_0^*}{2} - \frac{Y_1^*}{\phi}\right)\times\sin\left(\frac{\pi}{6}t\right)\right).
\end{aligned}$$

Since $\cos(\theta + 2\pi) = \cos(\theta)$ and $\sin(\theta + 2\pi) = \sin(\theta)$ the term

$$c\times\cos\left(\frac{\pi}{6}t\right) - d\times\sin\left(\frac{\pi}{6}t\right)$$

has a period P satisfying

$$\frac{\pi}{6}(t+P) = \frac{\pi}{6}t + 2\pi \Longrightarrow P = \frac{2\pi}{\frac{\pi}{6}} = 12.$$

7.2 Differential Equations with Complex Roots

Problem 7.28 *[D***] Consider*

$$Y''(t) + 3Y'(t) + 5Y(t) = 10 \textit{ with } Y(0) = 1,\ Y'(0) = 2$$

*If $Y(t) = Y^* + A_1 e^{r_1 t} + A_2 e^{r_2 t}$, calculate Y^* and explain what it is. Calculate r_1 and r_2 and determine whether or not this differential equation is stable. [B***] Find A_1 and A_2.*

Answer: We have $Y^* = \frac{10}{5} = 2$. As well

$$r^2 + 3r + 5 = 0 \Longrightarrow r_1 = -\frac{3}{2} + \frac{\sqrt{11}}{2}i, \; r_2 = -\frac{3}{2} - \frac{\sqrt{11}}{2}i.$$

Since the real part of r_1 and r_2, here $-\frac{3}{2}$, is negative, it follows that the differential equation is stable.

It follows that

$$\begin{bmatrix} 1 & 1 \\ -\frac{3}{2} + \frac{\sqrt{11}}{2}i & -\frac{3}{2} - \frac{\sqrt{11}}{2}i \end{bmatrix} \begin{bmatrix} A_1 \\ A_2 \end{bmatrix} = \begin{bmatrix} 1 \\ 2 \end{bmatrix}$$

so

$$\begin{bmatrix} A_1 \\ A_2 \end{bmatrix} = \begin{bmatrix} 1 & 1 \\ -\frac{3}{2} + \frac{\sqrt{11}}{2}i & -\frac{3}{2} - \frac{\sqrt{11}}{2}i \end{bmatrix}^{-1} \begin{bmatrix} 1 \\ 2 \end{bmatrix} = \begin{bmatrix} \frac{1}{2} - \frac{7\sqrt{11}}{22}i \\ \frac{1}{2} + \frac{7\sqrt{11}}{22}i \end{bmatrix}.$$

The exact solution is

$$\begin{aligned} Y(t) &= 2 + \left(\frac{1}{2} - \frac{7\sqrt{11}}{22}i\right) e^{\left(-\frac{3}{2} + \frac{\sqrt{11}}{2}i\right)t} + \left(\frac{1}{2} + \frac{7\sqrt{11}}{22}i\right) e^{\left(-\frac{3}{2} - \frac{\sqrt{11}}{2}i\right)t} \\ &= 2 - e^{-\frac{3}{2}t} \cos\left(\frac{\sqrt{11}}{2}t\right) + e^{-\frac{3}{2}t} \frac{1}{\sqrt{11}} \sin\left(\frac{\sqrt{11}}{2}t\right) \end{aligned}$$

which is shown below.

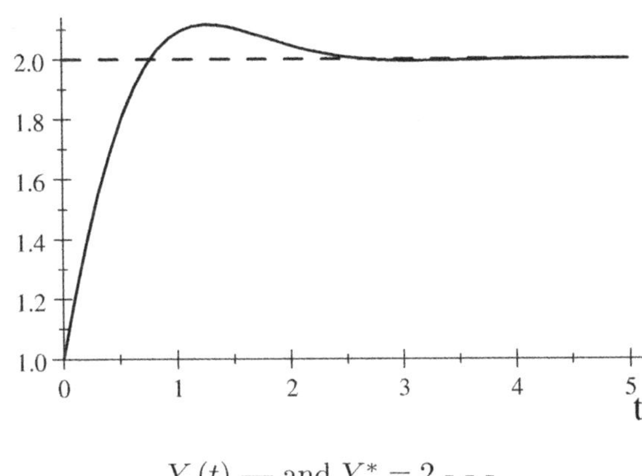

$Y(t)$ — and $Y^* = 2$ - - -

Problem 7.29 *[D***] Consider*

$$Y''(t) - 3Y'(t) + 5Y(t) = 10, \textit{ with } Y(0) = 3, \; Y'(0) = 2$$

*If $Y(t) = Y^* + A_1 e^{r_1 t} + A_2 e^{r_2 t}$, calculate Y^* and explain what it is. Calculate r_1 and r_2 and determine whether or not this differential equation is stable. [B***] Find A_1 and A_2 and the solution to $Y(t)$.*

Answer: We have $Y^* = \frac{10}{5} = 2$. As well

$$\begin{aligned} r^2 - 3r + 5 &= 0 \\ &\implies r_1 = \frac{3}{2} + \frac{\sqrt{11}}{2}i,\ r_2 = \frac{3}{2} - \frac{\sqrt{11}}{2}i. \end{aligned}$$

Since the real part of r_1 and r_2, here $\frac{3}{2}$, is positive, it follows that the differential equation is unstable.

It follows that

$$\begin{bmatrix} 1 & 1 \\ \frac{3}{2} + \frac{\sqrt{11}}{2}i & \frac{3}{2} - \frac{\sqrt{11}}{2}i \end{bmatrix} \begin{bmatrix} A_1 \\ A_2 \end{bmatrix} = \begin{bmatrix} 1 \\ 2 \end{bmatrix}$$

so

$$\begin{bmatrix} A_1 \\ A_2 \end{bmatrix} = \begin{bmatrix} 1 & 1 \\ \frac{3}{2} + \frac{\sqrt{11}}{2}i & \frac{3}{2} - \frac{\sqrt{11}}{2}i \end{bmatrix}^{-1} \begin{bmatrix} 1 \\ 2 \end{bmatrix} = \begin{bmatrix} \frac{1}{2} - \frac{\sqrt{11}}{22}i \\ \frac{1}{2} + \frac{\sqrt{11}}{22}i \end{bmatrix}.$$

The exact solution is

$$\begin{aligned} Y(t) &= 2 + \left(\frac{1}{2} - \frac{\sqrt{11}}{22}i\right) e^{\left(\frac{3}{2} + \frac{\sqrt{11}}{2}i\right)t} + \left(\frac{1}{2} + \frac{\sqrt{11}}{22}i\right) e^{\left(\frac{3}{2} - \frac{\sqrt{11}}{2}i\right)t} \\ &= 2 + e^{\frac{3}{2}t}\left(\cos\left(\frac{\sqrt{11}}{2}t\right) + \frac{1}{\sqrt{11}}\sin\left(\frac{\sqrt{11}}{2}t\right)\right) \end{aligned}$$

which is shown below.

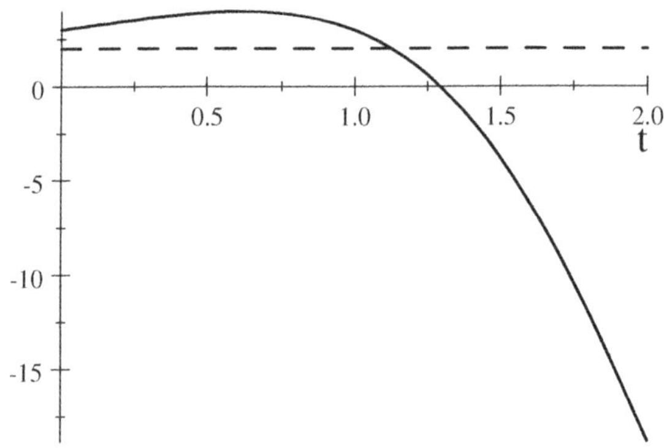

$Y(t)$ — and $Y^* = 2$ - - -

Problem 7.30 *[B***] Determine if*

$$Y''(t) - 4Y'(t) + 5Y(t) = 1$$

is stable or unstable

Answer: For

$$Y''(t) - 4Y'(t) + 5Y(t) = 1$$

the characteristic polynomial has roots as

$$\begin{aligned} r^2 - 4r + 5 &= 0 \Longrightarrow r = \frac{4 \pm \sqrt{(4)^2 - 4 \times 5}}{2} \\ &\Longrightarrow r_1 = 2 + i,\ r_2 = 2 - i. \end{aligned}$$

Since the real part of r_1 and r_2 is 2 which is positive, $Y(t)$ is unstable.

Problem 7.31 *[B***] Consider a second-order differential equation of the form*

$$Y''(t) + \phi Y'(t) + \frac{\phi^2}{2} Y(t) = \alpha$$

where $\phi > 0$ and for given starting values $Y(0), Y'(0)$. Find the equilibrium value Y^*. Show that the roots of the characteristic polynomial are given by

$$r_1 = -\frac{\phi}{2} + \frac{\phi}{2} i,\ r_2 = -\frac{\phi}{2} - \frac{\phi}{2} i.$$

Show that $Y(t)$ is stable. Show that

$$Y_t = Y^* + 2e^{-\frac{\phi}{2}t} \left(\frac{\tilde{Y}(0)}{2} \times \cos\left(\frac{\phi}{2}\right) - \left(\frac{\tilde{Y}(0)}{2} + \frac{Y'(0)}{\phi}\right) \times \sin\left(\frac{\phi}{2}\right) \right)$$

where $\tilde{Y}(0) = Y(0) - Y^*$. Show the term involving $\cos(\,)$ and $\sin(\,)$ repeats itself every $\frac{8\pi}{\phi}$ periods.

Answer: The equilibrium value Y^* is

$$Y^* = \frac{\alpha}{\frac{\phi^2}{2}} = \frac{2\alpha}{\phi^2}.$$

The characteristic polynomial has roots as

$$\begin{aligned} r^2 + \phi r + \frac{\phi^2}{2} &= 0 \\ &\Longrightarrow r_1, r_2 = \frac{-\phi \pm \sqrt{\phi^2 - 4 \times \frac{\phi^2}{2}}}{2} = -\frac{\phi}{2} \pm \frac{\phi}{2} i \\ &\Longrightarrow r_1 = -\frac{\phi}{2} + \frac{\phi}{2} i,\ r_2 = -\frac{\phi}{2} - \frac{\phi}{2} i. \end{aligned}$$

Since stability requires that the real part of r_1 and r_2 be negative. Since $\phi > 0$ it follows that

$$\operatorname{Re}[r_1] = \operatorname{Re}[r_2] = -\frac{\phi}{2} < 0$$

and so $Y(t)$ is stable.

The general solution is then

$$Y(t) = Y^* + A_1 e^{r_1 t} + A_2 e^{r_2 t} = Y^* + A_1 e^{\left(-\frac{\phi}{2}+\frac{\phi}{2}i\right)t} + A_2 e^{\left(-\frac{\phi}{2}-\frac{\phi}{2}i\right)t}$$

where we also will need

$$Y'(t) = \left(-\frac{\phi}{2}+\frac{\phi}{2}i\right) A_1 e^{\left(-\frac{\phi}{2}+\frac{\phi}{2}i\right)t} + \left(-\frac{\phi}{2}-\frac{\phi}{2}i\right) A_2 e^{\left(-\frac{\phi}{2}-\frac{\phi}{2}i\right)t}.$$

A_1 and A_2 depend on the starting values $\tilde{Y}(0) = Y(0) - Y^*$ and $Y'(0)$ as

$$\begin{aligned} Y(0) &= Y^* + A_1 e^{\left(-\frac{\phi}{2}+\frac{\phi}{2}i\right)\times 0} + A_2 e^{\left(-\frac{\phi}{2}-\frac{\phi}{2}i\right)\times 0} = A_1 + A_2 \\ Y'(0) &= \left(-\frac{\phi}{2}+\frac{\phi}{2}i\right) A_1 e^{\left(-\frac{\phi}{2}+\frac{\phi}{2}i\right)\times 0} + \left(-\frac{\phi}{2}-\frac{\phi}{2}i\right) A_2 e^{\left(-\frac{\phi}{2}-\frac{\phi}{2}i\right)\times 0} \\ &= \left(-\frac{\phi}{2}+\frac{\phi}{2}i\right) A_1 + \left(-\frac{\phi}{2}-\frac{\phi}{2}i\right) A_2 \end{aligned}$$

so that in matrix notation

$$\begin{bmatrix} 1 & 1 \\ -\frac{\phi}{2}+\frac{\phi}{2}i & -\frac{\phi}{2}-\frac{\phi}{2}i \end{bmatrix} \begin{bmatrix} A_1 \\ A_2 \end{bmatrix} = \begin{bmatrix} \tilde{Y}(0) \\ Y'(0) \end{bmatrix}$$

or

$$\begin{aligned} \begin{bmatrix} A_1 \\ A_2 \end{bmatrix} &= \begin{bmatrix} 1 & 1 \\ -\frac{\phi}{2}+\frac{\phi}{2}i & -\frac{\phi}{2}-\frac{\phi}{2}i \end{bmatrix}^{-1} \begin{bmatrix} \tilde{Y}(0) \\ Y'(0) \end{bmatrix} \\ &= \begin{bmatrix} \frac{\tilde{Y}(0)}{2} - \left(\frac{\tilde{Y}(0)}{2} + \frac{Y'(0)}{\phi}\right) i \\ \frac{\tilde{Y}(0)}{2} + \left(\frac{\tilde{Y}(0)}{2} + \frac{Y'(0)}{\phi}\right) i \end{bmatrix} \end{aligned}$$

and so

$$\begin{aligned} A_1 &= c + di, \ A_2 = c - di \\ c &= \frac{\tilde{Y}(0)}{2}, \ d = -\left(\frac{\tilde{Y}(0)}{2} + \frac{Y'(0)}{\phi}\right). \end{aligned}$$

We thus have the solution

$$\begin{aligned}
Y_t &= Y^* + A_1 e^{\left(-\frac{\phi}{2}+\frac{\phi}{2}i\right)t} + A_2 e^{\left(-\frac{\phi}{2}-\frac{\phi}{2}i\right)t} \\
&= Y^* + e^{-\frac{\phi}{2}t}\left(A_1 e^{\frac{\phi}{2}it} + A_2 e^{-\frac{\phi}{2}it}\right) \\
&= Y^* + e^{-\frac{\phi}{2}t}\left((c+di)\, e^{\frac{\phi}{2}it} + (c-di)\, e^{-\frac{\phi}{2}it}\right) \\
&= Y^* + e^{-\frac{\phi}{2}t}\left((c+di)\left(\cos\left(\frac{\phi}{2}\right) + i\sin\left(\frac{\phi}{2}\right)\right) + (c-di)\left(\cos\left(-\frac{\phi}{2}\right) + i\sin\left(-\frac{\phi}{2}\right)\right)\right) \\
&= Y^* + e^{-\frac{\phi}{2}t}\left(2c \times \cos\left(\frac{\phi}{2}\right) - 2d\sin\left(\frac{\phi}{2}\right)\right) \\
&= Y^* + 2e^{-\frac{\phi}{2}t}\left(c \times \cos\left(\frac{\phi}{2}\right) - d\sin\left(\frac{\phi}{2}\right)\right) \\
&= Y^* + 2e^{-\frac{\phi}{2}t}\left(\frac{\tilde{Y}(0)}{2} \times \cos\left(\frac{\phi}{2}\right) + \left(\frac{\tilde{Y}(0)}{2} + \frac{Y'(0)}{\phi}\right) \times \sin\left(\frac{\phi}{2}\right)\right)
\end{aligned}$$

or

$$Y(t) = Y^* + 2e^{-\frac{\phi}{2}t}\left(\frac{\tilde{Y}(0)}{2} \times \cos\left(\frac{\phi}{2}\right) + \left(\frac{\tilde{Y}(0)}{2} + \frac{Y'(0)}{\phi}\right) \times \sin\left(\frac{\phi}{2}\right)\right).$$

Since $\cos(\theta + 2\pi) = \cos(\theta)$ and $\sin(\theta + 2\pi) = \sin(\theta)$ the term

$$c \times \cos\left(\frac{\phi}{2}\right) - d \times \sin\left(\frac{\phi}{2}\right)$$

has a period P satisfying

$$\frac{\phi}{4}(t+P) = \frac{\phi}{4}t + 2\pi \Longrightarrow P = \frac{2\pi}{\frac{\phi}{4}} = \frac{8\pi}{\phi}.$$

Problem 7.32 *[B***] Consider a second-order differential equation of the form*

$$Y''(t) + \phi Y'(t) + \phi^2 Y(t) = \alpha$$

where $\phi > 0$ and for given starting values $Y(0), Y'(0)$. Find the equilibrium value Y^*. Show that the roots of the characteristic polynomial are given by

$$r_1 = -\frac{\phi}{2} + \frac{\sqrt{3}\phi}{2}i,\ r_2 - \frac{\phi}{2} - \frac{\sqrt{3}\phi}{2}i.$$

*Show that $Y(t)$ is stable. [A***] Show that*

$$Y(t) = Y^* + 2e^{-\frac{\phi}{2}t}\left(\frac{\tilde{Y}(0)}{2} \times \cos\left(\frac{\sqrt{3}\phi}{2}\right) + \frac{1}{\sqrt{3}}\left(\frac{\tilde{Y}(0)}{2} + \frac{Y'(0)}{\phi}\right) \times \sin\left(\frac{\sqrt{3}\phi}{2}\right)\right)$$

where $\tilde{Y}(0) = Y(0) - Y^*$. Show the term involving $\cos(\ )$ and $\sin(\ )$ repeats itself every $\frac{4\pi}{\sqrt{3}\phi}$ periods.

Answer: The equilibrium value Y^* is $Y^* = \frac{\alpha}{\phi^2}$. The characteristic polynomial has roots as

$$\begin{aligned}
r^2 + \phi r + \phi^2 &= 0 \\
&\implies r_1, r_2 = \frac{-\phi \pm \sqrt{\phi^2 - 4 \times \phi^2}}{2} = -\frac{\phi}{2} \pm \frac{\sqrt{3}\phi}{2} i \\
&\implies r_1 = -\frac{\phi}{2} + \frac{\sqrt{3}\phi}{2} i, r_2 = -\frac{\phi}{2} - \frac{\sqrt{3}\phi}{2} i.
\end{aligned}$$

Since $\phi > 0$ it follows that

$$\operatorname{Re}[r_1] = \operatorname{Re}[r_2] = -\frac{\phi}{2} < 0$$

and so $Y(t)$ is stable.

The general solution is then

$$Y(t) = Y^* + A_1 e^{r_1 t} + A_2 e^{r_2 t} = Y^* + A_1 e^{\left(-\frac{\phi}{2} + \frac{\sqrt{3}\phi}{2} i\right) t} + A_2 e^{\left(-\frac{\phi}{2} - \frac{\sqrt{3}\phi}{2} i\right) t}$$

where we also need

$$Y'(t) = \left(-\frac{\phi}{2} + \frac{\sqrt{3}\phi}{2} i\right) A_1 e^{\left(-\frac{\phi}{2} + \frac{\sqrt{3}\phi}{2} i\right) t} + \left(-\frac{\phi}{2} - \frac{\sqrt{3}\phi}{2} i\right) A_2 e^{\left(-\frac{\phi}{2} - \frac{\sqrt{3}\phi}{2} i\right) t}.$$

Here A_1 and A_2 depend on the starting values $\tilde{Y}(0) = Y(0) - Y^*$ and $Y'(0)$ as

$$\begin{aligned}
Y(0) &= Y^* + A_1 e^{\left(-\frac{\phi}{2} + \frac{\sqrt{3}\phi}{2} i\right) \times 0} + A_2 e^{\left(-\frac{\phi}{2} - \frac{\sqrt{3}\phi}{2} i\right) \times 0} = A_1 + A_2 \\
Y'(0) &= \left(-\frac{\phi}{2} + \frac{\sqrt{3}\phi}{2} i\right) A_1 e^{\left(-\frac{\phi}{2} + \frac{\sqrt{3}\phi}{2} i\right) \times 0} + \left(-\frac{\phi}{2} - \frac{\sqrt{3}\phi}{2} i\right) A_2 e^{\left(-\frac{\phi}{2} - \frac{\sqrt{3}\phi}{2} i\right) \times 0} \\
&= \left(-\frac{\phi}{2} + \frac{\sqrt{3}\phi}{2} i\right) A_1 + \left(-\frac{\phi}{2} - \frac{\sqrt{3}\phi}{2} i\right) A_2
\end{aligned}$$

so that in matrix notation

$$\begin{bmatrix} 1 & 1 \\ -\frac{\phi}{2} + \frac{\sqrt{3}\phi}{2} i & -\frac{\phi}{2} - \frac{\sqrt{3}\phi}{2} i \end{bmatrix} \begin{bmatrix} A_1 \\ A_2 \end{bmatrix} = \begin{bmatrix} \tilde{Y}(0) \\ Y'(0) \end{bmatrix}$$

or

$$\begin{aligned}
\begin{bmatrix} A_1 \\ A_2 \end{bmatrix} &= \begin{bmatrix} 1 & 1 \\ -\frac{\phi}{2} + \frac{\sqrt{3}\phi}{2} i & -\frac{\phi}{2} - \frac{\sqrt{3}\phi}{2} i \end{bmatrix}^{-1} \begin{bmatrix} \tilde{Y}(0) \\ Y'(0) \end{bmatrix} \\
&= \begin{bmatrix} \frac{\tilde{Y}(0)}{2} - \frac{1}{\sqrt{3}} \left(\frac{\tilde{Y}(0)}{2} + \frac{Y'(0)}{\phi} \right) i \\ \frac{\tilde{Y}(0)}{2} + \frac{1}{\sqrt{3}} \left(\frac{\tilde{Y}(0)}{2} + \frac{Y'(0)}{\phi} \right) i \end{bmatrix}
\end{aligned}$$

and so

$$\begin{aligned} A_1 &= c + di,\ A_2 = c - di \\ c &= \frac{\tilde{Y}(0)}{2},\ d = -\frac{1}{\sqrt{3}}\left(\frac{\tilde{Y}(0)}{2} + \frac{Y'(0)}{\phi}\right). \end{aligned}$$

We thus have the solution

$$\begin{aligned} Y_t &= Y^* + A_1 e^{\left(-\frac{\phi}{2} + \frac{\sqrt{3}\phi}{2} i\right)t} + A_2 e^{\left(-\frac{\phi}{2} - \frac{\sqrt{3}\phi}{2} i\right)t} \\ &= Y^* + e^{-\frac{\phi}{2}t}\left(A_1 e^{\frac{\sqrt{3}\phi}{2} it} + A_2 e^{-\frac{\sqrt{3}\phi}{2} it}\right) \\ &= Y^* + e^{-\frac{\phi}{2}t}\left((c+di)\, e^{\frac{\sqrt{3}\phi}{2} it} + (c-di)\, e^{-\frac{\sqrt{3}\phi}{2} it}\right) \\ &= Y^* + e^{-\frac{\phi}{2}t}\left(\begin{array}{c} (c+di)\left(\cos\left(\frac{\sqrt{3}\phi}{2}\right) + i\sin\left(\frac{\sqrt{3}\phi}{2}\right)\right) \\ +(c-di)\left(\cos\left(-\frac{\sqrt{3}\phi}{2}\right) + i\sin\left(-\frac{\sqrt{3}\phi}{2}\right)\right) \end{array}\right) \\ &= Y^* + e^{-\frac{\phi}{2}t}\left(2c \times \cos\left(\frac{\sqrt{3}\phi}{2}\right) - 2d\sin\left(\frac{\sqrt{3}\phi}{2}\right)\right) \\ &= Y^* + 2e^{-\frac{\phi}{2}t}\left(c \times \cos\left(\frac{\sqrt{3}\phi}{2}\right) - d\sin\left(\frac{\sqrt{3}\phi}{2}\right)\right) \\ &= Y^* + 2e^{-\frac{\phi}{2}t}\left(\frac{\tilde{Y}(0)}{2} \times \cos\left(\frac{\sqrt{3}\phi}{2}\right) + \frac{1}{\sqrt{3}}\left(\frac{\tilde{Y}(0)}{2} - \frac{Y'(0)}{\phi}\right) \times \sin\left(\frac{\sqrt{3}\phi}{2}\right)\right) \end{aligned}$$

or

$$Y(t) = Y^* + 2e^{-\frac{\phi}{2}t}\left(\frac{\tilde{Y}(0)}{2} \times \cos\left(\frac{\sqrt{3}\phi}{2}\right) + \frac{1}{\sqrt{3}}\left(\frac{\tilde{Y}(0)}{2} + \frac{Y'(0)}{\phi}\right) \times \sin\left(\frac{\sqrt{3}\phi}{2}\right)\right).$$

Since $\cos(\theta + 2\pi) = \cos(\theta)$ and $\sin(\theta + 2\pi) = \sin(\theta)$ the term

$$c \times \cos\left(\frac{\sqrt{3}\phi}{2}\right) - d \times \sin\left(\frac{\sqrt{3}\phi}{2}\right)$$

has a period P satisfying

$$\frac{\sqrt{3}\phi}{2}(t + P) = \frac{\sqrt{3}\phi}{2}t + 2\pi \Longrightarrow P = \frac{2\pi}{\frac{\sqrt{3}\phi}{2}} = \frac{4\pi}{\sqrt{3}\phi}.$$

Problem 7.33 *[B***] Consider a second-order differential equation of the form*

$$Y''(t) + \phi Y'(t) + \frac{\phi^2}{3} Y(t) = \alpha$$

where $\phi > 0$ and for given starting values $Y(0), Y'(0)$. Find the equilibrium value Y^*. Show that the roots of the characteristic polynomial are given by

$$r_1 = -\frac{\phi}{2} + \frac{\phi}{2\sqrt{3}} i,\ r_2 = -\frac{\phi}{2} - \frac{\phi}{2\sqrt{3}} i.$$

*Show that $Y(t)$ is stable. [A***] Show that*

$$Y(t) = Y^* + 2e^{-\frac{\phi}{2}t}\left(\frac{\tilde{Y}(0)}{2}\times\cos\left(\frac{\phi}{2\sqrt{3}}\right) + \sqrt{3}\left(\frac{\tilde{Y}(0)}{2} + \frac{Y'(0)}{\phi}\right)\times\sin\left(\frac{\phi}{2\sqrt{3}}\right)\right).$$

where $\tilde{Y}(0) = Y(0) - Y^*$. Show the term involving $\cos(\;)$ *and* $\sin(\;)$ *repeats itself every $\frac{4\sqrt{3}\pi}{\phi}$ periods.*

Answer: The equilibrium value is $Y^* = \frac{3\alpha}{\phi^2}$.The characteristic polynomial has roots as

$$\begin{aligned} r^2 + \phi r + \frac{\phi^2}{3} &= 0 \\ \implies r_1, r_2 &= \frac{-\phi \pm \sqrt{\phi^2 - 4\times\frac{\phi^2}{3}}}{2} = -\frac{\phi}{2} \pm \frac{\phi}{2\sqrt{3}}i \\ \implies r_1 &= -\frac{\phi}{2} + \frac{\phi}{2\sqrt{3}}i, r_2 = -\frac{\phi}{2} - \frac{\phi}{2\sqrt{3}}i. \end{aligned}$$

Since $\phi > 0$ it follows that

$$\text{Re}[r_1] = \text{Re}[r_2] = -\frac{\phi}{2} < 0$$

and so $Y(t)$ is stable.

The general solution is then

$$Y(t) = Y^* + A_1 e^{r_1 t} + A_2 e^{r_2 t} = Y^* + A_1 e^{\left(-\frac{\phi}{2}+\frac{\phi}{2\sqrt{3}}i\right)t} + A_2 e^{\left(-\frac{\phi}{2}-\frac{\phi}{2\sqrt{3}}i\right)t}$$

where we will also need

$$Y'(t) = \left(-\frac{\phi}{2} + \frac{\phi}{2\sqrt{3}}i\right)A_1 e^{\left(-\frac{\phi}{2}+\frac{\phi}{2\sqrt{3}}i\right)t} + \left(-\frac{\phi}{2} - \frac{\phi}{2\sqrt{3}}i\right)A_2 e^{\left(-\frac{\phi}{2}-\frac{\phi}{2\sqrt{3}}i\right)t}.$$

A_1 and A_2 depend on the starting values $\tilde{Y}(0) = Y(0) - Y^*$ and $Y'(0)$ as

$$\begin{aligned} Y(0) &= Y^* + A_1 e^{\left(-\frac{\phi}{2}+\frac{\phi}{2\sqrt{3}}i\right)\times 0} + A_2 e^{\left(-\frac{\phi}{2}-\frac{\phi}{2\sqrt{3}}i\right)\times 0} = A_1 + A_2 \\ Y'(0) &= \left(-\frac{\phi}{2} + \frac{\phi}{2\sqrt{3}}i\right)A_1 e^{\left(-\frac{\phi}{2}+\frac{\phi}{2\sqrt{3}}i\right)\times 0} + \left(-\frac{\phi}{2} - \frac{\phi}{2\sqrt{3}}i\right)A_2 e^{\left(-\frac{\phi}{2}-\frac{\phi}{2\sqrt{3}}i\right)\times 0} \\ &= \left(-\frac{\phi}{2} + \frac{\phi}{2\sqrt{3}}i\right)A_1 + \left(-\frac{\phi}{2} - \frac{\phi}{2\sqrt{3}}i\right)A_2 \end{aligned}$$

so that in matrix notation

$$\begin{bmatrix} 1 & 1 \\ -\frac{\phi}{2} + \frac{\phi}{2\sqrt{3}}i & -\frac{\phi}{2} - \frac{\phi}{2\sqrt{3}}i \end{bmatrix}\begin{bmatrix} A_1 \\ A_2 \end{bmatrix} = \begin{bmatrix} \tilde{Y}(0) \\ Y'(0) \end{bmatrix}$$

or

$$\begin{aligned}\begin{bmatrix} A_1 \\ A_2 \end{bmatrix} &= \begin{bmatrix} 1 & 1 \\ -\frac{\phi}{2}+\frac{\phi}{2\sqrt{3}}i & -\frac{\phi}{2}-\frac{\phi}{2\sqrt{3}}i \end{bmatrix}^{-1} \begin{bmatrix} \alpha \\ \beta \end{bmatrix} \begin{bmatrix} \tilde{Y}(0) \\ Y'(0) \end{bmatrix} \\ &= \begin{bmatrix} \frac{\tilde{Y}(0)}{2} - \sqrt{3}\left(\frac{\tilde{Y}(0)}{2}+\frac{Y'(0)}{\phi}\right)i \\ \frac{\tilde{Y}(0)}{2} + \sqrt{3}\left(\frac{\tilde{Y}(0)}{2}+\frac{Y'(0)}{\phi}\right)i \end{bmatrix}\end{aligned}$$

and so

$$A_1 = c+di,\ A_2 = c-di,\ c = \frac{\tilde{Y}(0)}{2},\ d = -\sqrt{3}\left(\frac{\tilde{Y}(0)}{2}+\frac{Y'(0)}{\phi}\right).$$

The general solution is then

$$Y(t) = Y^* + A_1 e^{r_1 t} + A_2 e^{r_2 t} = Y^* + A_1 e^{\left(-\frac{\phi}{2}+\frac{\phi}{2\sqrt{3}}i\right)t} + A_2 e^{\left(-\frac{\phi}{2}-\frac{\phi}{2\sqrt{3}}i\right)t}$$

where we will also need

$$Y'(t) = \left(-\frac{\phi}{2}+\frac{\phi}{2\sqrt{3}}i\right)A_1 e^{\left(-\frac{\phi}{2}+\frac{\phi}{2\sqrt{3}}i\right)t} + \left(-\frac{\phi}{2}-\frac{\phi}{2\sqrt{3}}i\right)A_2 e^{\left(-\frac{\phi}{2}-\frac{\phi}{2\sqrt{3}}i\right)t}.$$

A_1 and A_2 depend on the starting values $\tilde{Y}(0) = Y(0) - Y^*$ and $Y'(0)$ as

$$\begin{aligned} Y(0) &= Y^* + A_1 e^{\left(-\frac{\phi}{2}+\frac{\phi}{2\sqrt{3}}i\right)\times 0} + A_2 e^{\left(-\frac{\phi}{2}-\frac{\phi}{2\sqrt{3}}i\right)\times 0} = A_1 + A_2 \\ Y'(0) &= \left(-\frac{\phi}{2}+\frac{\phi}{2\sqrt{3}}i\right)A_1 e^{\left(-\frac{\phi}{2}+\frac{\phi}{2\sqrt{3}}i\right)\times 0} + \left(-\frac{\phi}{2}-\frac{\phi}{2\sqrt{3}}i\right)A_2 e^{\left(-\frac{\phi}{2}-\frac{\phi}{2\sqrt{3}}i\right)\times 0} \\ &= \left(-\frac{\phi}{2}+\frac{\phi}{2\sqrt{3}}i\right)A_1 + \left(-\frac{\phi}{2}-\frac{\phi}{2\sqrt{3}}i\right)A_2 \end{aligned}$$

so that in matrix notation

$$\begin{bmatrix} 1 & 1 \\ \frac{\phi}{2}+\frac{\phi}{2\sqrt{3}}i & \frac{\phi}{2}-\frac{\phi}{2\sqrt{3}}i \end{bmatrix} \begin{bmatrix} A_1 \\ A_2 \end{bmatrix} = \begin{bmatrix} \tilde{Y}(0) \\ Y'(0) \end{bmatrix}$$

or

$$\begin{aligned}\begin{bmatrix} A_1 \\ A_2 \end{bmatrix} &= \begin{bmatrix} 1 & 1 \\ \frac{\phi}{2}+\frac{\phi}{2\sqrt{3}}i & \frac{\phi}{2}-\frac{\phi}{2\sqrt{3}}i \end{bmatrix}^{-1} \begin{bmatrix} \tilde{Y}(0) \\ Y'(0) \end{bmatrix} \\ &= \begin{bmatrix} \frac{\tilde{Y}(0)}{2} + \frac{1}{\sqrt{3}}\left(\frac{\tilde{Y}(0)}{2}-\frac{Y'(0)}{\phi}\right)i \\ \frac{Y(0)}{2} - \frac{1}{\sqrt{3}}\left(\frac{\tilde{Y}(0)}{2}-\frac{Y'(0)}{\phi}\right)i \end{bmatrix}\end{aligned}$$

where

$$\begin{aligned} A_1 &= c+di = \frac{\tilde{Y}(0)}{2} + \frac{1}{\sqrt{3}}\left(\frac{\tilde{Y}(0)}{2}-\frac{Y'(0)}{\phi}\right)i \\ A_2 &= c-di = \frac{\tilde{Y}(0)}{2} - \frac{1}{\sqrt{3}}\left(\frac{\tilde{Y}(0)}{2}-\frac{Y'(0)}{\phi}\right)i. \end{aligned}$$

We thus have the solution

$$\begin{aligned}
Y_t &= Y^* + A_1 e^{\left(-\frac{\phi}{2}+\frac{\phi}{2\sqrt{3}}i\right)t} + A_2 e^{\left(-\frac{\phi}{2}-\frac{\phi}{2\sqrt{3}}i\right)t} \\
&= Y^* + e^{-\frac{\phi}{2}t}\left(A_1 e^{\frac{\phi}{2\sqrt{3}}it} + A_2 e^{-\frac{\phi}{2\sqrt{3}}it}\right) \\
&= Y^* + e^{-\frac{\phi}{2}t}\left((c+di)\, e^{\frac{\phi}{2\sqrt{3}}it} + (c-di)\, e^{-\frac{\phi}{2\sqrt{3}}it}\right) \\
&= Y^* + e^{-\frac{\phi}{2}t}\left(\begin{array}{c} (c+di)\left(\cos\left(\frac{\phi}{2\sqrt{3}}\right) + i\sin\left(\frac{\phi}{2\sqrt{3}}\right)\right) \\ +(c-di)\left(\cos\left(-\frac{\phi}{2\sqrt{3}}\right) + i\sin\left(-\frac{\phi}{2\sqrt{3}}\right)\right) \end{array}\right) \\
&= Y^* + e^{-\frac{\phi}{2}t}\left(2c \times \cos\left(\frac{\phi}{2\sqrt{3}}\right) - 2d\sin\left(\frac{\phi}{2\sqrt{3}}\right)\right) \\
&= Y^* + 2e^{-\frac{\phi}{2}t}\left(c \times \cos\left(\frac{\phi}{2\sqrt{3}}\right) - d\sin\left(\frac{\phi}{2\sqrt{3}}\right)\right) \\
&= Y^* + 2e^{-\frac{\phi}{2}t}\left(\frac{\tilde{Y}(0)}{2}\cos\left(\frac{\phi}{2\sqrt{3}}\right) + \sqrt{3}\left(\frac{\tilde{Y}(0)}{2} - \frac{Y'(0)}{\phi}\right)\sin\left(\frac{\phi}{2\sqrt{3}}\right)\right)
\end{aligned}$$

or

$$Y(t) = Y^* + 2e^{-\frac{\phi}{2}t}\left(\frac{\tilde{Y}(0)}{2}\cos\left(\frac{\phi}{2\sqrt{3}}\right) + \sqrt{3}\left(\frac{\tilde{Y}(0)}{2} + \frac{Y'(0)}{\phi}\right)\sin\left(\frac{\phi}{2\sqrt{3}}\right)\right).$$

Since $\cos(\theta + 2\pi) = \cos(\theta)$ and $\sin(\theta + 2\pi) = \sin(\theta)$ the term

$$c \times \cos\left(\frac{\phi}{2\sqrt{3}}\right) - d \times \sin\left(\frac{\phi}{2\sqrt{3}}\right)$$

has a period P satisfying

$$\begin{aligned}
\frac{\phi}{2\sqrt{3}}(t+P) &= \frac{\phi}{2\sqrt{3}}t + 2\pi \\
&\implies P = \frac{2\pi}{\frac{\phi}{2\sqrt{3}}} = \frac{4\sqrt{3}\pi}{\phi}.
\end{aligned}$$

Problem 7.34 *[B***] Consider a second-order differential equation of the form*

$$Y''(t) + 2aY'(t) + \left(a^2 + b^2\right)Y(t) = \alpha$$

where $b > 0$ and for given starting values $Y(0), Y'(0)$. Find the equilibrium value Y^*. Show that the roots of the characteristic polynomial are given by

$$r_1 = -a + bi,\ r_2 = -a - bi.$$

*Show that $Y(t)$ is stable if $a > 0$. [A***] Show that*

$$Y_t = Y^* + 2e^{-at}\left(\frac{\tilde{Y}(0)}{2} \times \cos(bt) + \frac{1}{2}\left(\frac{\tilde{Y}(0)\,a}{b} + \frac{Y'(0)}{b}\right) \times \sin(bt)\right).$$

where $\tilde{Y}(0) = Y(0) - Y^*$. Show the term involving $\cos(\ )$ and $\sin(\ )$ repeats itself every $\frac{2\pi}{b}$ periods.

Answer: The equilibrium value is $Y^* = \frac{\alpha}{a^2+b^2}$. The characteristic polynomial has roots as

$$\begin{aligned} r^2 + 2ar + (a^2+b^2) &= 0 \\ &\Longrightarrow r_1, r_2 = \frac{-2a \pm \sqrt{(2a)^2 - 4\times(a^2+b^2)}}{2} \frac{-2a \pm \sqrt{-4b^2}}{2} = -a \pm bi \\ &\Longrightarrow r_1 = -a + bi,\ r_2 = -a - bi. \end{aligned}$$

Stability requires that the real part of r_1 and r_2 given by $-a$ be negative or

$$-a < 0 \Longrightarrow a > 0.$$

The general solution is then

$$Y(t) = Y^* + A_1 e^{(-a+bi)t} + A_2 e^{(-a-bi)t}$$

where we also will need

$$Y'(t) = (-a+bi) A_1 e^{(-a+bi)t} + (-a-bi) A_1 e^{(-a-bi)t}.$$

where A_1 and A_2 depend on the starting values $\tilde{Y}(0) = Y(0) - Y^*$ and $Y'(0)$ as

$$\begin{aligned} Y(0) &= Y^* + A_1 e^{(-a+bi)\times 0} + A_2 e^{(-a-bi)\times 0} = A_1 + A_2 \\ Y'(0) &= (-a+bi) A_1 e^{(-a+bi)\times 0} + (-a-bi) A_1 e^{(-a-bi)\times 0} \\ &= (-a+bi) A_1 + (-a-bi) A_1 \end{aligned}$$

so that in matrix notation

$$\begin{bmatrix} 1 & 1 \\ -a+bi & -a-bi \end{bmatrix} \begin{bmatrix} A_1 \\ A_2 \end{bmatrix} = \begin{bmatrix} \tilde{Y}(0) \\ Y'(0) \end{bmatrix}$$

or

$$\begin{aligned} \begin{bmatrix} A_1 \\ A_2 \end{bmatrix} &= \begin{bmatrix} 1 & 1 \\ -a+bi & -a-bi \end{bmatrix}^{-1} \begin{bmatrix} \tilde{Y}(0) \\ Y'(0) \end{bmatrix} \\ &= \begin{bmatrix} \frac{\tilde{Y}(0)}{2} - \frac{1}{2}\left(\frac{\tilde{Y}(0)a}{b} + \frac{Y'(0)}{b}\right) i \\ \frac{\tilde{Y}(0)}{2} + \frac{1}{2}\left(\frac{\tilde{Y}(0)a}{b} + \frac{Y'(0)}{b}\right) i \end{bmatrix} \end{aligned}$$

and so

$$A_1 = c + di,\ A_2 = c - di, c = \frac{\tilde{Y}(0)}{2},\ d = -\frac{1}{2}\left(\frac{\tilde{Y}(0)\,a}{b} + \frac{Y'(0)}{b}\right).$$

We thus have the solution

$$
\begin{aligned}
Y_t &= Y^* + Y^* + A_1 e^{(-a+bi)t} + A_2 e^{(-a-bi)t} \\
&= Y^* + e^{-at}\left((c+di)\, e^{bit} + (c-di)\, e^{-bit}\right) \\
&= Y^* + e^{-at}\left((c+di)\left(\cos(bt) + i\sin(bt)\right) + (c-di)\left(\cos(-bt) + i\sin(-bt)\right)\right) \\
&= Y^* + e^{-at}\left(2c \times \cos(bt) - 2d \times \sin(bt)\right) \\
&= Y^* + 2e^{-at}\left(\frac{\tilde{Y}(0)}{2}\cos(bt) + \frac{1}{2}\left(\frac{\tilde{Y}(0)\, a}{b} + \frac{Y'(0)}{b}\right)\sin(bt)\right)
\end{aligned}
$$

or

$$
Y(t) = Y^* + 2e^{-at}\left(\frac{\tilde{Y}(0)}{2}\cos(bt) + \frac{1}{2}\left(\frac{\tilde{Y}(0)\, a}{b} + \frac{Y'(0)}{b}\right)\sin(bt)\right).
$$

Since $\cos(\theta + 2\pi) = \cos(\theta)$ and $\sin(\theta + 2\pi) = \sin(\theta)$ the term

$$
c \times \cos(bt) - d \times \sin(bt)
$$

has a period P satisfying

$$
b(t + P) = bt + 2\pi \Longrightarrow P = \frac{2\pi}{b}.
$$

Made in the USA
Las Vegas, NV
10 May 2024

89772655R00149